Lecture Notes in Physics

Volume 850

For further volumes:
http://www.springer.com/series/5304

The Lecture Notes in Physics

The series Lecture Notes in Physics (LNP), founded in 1969, reports new developments in physics research and teaching—quickly and informally, but with a high quality and the explicit aim to summarize and communicate current knowledge in an accessible way. Books published in this series are conceived as bridging material between advanced graduate textbooks and the forefront of research and to serve three purposes:

- to be a compact and modern up-to-date source of reference on a well-defined topic
- to serve as an accessible introduction to the field to postgraduate students and nonspecialist researchers from related areas
- to be a source of advanced teaching material for specialized seminars, courses and schools

Both monographs and multi-author volumes will be considered for publication. Edited volumes should, however, consist of a very limited number of contributions only. Proceedings will not be considered for LNP.

Volumes published in LNP are disseminated both in print and in electronic formats, the electronic archive being available at springerlink.com. The series content is indexed, abstracted and referenced by many abstracting and information services, bibliographic networks, subscription agencies, library networks, and consortia.

Proposals should be sent to a member of the Editorial Board, or directly to the managing editor at Springer:

Christian Caron
Springer Heidelberg
Physics Editorial Department I
Tiergartenstrasse 17
69121 Heidelberg/Germany
christian.caron@springer.com

Ángel S. Sanz · Salvador Miret-Artés

A Trajectory Description of Quantum Processes. I. Fundamentals

A Bohmian Perspective

 Springer

Dr. Ángel S. Sanz
Instituto de Física Fundamental
Consejo Superior de Investigaciones
 Científicas
Serrano 123
28006 Madrid
Spain

Prof. Salvador Miret-Artés
Instituto de Física Fundamental
Consejo Superior de Investigaciones
 Científicas
Serrano 123
28006 Madrid
Spain

ISSN 0075-8450
ISBN 978-3-642-18091-0
DOI 10.1007/978-3-642-18092-7
Springer Heidelberg New York Dordrecht London

e-ISSN 1616-6361
e-ISBN 978-3-642-18092-7

Library of Congress Control Number: 2011944205

Printed on acid-free paper

Springer is part of Springer Science+Business Media (www.springer.com)

To Ruth
S. A. S.
and
To Virginia, Jessica, David, my fathers
and brother[†]
S. M. A.

Preface

Trajectory-based formalisms used to describe non-relativistic quantum processes are being continuously developed. Perhaps with increased emphasis in the last 15 years or so, since dealing with "classical" concepts is very appealing owing to the physical insight or intuition one gains into the process under study. The three main theoretical frameworks in use nowadays—apart from classical mechanics—but with significant advances since their initial formulation are, in chronological order, the Jeffreys–Wentzel–Kramers–Brillouin (JWKB) approach (1926), the Feynman path integral approach (1948) based on earlier remarks by Dirac (1935), and the Bohmian approach (1952) with its roots in the pioneering works of Madelung's hydrodynamic formulation of quantum mechanics (1926) and de Broglie's pilot wave theory (1927). Since then, many hybrid methods combining classical and quantum mechanics have been developed, mainly to tackle an accurate description of many–degrees-of-freedom systems.

The semiclassical JWKB approximation is a short-wavelength description of quantum mechanics. The idea behind this approach is to build wave functions from classical trajectories and, in a more pictorial way, to "sew quantum mechanical flesh onto classical bones", quoting Berry and Mount (1972). From a mathematical viewpoint, this treatment is based on asymptotic series. For simple bound systems, quantization schemes are usually based on the JWKB method, the Bohr-Sommerfeld quantization rule, or the multidimensional generalization of the latter, namely the Einstein-Brillouin-Keller quantization rule. Near turning points this approximation breaks down, giving rise to caustics or coalescence of classical trajectories. In order to solve this problem, uniform approximations were developed by linearizing the interaction potential in the vicinity of turning points. Taking into account this theoretical scheme, one of the processes that has been more intensively studied is that of tunneling through a barrier. In classically forbidden regions, trajectories are analytically continued in the complex plane in order to account for a quantum problem (for example, tunneling) that has no classical analog. Furthermore, this approach has also been exploited within semiclassical scattering, starting with Ford and Wheeler (1959), who explained the rainbow effect observed in the gas phase. Nonetheless, from a practical point of view, it is easier to solve an initial-value problem than a boundary-value one. This is the reason why real-time propagators are usually based on the so-called initial-value representation in phase space. Alternatively, a very powerful and elegant route to the semiclassical approach is Feynman's path integral, which is another formulation of quantum mechanics. In this formulation, the time propagator arising from the integral representation of the Schrödinger equation is written in terms of a path integral—or sum over classical paths—which is dominated by those trajectories extremalizing the action. At present, one of the most important applications of this approach is the calculation of the density matrix for many-body (or many–degrees-of-freedom) systems—actually, the path-integral Monte Carlo method used to deal with particle clusters is based on it.

Bohmian mechanics mainly arose as a result of the unsatisfactory interpretation of standard quantum mechanics, which claimed that the wave function provides the most general and complete physical information about a quantum system. This led to a very exciting, never-ending debate focused on the completeness of the wave function and the quantum theory of measurement. Within Bohmian mechanics, a quantum system is described by a well-defined (in space and time) trajectory, namely a quantum or Bohmian trajectory; the evolution of this trajectory is determined by the wave function associated with the system. Quite recently, a revival of the debate about the role played by this mechanics in quantum physics can be found in the specialized literature. There are several groups for whom this theory constitutes the natural framework of quantum mechanics, whereas other groups consider it as an alternative and exact formulation that enables us to characterize, interpret and predict quantum processes, standing on equal footing with the standard theory. The central topic of this monograph is Bohmian mechanics. This formulation has also received an important impulse over the last 15 years from different communities, which translates into an impressive and fruitful theoretical development.

At present, there are several books on Bohmian mechanics, which somehow summarize the trends mentioned above. *The Quantum Theory of Motion* (1993) by Holland, *The Undivided Universe: an Ontological Interpretation of Quantum Theory* (1993) by Bohm and Hiley, and *Bohmsche Mechanik als Grundlage der Quantenmechanik* (2001) by Dürr—with its recently published English's version, *Bohmian Mechanics* (2009), in collaboration with Teufel—mainly deal with the conceptual grounds and foundations of this theory as well as epistemological problems. On the other hand, Wyatt's monograph, *Quantum Dynamics with Trajectories* (2005), tackles a more practical side of this theory, dealing with the potential application of quantum trajectories in applied problems and stressing their computational aspect as a means of solving the time-dependent Schrödinger equation. A collection of chapters published very recently in two monographs, *Quantum Trajectories* (2010) edited by Chattaraj and *Applied Bohmian Mechanics: From Nanoscale Systems to Cosmology* (2012) edited by Oriols and Mompart, give an ample overview of Bohmian mechanics in which the corresponding theory has been successfully applied. However, in spite of the wide range of applications within Bohmian mechanics covered by these books, from the foundations to computation, its interpretational importance is, somehow, lacking. It is missing in the sense that one cannot find many applications and discussions of this trajectory-based viewpoint within the context of realistic quantum phenomena, that are of broad interest to different scientific communities.

Moreover, explaining the dynamics of quantum phenomena in terms of trajectories has always attracted many physicists and chemists. The interpretations arising from Bohmian mechanics are very intuitive, powerful, and simpler than those provided by the standard version of quantum mechanics. Taking this into account, the main purpose of this monograph, and what justifies its publication is to provide and promote the interpretational aspects of Bohmian mechanics as an alternative way of understanding quantum physics and gaining more physical intuition, in particular, with regard to the visualization of the evolution of individual systems (at the same level as Newtonian mechanics with respect to classical statistical mechanics). Furthermore, and from our own longstanding experience in the field, Bohmian mechanics can tackle any quantum problem that standard quantum mechanics does, providing an alternative way of interpreting the phenomenon under analysis. Obviously, the effort invested by many researchers in standard quantum mechanics completely outweighs that invested in Bohmian mechanics. However, we think that this situation will be corrected in the near future owing to the fact that this theory appears in more and more modern quantum mechanics textbooks at the introductory level (indeed, as John Bell suggested, quantum mechanics should be studied from a Bohmian perspective in order to make clear the most striking and strange features of that theory).

With this goal, and in order to be as self-contained as possible, this monograph has been divided into two volumes. The first volume is focused on the classical and quantum theoretical background, whereas the second volume is devoted to simple and basic quantum processes to provide a new and alternative interpretation in terms of quantum trajectories. The chapters of this first volume, which are intended to be as self-conatined as possible, are organized as follows.

In Chap. 1, a brief survey of classical mechanics is presented ranging from trajectories to ensembles of trajectories, paying attention to the dynamics or time evolution of micro-objects when interacting with other micro-particles or with some external potential function. Newtonian physics is based on the idea of a first cause behind the motion of objects. However, perhaps one of the most elegant ways of rationalizing physical laws arises through the *calculus of variations*, from which such laws emerge as a consequence of the application of a *variational principle* (in this sense, Appendix A reminds the reader of the essentials of the calculus of variations for variables and fields). The three main formulations of classical mechanics, that is, Lagrangian, Hamiltonian, and Hamilton–Jacobi formulations, are briefly set out, since they represent the fundamental building blocks of any dynamical theory in terms of time or energy as primary parameters. Very often a first classical approach to a given quantum problem provides us with a complementary understanding of the corresponding dynamics, which results in a considerable gain in intuition—in particular, if the phase-space formulation is used. When dealing with ensembles of trajectories, we expect a natural transition from regular to chaotic motion owing to the underlying stochasticity present in dynamical (Hamiltonian) problems with two or higher dimensions. Furthermore, the extension to classical statistical mechanics, where the motion is deterministic but unpredictable, is analyzed in terms of the Liouville equation and a field theory. Several important aspects of continuum mechanics are very briefly commented on owing to its basic importance in quantum fluid dynamics.

In Chap. 2, the dynamics of open classical systems are introduced. Open classical systems are usually defined as those where the system of interest is surrounded by an environment at a certain temperature (heat bath or reservoir), exchanging energy in both directions. Strictly speaking real physical systems do not exist in complete isolation in Nature; all physical systems are open systems since the interaction with their environment can never be totally neglected. The mathematics required to understand this dynamics is the theory of probability and stochastic processes. This theory is briefly described in Appendix B since it plays a fundamental role in any classical or quantum dynamics. When a "coarse-grained" description is used, where we focus only on the dynamics of the system of interest, neglecting the details of the time evolution of the environment, two types of mechanics arise naturally: the dissipative and stochastic mechanics. In both types of mechanics, there are three standard routes to introduce dissipation and/or stochasticiy. First, from a phenomenological viewpoint, empirical equations are introduced, such as the standard Langevin equation, where a few parameters are required to describe the system–environment interaction. Second, starting from the Liouville equation, which is satisfied by any dynamical variable in phase space, projection operator techniques are applied until a generalized Langevin equation is finally reached. Third, when the starting point is a conservative many-body problem (system plus environment is an isolated system), dissipative forces can be obtained as well as an external stochasticity owing to the fluctuations or noise of the heat bath. A clear distinction between the two mechanics and some links between these three different approaches are presented and discussed.

In Chap. 3, some elements of quantum mechanics are presented. Time-independent and time-dependent Schrödinger equations are derived from the so-called Hamiltonian analogy through the calculus of variation together with de Broglie's ideas of associating a wavelength with matter particles. Some basic notions of wave mechanics, current densities, ensemble distributions and density matrix in phase space are also reviewed. Special emphasis is placed on some approaches to quantum mechanics in which classical concepts and/or trajectories are the main ingredients such as the path integral formulation, semiclassical mechanics and, the eikonal approach.

Chapter 4 is devoted to wave optics in connection to quantum mechanics. The issues covered in this chapter are almost entirely based on the physics described by the *wave equation*. This allows us to understand and offer an alternative optical perspective of many of the basic elements and concepts found in quantum mechanics within the context of any wave theory, and not as something purely specific to quantum physics. According to Ballentine, quantum phenomena can be illustrated by means of three traits: *discreteness*, *diffraction*, and *coherence*. Thus, the chapter is organized in such a way that shows how such features and related concepts are already present in wave optics, though in a general way. Thus, starting from the main ingredients of wave optics, namely Maxwell's equations and the wave equation, we will move into the superposition or Huygens–Fresnel principle, very closely connected to the notion of *coherence* and the appearance of interference and diffraction phenomena. Regarding discreteness, it is not necessary to go as far as the photoelectric effect, but we already find it in *optical waveguides*, which are the optical analogs of quantum "bound" systems. In order to cover the full spectrum of phenomena that can be found in quantum mechanics, we also revisit the *Goos–Hänchen effect* or the *Hartman effect*, which are good examples related to *optical tunneling*. Furthermore, a direct link to the language of quantum mechanics can be established through the *hydrodynamical formulation of electromagnetism*, a generalized formulation based on the so-called *Riemann–Silberstein vector*.

The dynamics of open quantum systems is briefly treated in Chap. 5. The system-plus-reservoir model used in the classical context is also followed here in quantum mechanics, and dissipation and stochasticity are easier to tackle and understand. Both system and reservoir are in continuous interaction and the effects—coherence loss or decoherence, population transfer, and/or (system–environment) energy exchange—arising from that interaction will depend to a greater or lesser extent on the coupling strength and its intrinsic nature. The system time-evolution is not unitary and therefore cannot be described in terms of the Schrödinger equation. In these cases, it is then necessary to resort to statistical quantum methods invoking, for example, the density matrix and Langevin formalisms and/or introducing, in general, quantum stochasticity into the time-evolution equations: the Linblad equation, quantum Langevin-type equations, and so on. The energy transfer from the system to the environment is termed quantum *relaxation* or *damping*. If there is no chance for the energy to move backwards into the system, the unidirectional energy flow into the reservoir is then called quantum *dissipation*. On short time scales, the distinction between quantum relaxation and dissipation is obviously unclear. Under certain conditions the duration of the reservoir correlations is very short compared to the dynamical evolution of the system. This leads to a total memory loss of the bath dynamics, which gives rise to a subsequent irreversible loss of coherence and energy (or population) relaxation in the system. This is called a *Markovian regime*. Within this regime, the time evolution of the system depends only on the present state of the system; this is called a *Markovian process*. As will be seen, when this happens, the system dynamics can be characterized by (relatively) simple Markovian master equations, where one does not need to take into account the reservoir dynamics, and its effects on the system are described by means of certain operators. In analogy to open classical systems, there are also three main different approaches to dealing with quantum dissipative dynamics: (i) effective time-dependent Hamiltonians, (ii) the nonlinear Schrödinger equation, and (iii) the system-plus-reservoir model within a conservative scenario. In particular, the so-called stochastic Schrödinger equation, written in terms of an Itô differential equation, gives rise to quantum trajectories, not to be confused with those coming from Bohmian mechanics. Finally, in this chapter as well as in Appendix B the measurement process is also very briefly discussed through the introduction of the so-called weak measurement due to Aharonov, Albert and Vaidman, in distinction to the more standard von Neumann strong measurement, since observing very weak effects is becoming more and more important at present.

Chapter 6 can be considered the main chapter of this monograph; to some extent, the previous chapters have been written for the purpose of providing the reader, as far as possible, with the background necessary to better understand the approach developed by David Bohm, nowadays known as Bohmian mechanics. He essentially based his theory on the assumption that a quantum system consists, at the same time, of a wave and a particle. The wave evolves according to Schrödinger's equation and the particle moves according to a certain guidance condition (quantum trajectories), which makes the particle motion dependent on the wave evolution. Although Bohmian mechanics is usually regarded as a "reinterpretation" or an alternative picture of standard quantum mechanics, it is also common to refer to it as a "theory" in order to stress the conceptual difference between the two approaches to the microscopic world. Bohm's ideas were applied to different prototypical models of quantum mechanics during the late 1970s and, particularly, the 1980s and early 1990s, and the attention paid by the scientific community was not very great. However, in the last ten years or so, Bohmian mechanics has passed from being merely a way to formulate a quantum mechanics "without observers" to become a well-known (and increasingly accepted) theoretical framework used as a source of new quantum computational methods as well as new quantum interpretations. This chapter ends by considering open quantum systems from this point of view.

Finally, Chap. 7 has been organized to take into account a gradual transition from simple (light) rays to hydrodynamic (photon) trajectories/paths, i.e., from geometric optics to what we shall denote as *hydrodynamic optics*. This trajectory-based description is analyzed for the propagation of plane waves and Young-type experiments with polarized light, the latter being intimately related to the so-called Arago–Fresnel laws of diffraction for polarized light. Afterwards, a brief account on the relation between hydrodynamic optics and the formulation based on the Riemann–Silberstein vector is given. The reported diffraction pattern for the two-slit experiment has been very recently inferred and explained in terms of photon paths from experiment. The weak measurement of an observable (position and/or momentum) for a quantum system is preselected in an initial state and postselected by a strong measurement in a final different state. Experiments of this kind have also led to a direct measurement of the photon quantum wave function. In our opinion, Bohmian mechanics can again undergo a new revival thanks to these experiments (where weak measurements are carried out), which can provide information on quantum trajectories of the underlying dynamics of any quantum process.

This monograph is the result of more than 15 years working on trajectory-based formalisms, in particular, on Bohmian mechanics. Concerning citations, we have tried to furnish a historical development of the different topics presented here. However, to provide a selection of the very last references in very active fields is really difficult. We apologize to those who think they should be cited and are not. During this long but exciting time, we have benefitted from discussions with many colleagues from abroad and from Spain. In this sense, we would like to acknowledge fruitful discussions and collaborations with J. A. Deswick, J. M. Bofill, F. Borondo, M. Božić, P. Brumer, J. Campos-Martínez, P. K. Chattaraj, C. C. Chou, M. Davidović, D. Dürr, E. R. Floyd, X. Giménez, S. Goldstein, T. González-Lezana, B. J. Hiley, B. K. Kendrick, J. Margalef-Roig, B. Poirier, E. Pollak, O. Roncero, J. S. Sánchez-Gómez, D. J. Tannor, T. Uzer and R. E. Wyatt. Also, we would like to thank all members (past and present) of the *Departamento de Física Atómica, Molecular y de Agregados* of the *Instituto de Física Fundamental* (CSIC) in Madrid, where this work has been carried out from its inception, benefiting support from the projects FIS2007-62006, FIS2010-18132, FIS2010-22082 and FIS2011-29596-C02-01 from the Ministerio de Ciencia e Innovación (Spain), a "Ramón y Cajal" Research Fellowship (A. S. S.), and the COST Action MP1006 "Fundamental Problems in Quantum Physics". Special thanks go to Gerardo Delgado-Barrio and Pablo Villarreal, *founding fathers* of this department, for their continuous support of and enthusiasm for our work. Finally, we thank A. Lahee, our Editor, for her enthusiasm when we proposed the monograph to her, as well as her patience and for extending —several times— the deadline for finishing this project.

Madrid, October 2011   Ángel S. Sanz
                                                                                                                             Salvador Miret-Artés

Contents

Chapter 1
From Trajectories to Ensembles in Classical Mechanics

1.1 Introduction

Our physical intuition is based on our everyday experience. From this viewpoint, Newton's mechanics seems to be the natural theoretical framework to describe the objects of classical mechanics. This framework, grounded on the mechanical laws for celestial bodies discovered by Copernicus [1] and Kepler [2], as well as in the rational "thought" systems formerly established by Galileo [3] and Descartes [4], is summarized by the well-known Newton's laws of motion [5]. The motion of (classical) objects is explained in terms of their response to the interactions among themselves or other external interactions. Regardless of whether such interactions are described in terms of forces at a distance, (potential) fields or particle exchanges, the evolution of a body is thus seen as the consequence or *effect* of an interaction on it (the *cause*). The *causality principle* enjoys of a wide acceptance, which has led to also consider that physical processes are *local*, i.e., what happens in certain space region does not affect what may happen in other distant space regions out of the corresponding light cone. This is not the case, though, in quantum mechanics (see Chap. 3)—neither in Bohmian mechanics (see Chap. 6)—, where *nonlocality* constitutes a very distinctive feature, as shown by Bell in the 1960s [6, 7].

Newtonian physics is therefore based on the idea of a first cause behind the motion of objects. Actually, Newton postulated a relationship between the acting force (cause) and the rate of change with time of the momentum (effect) rather than its velocity. This approach, however, is not unique when formulating classical mechanics, as mentioned before. An alternative and conceptually different approach, possibly one of the most elegant ways to rationalize physical laws, arises through the *calculus of variations*. Within this formulation, such laws emerge as a consequence of the application of a *variational principle* [8–10]. The calculus of variations allows us to determine the system dynamics in terms of some characteristic quantities, which are found to be either *extremal* (maximum, minimum or saddle-point values) or *stationary* (their rate of change is zero). This constitutes a very important conceptual difference with respect to the Newtonian formulation: now the dynamics arises

A. S. Sanz and S. Miret-Artés, *A Trajectory Description of Quantum Processes.* 1
I. Fundamentals, Lecture Notes in Physics 850, DOI: 10.1007/978-3-642-18092-7_1,
© Springer-Verlag Berlin Heidelberg 2012

as a consequence of a *need*—the physical solution is obtained via extremization of a quantity—rather than an effect from external causes. The problem of finding *geodesics* [11], i.e., the curves describing the shortest distance between two points on a given space surface constitutes a well-known application of the calculus of variations. Another important application is that of finding the fastest descendent curve or *brachistochrone curve* [12, 13] described by a body which is falling under gravity and constrained to move along the curve joining two points. These problems summarize the essence of two of the most fundamental principles in physics, which precisely arise from the calculus of variations. On the one hand, the *least time principle* or *Fermat's principle* (see Chap. 4), which states that light always follows the path of shortest optical length connecting two points (where the optical length depends on the material constituting the medium along which it propagates) [14], and was formerly proposed to describe reflection by Hero of Alexandria in his *Catoptrica* (circa 60 AC) [15] and later on expanded to refraction by Ibn al-Haytham "Alhazen" in 1021 in his *Book of Optics* [16]. On the other hand, its mechanical counterpart, the *least action principle* [8], which is generally known in the form given formerly by Maupertuis [17, 18] and, about a century later, by Hamilton [19, 20] for the mechanical action.

When going through the different widely accepted (and used) formulations of classical mechanics, the principles upon which they have been risen become very apparent. For example, though different in their starting point, Newtonian and Lagrangian mechanics make emphasis on the concept of *time evolution*. Indeed, the former arises very nicely from the latter after applying the calculus of variations via extremization with respect to time. The leading role of time in these formulations is also noticeable in their own mathematical structure, which relies on sets of second-order differential equations. The solutions to these equations are determined by specifying the value of positions[1] and velocities at a certain time—although the velocities are not totally independent of the positions, since the former are just the first time-derivative of the latter. This mechanistic view of physics is well summarized by Laplace's famous words [21]:

> We may regard the present state of the universe as the effect of the past and the cause of the future. Given for one instant an intelligence which could comprehend all the forces by which nature is animated and the respective positions of the beings which compose it, if moreover this intelligence were vast enough to submit these data to analysis, it would embrace in the same formula both the movements of the largest bodies in the universe and those of the lightest atom: to it nothing would be uncertain, and the future as the past would be present to its eyes.

On the other hand, the Hamiltonian and Hamilton–Jacobi formulations are built upon the concept of *energy conservation*: the motion laws arise via extremization with respect to energy, a property intrinsic to a given system (regardless of whether it can also be dissipated or absorbed). In this way, instead of positions and velocities,

---

[1] In general, from now on, any variable x (or its associated generalized coordinate, q) will be regarded as a "position", independently of whether it refers to a "true" position or to any other coordinate describing a physical system. The evolution of this variable will be given by a differential equation.

the physical system is described by a set of generalized coordinates and momenta, which are independent and obey first-order differential equations.

The Hamilton–Jacobi formulation of classical mechanics establishes a very direct link between energy and time, which are considered as conjugated. This idea has a particular relevance in quantum mechanics. For example, the passage from the time-independent Schrödinger equation to the time dependent one, or the time-energy uncertainty relation are based on it. Furthermore, this formulation allows us to establish the direct link between classical and quantum mechanics [8, 10] (actually, Schrödinger's wave equation was formerly derived starting from this formulation, as will be seen in Chap. 3) as well as Bohmian mechanics (see Chap. 6). The Hamilton–Jacobi formulation of classical mechanics (as well as the Hamiltonian one) is also very close to optics, this connection constituting a fundamental step to reach a deeper understanding of the meaning of the concept of *duality* that appears when dealing with both matter waves and radiation. The relationship comes from the possibility of describing motion by means of a similar wave-like language; in optics waves are surfaces of constant phase (see Chap. 4), while in classical mechanics the role of the wave is played by surfaces of constant action. Particle trajectories are then "rays" perpendicular to such surfaces at each time in the same way as optical rays are perpendicular to constant phase surfaces—later on, in Chap. 6, Bohmian trajectories are seen to be perpendicular to the surfaces of constant quantum phase.

Classical mechanics is also used to describe the dynamics or time evolution of micro-objects when interacting with other micro-particles or with some external potential function. This is mainly the issue under discussion in this Chapter, since many times it happens that a first classical approach to a given quantum problem provides a better understanding of its dynamical behavior. Following this scheme, the transition from regular to chaotic motion and to intrinsic stochasticity will be also analyzed. Extension to classical statistical mechanics where the motion is deterministic but unpredictable presented in terms of the Liouvillian dynamics will be carried out. Certain criteria to discern if a system has to be considered as classical or quantum will also be presented. In order to be self-contained and pay special attention to the ensemble of trajectories, certain aspects of continuum mechanics will be briefly exposed to better understand quantum fluid dynamics.

1.2 Fundamental Grounds of Classical Mechanics

1.2.1 Hamilton's Principle and Equations of Motion

One of the most important application of the calculus of variations (see Appendix A) is the derivation of the mechanical equations of motion avoiding the use of the concept of Newtonian force. This leads to the least action principle, also known as *Hamilton's variational principle* or *Hamilton's principle*, which states [22] that

[...] of all the possible paths along which a dynamical system may move from one point to another within a specified time interval (consistent with any constraints), the actual path followed is that which minimizes the time integral of the difference between the kinetic and potential energies.

This simple idea constitutes one of the most general and fundamental postulates of physics, not only because all of classical mechanics can be derived from it, but also quantum mechanics (see Chap. 3).

Mathematically, for a classical N-dimensional system (i.e., a system characterized by N degrees of freedom regardless of what they may represent), Hamilton's principle applies as follows. Consider the functional

$$S[L] \equiv \int_{t_a}^{t_b} L(q, \dot{q}, t)dt, \tag{1.1}$$

where t denotes time, $q = (q_1, q_2, \ldots, q_N)$ stands for a set of generalized coordinates describing the system and $\dot{q} = (\dot{q}_1, \dot{q}_2, \ldots, \dot{q}_N)$, with $\dot{q}_i \equiv dq_i/dt$, are the associated (generalized) velocities [8, 10]. In (1.1),

$$L(q, \dot{q}, t) = T(q, \dot{q}, t) - V(q, t) \tag{1.2}$$

is the so-called *Lagrangian function*, with T and V being the kinetic and potential energies, respectively. In generalized coordinates, the kinetic energy may depend on both q and $\dot{q}$—for example, in Cartesian coordinates it only depends on the velocities and not on the positions, but in spherical coordinates the dependence is on both—, while the potential energy always depends on q. Moreover, these energies may also depend explicitly on time, this being the case of *dissipative systems* (see Chap. 2).

Applying the calculus of variations to (1.1), one obtains

$$\frac{\partial L}{\partial q_i} - \frac{d}{dt}\frac{\partial L}{\partial \dot{q}_i} = 0, \quad i = 1, 2, \ldots, N. \tag{1.3}$$

This set of equations are the so-called *Lagrange equations of motion* of classical mechanics (within its Lagrangian formulation). They can also be expressed as

$$\frac{\partial T}{\partial q_i} - \frac{d}{dt}\frac{\partial T}{\partial \dot{q}_i} = \frac{\partial V}{\partial q_i}, \quad i = 1, 2, \ldots, N, \tag{1.4}$$

in terms of the kinetic and potential energies. If T is given in Cartesian coordinates (i.e., $T = \sum_i m_i \dot{x}_i^2/2$), then (1.4) acquires the more familiar form

$$\frac{d}{dt}(m_i \dot{x}_i) = -\frac{\partial V}{\partial x_i}, \quad i = 1, 2, \ldots, N, \tag{1.5}$$

where *Newton's equations of motion* or Newton's second law is readily recognized. According to Hamilton's principle, *classical trajectories* are the curves for which

(1.1) becomes an extremal when going from $q_a = q(t_a)$ to $q_b = q(t_b)$. The functional $S[q_b; q_a]$ is then said to be a first-order invariant under small perturbations of the trajectory $q(t)$. This is equivalent to find the curves $q_i(t)$ satisfying the set of equations (1.3) or, equivalently, (1.5). Since (1.3) consists of a set of N second-order differential equations, in order to obtain a complete solution the value of the q_i and that of their first derivatives, $\dot{q}_i$, have to be specified at a certain time, usually the initial time (t_0).

After Jacobi, the nature of the extrema found in the calculus of variations can be determined from an eigenvalue analysis of the matrix associated with the quadratic form that comes from the so-called second variation or third term of the action integral expansion in terms of the displacements. Furthermore, it can be shown that this quadratic form in the displacements of all possible paths around a given trajectory has as many negative eigenvalues as conjugate points can be found along the trajectory. These points are defined when one of the eigenvalues vanishes.

The system dynamics can be equivalently described within the Hamiltonian framework. In this case, first the *canonical momentum* along a particular trajectory $q = q(t)$ is determined by means of the relation

$$p_i = \frac{\partial L}{\partial \dot{q}_i}, \quad i = 1, 2, \ldots, N. \tag{1.6}$$

Then, the system Hamiltonian is obtained by means of the *Euler–Legendre transformation*

$$H(q, p, t) = \sum_{i=1}^{N} p_i \dot{q}_i - L(q, \dot{q}, t). \tag{1.7}$$

Substituting (1.7) into (1.1) and then applying Hamilton's principle, a set of $2N$ first-order differential equations is obtained, namely the *Hamilton* or *canonical equations of motion* of the Hamiltonian formulation of classical mechanics,

$$\dot{q}_i = \frac{\partial H}{\partial p_i}, \tag{1.8a}$$

$$\dot{p}_i = -\frac{\partial H}{\partial q_i}. \tag{1.8b}$$

In order to solve these equations of motion, a set of initial conditions $(q_0, p_0) = (q(t_0), p(t_0))$ has to be specified. Equations (1.8) can be regarded as a vector field in phase space,

$$\mathbf{F} \equiv (\dot{q}, \dot{p}) = (\nabla_p H, -\nabla_q H), \tag{1.9}$$

where $\nabla_p \equiv (\partial/\partial p_1, \partial/\partial p_2, \cdots, \partial/\partial p_N)$ and $\nabla_q \equiv (\partial/\partial q_1, \partial/\partial q_2, \cdots, \partial/\partial q_N)$ (from now on, the notation $\nabla \equiv \nabla_q$ will be considered). That is, the evolution of the physical system can be related to that of a *flow* in phase space, which is strongly connected with the also often used matrix or symplectic notation of Hamilton's equations of motion [23] (see Sect. 1.3.1). The opposite signs in the two terms of (1.9) have some important consequences when analyzed in the corresponding symplectic geometry. A system is said to be a *Hamiltonian system* when the vector field is defined by the above Hamilton equations. Hamiltonian systems constitute a special class of dynamical systems characterized by a smooth vector field in phase space.

Due to the symmetry displayed by the canonical equations when compared to the Euler–Lagrange ones as well as the fact that they are first-order differential equations for a set of $2N$ independent variables (the Lagrangian coordinates and momenta do not constitute a set of independent variables), Hamilton's formulation results simpler to study the dynamics of classical systems. Nevertheless, there are circumstances where working with them is not appropriate or comfortable, but it is more convenient to carry out a variable transformation from (q, p) to a set of new variables (q', p'), whose equations of motion are simpler and more insightful (physically). In this regard, *canonical* or *contact transformations* constitute a suitable type of transformation, for the new variables will also be canonical. That is, given the canonical transformation $q = q(q', p')$ and $p = p(q', p')$, then

$$\dot{q}_i' = \frac{\partial H'}{\partial p_i'}, \tag{1.10a}$$

$$\dot{p}_i' = -\frac{\partial H'}{\partial q_i'}, \tag{1.10b}$$

where H' is the Hamiltonian associated with the (new) canonical variables q' and p'. Actually, (1.10a) and (1.10b) can also be derived by using Hamilton's principle provided that the relation

$$\sum_{i=1}^{N} p_i \dot{q}_i - H = \sum_{i=1}^{N} p_i' \dot{q}_i' - H' + \frac{dG}{dt} \tag{1.11}$$

is satisfied. Here G is an arbitrary differentiable function depending on q, p, q' and p' (eventually it might also depend on t). A sufficient and necessary condition for the transformation to be canonical is that at least G depends on one of the old variables and another of the new ones. Then, G is called the transformation *generating function*, which might also be accompanied by a term describing an Euler–Legendre transformation required to change accordingly the right-hand side of (1.11).

Consider the generating function $G = S(q, p', t)$ together with the Legendre transformation $-p'q'$. Substituting them into the right-hand side of (1.11) and then rearrange terms, leads to

$$p_i = \frac{\partial S}{\partial q_i}, \tag{1.12a}$$

$$q_i' = \frac{\partial S}{\partial p_i'}, \tag{1.12b}$$

$$H' = H + \frac{\partial S}{\partial t}. \tag{1.12c}$$

As can be noticed, if H' is forced to be identically zero, the Hamilton equations corresponding to the new variables also vanish, i.e.,

$$\dot{q}_i' = \frac{\partial H'}{\partial p_i'} = 0, \tag{1.13a}$$

$$\dot{p}_i' = -\frac{\partial H'}{\partial q_i'} = 0, \tag{1.13b}$$

and therefore the q_i' and the p_i' constitute a set of $2N$ non-independent constants ($q_i' = \beta_i$, $p_i' = \alpha_i$), which are related through (1.12b) (i.e., $\beta_i = \partial S / \partial \alpha_i$). The problem then consists of solving the partial differential equation

$$H\left(q, \frac{\partial S}{\partial q}, t\right) + \frac{\partial S}{\partial t} = 0, \tag{1.14}$$

where S is a function that only depends on the old coordinates. Equation (1.14) is the so-called *Hamilton–Jacobi equation* [8–10], more familiar when it is expressed as

$$-\frac{\partial S}{\partial t} = \frac{(\nabla S)^2}{2m} + V, \tag{1.15}$$

with $\nabla S \equiv \partial S / \partial q = (\partial S / \partial q_1, \partial S / \partial q_2, \cdots, \partial S / \partial q_N)$. Note that, computing the total derivative dS/dt by means of (1.12a) and (1.12b), and then integrating over time, the resulting expression is (1.1). Therefore, the function S, the so-called *Hamilton's principal function*, is the *classical action* of the system. The lack of uniqueness of the classical function is well known. Pertinent discussions about this issue can be found in any standard textbook about classical mechanics.

Within this Hamilton–Jacobi formulation of classical mechanics, the particle equation of motion is given by (1.12a), the so-called *Jacobi's law*, as a function of the canonical coordinates (q, p) and Hamilton's principal function, S. The motions generated by (1.12a) present the particularity that any trajectory is always perpendicular to the surfaces of constant S, which allows us to construct in a very simple fashion families of trajectories by considering a starting point and taking the normal to a series of surfaces with constant S. This is the so-called *Huygens' construction* (see Sect. 4.3.1). More formally, the trajectories are the characteristics of the Cauchy problem associated with the Hamilton–Jacobi equation. The orthogonality condition between constant S surfaces and trajectories remains valid even in those cases where (1.12a) is modified by the presence of electromagnetic fields or other more complicated situations, such as arbitrary many-body systems acted by time-dependent potentials. In such cases, it is always possible to define a new function S' by means of non-Euclidean metrics, so that the corresponding trajectories will be orthogonal to surfaces with constant S'. The speed of the wavefront or constant S surface is given by $-(\partial S/\partial t)/|\nabla S|$, which can be related to particle velocities. This way of interpreting S allows us to establish a narrow relationship between classical mechanics and geometric optics (see Chap. 7), where the rays are orthogonal to the wavefronts or can be treated as such by means of a generalized geometry [24, 25], thus anticipating the possibility of a wave theory of matter, materialized in the formulation of the wave equation by Schrödinger. This connection between mechanics and optics becomes more apparent in the case of *conservative systems*, where the total system energy, E, evaluated along a trajectory remains constant in time. In this case, from the total time-derivative of (1.1), $dS/dt = L$, one reaches

$$\frac{\partial S}{\partial t} = L - \sum_{i=1}^{N} p_i \dot{q}_i = -H. \tag{1.16}$$

Now, since H describes the system energy, in the case of a conservative system $H = E$ and therefore

$$-\frac{\partial S}{\partial t} = E = \text{constant}. \tag{1.17}$$

Equation (1.15) then becomes

$$E = \frac{(\nabla S)^2}{2m} + V, \tag{1.18}$$

which is the so-called *time-independent Hamilton–Jacobi equation*. Because of (1.17), it is also common to express the classical action by separating its space and time dependent parts as

$$S = \mathcal{W}(q) - Et, \tag{1.19}$$

where $\mathcal{W}$ is the so-called *reduced Jacobi function* or Hamilton's characteristic function. Thus, for conservative systems, (1.12a) can also be expressed as $p = \nabla \mathcal{W}/m$, which does not depend explicitly on time.

The Hamilton–Jacobi formulation results very useful in one-dimensional problems and separable higher-dimensional ones. With no loss of generality, consider the former. From (1.18) together with (1.12a),

$$p = \pm\sqrt{2m[E - V(q)]}. \tag{1.20}$$

It is straightforward now to obtain the corresponding classical trajectories from this expression by integrating in time and once an initial condition q_0 is specified (indeed, two trajectories are obtained with opposite momentum). Also note that the initial value of the momentum, p_0, is fixed by the condition of the energy conservation—more specifically, it is $p_0 = \pm\sqrt{2m[E - V(q_0)]}$—and, therefore, needs not to be supplied initially. This constitutes an important reduction of the dimensionality of the problem in the sense that it will avoid it to solve two differential equations, as it happens within the Hamiltonian formulation. Accordingly, the trajectories arise from the integral

$$\int_{q_0=q(t_0)}^{q=q(t)} \frac{dq'}{\sqrt{2m[E - V(q')]}} = t - t_0, \tag{1.21}$$

where the left-hand side is linear with time. These trajectories are allowed to cross at the same time in the configuration where they are defined (e.g., two trajectories with opposite momentum, as given by (1.20)). However, they cannot cross in phase space, this being called a *congruence*.

The Lagrangian, Hamiltonian and Hamilton–Jacobi formulations of a given problem are all non-unique, as it has been extensively discussed in the literature. This becomes apparent through the so-called *inverse problem*, which consists of finding out the Lagrangian or Hamiltonian associated with the equations of motion describing a system. In principle, there is an infinite set of Lagrangians, Hamiltonians or classical actions that can be assigned to a dynamical system. Not all of them, of course, will be physically acceptable, since they may lead to singular or pathological behaviors, or they could violate some physical requirements. In spite of this, there is an equivalence among them, which arises from the fact that the equations of motion they generate are identical. In this regard, a set of Lagrangians or Hamiltonians are said to be *q-equivalent* if they generate the same equations of motion in the coordinate space—similarly, p and (q, p)-equivalences can also be defined.

The integrability (regularity) or non-integrability (chaoticity) of a dynamical system depends on the constant of motion characterizing such a system. The formal condition for a function A to be a constant of motion is

$$\frac{dA(q, p)}{dt} = \frac{\partial A}{\partial t} + \{A, H\} = 0, \tag{1.22}$$

with $\partial A/\partial t = 0$ (no explicit dependence on time) and where the Poisson bracket for any two functions A and B in phase space is defined as

$$\{A, B\} \equiv \sum_{i=1}^{N} \left(\frac{\partial A}{\partial q_i} \frac{\partial B}{\partial p_i} - \frac{\partial A}{\partial p_i} \frac{\partial B}{\partial q_i} \right). \tag{1.23}$$

In this way, for *integrable* or *regular systems* more than one constant of motion can be found apart from the total energy, while the latter is the only constant of motion in the case of *non-integrable* or *chaotic systems*. Thus, if more than one constant of motion exist, they will be independent, i.e., they cannot be expressed as a linear combination of the remaining ones. Furthermore, if the Poisson bracket of two constants of motion vanishes, they are *in involution*. If the number of constants of motion in involution is the same as the number of degrees of freedom describing the system, the corresponding trajectories will then be confined to an N-dimensional manifold or N-dimensional torus. This is the formal condition of integrability of a dynamical system. For this kind of systems, it is common to change to the so-called *action-angle variables*. The angle variables, usually denoted by $w = (w_1, \cdots , w_N)$, vary from 0 to 2π and play the role of position coordinates. Their corresponding canonically conjugated coordinates are the action variables, denoted by $I = (I_1, \cdots , I_N)$, and that play the role of a momentum with respect to w. Taking into account these new variables, Hamilton's equations of motion (1.8) read as

$$\dot{w}_i = \frac{\partial H}{\partial I_i} = \omega_i, \tag{1.24a}$$

$$\dot{I}_i = -\frac{\partial H}{\partial w_i} = 0. \tag{1.24b}$$

where ω_i are the frequencies characterizing the dynamical system. The complete solution of these equations is given by the quantities $w_i = \omega_i t + \phi_i$, with arbitrarily chosen phases ϕ_i. In general, the ratios between frequencies for a particular torus are irrational numbers. However, these ratios become rational numbers for special values of the constants of motion, giving rise to classical resonances. Because angle variables are periodic, with period 2π in each one of the angles, it is quite natural to make Fourier expansions of q and p as a function of w. The motion is thus called multiperiodic. Alternatively, this frequency analysis can also be obtained from the the power spectrum of the autocorrelation function of the dynamical variables (see Appendix B).

Periodic orbits, or solutions of the equations of motion with final positions and momenta coinciding with their initial counterparts, play a fundamental role in classical as well as quantum mechanics. They are densely distributed in phase space for a given mechanical system. The study of their stability is crucial to have a complete understanding of the underlying (regular or chaotic) dynamics. Furthermore, in general, a mechanical system can only be quantized if it is integrable, for there is no method to date to quantize non-integrable (chaotic) systems in a consistent manner—except for those cases where periodic unstable trajectories can be found

It is also worth mentioning that integrable systems are not always separable like, for example, the Toda lattice problem consisting of a one-dimensional lattice or chain of many particles coupled by nonlinear springs, the soliton being a special solution.

1.2.2 Classical Dynamics in Complex Phase Space

Usually, the physical properties of a dynamical system are studied by considering real-valued Hamiltonians (within the Hamilton formulation of classical mechanics), which are functions of a certain set of (also real-valued) coordinates and momenta. However, when dealing with microscopic processes, quantum particle dynamics will develop in regions of the configuration and/or phase space that can be either allowed or forbidden classically depending on the value of the (classical) total energy. For example, beyond the turning points where the total energy is equal to the potential energy function, particles may explore regions that are classically forbidden (i.e., unaccessible for a classical trajectory). In such cases, if one is interested in a classical description of the corresponding dynamics, an analytical continuation of the dynamical variables is required. Also, there are situations where a complexification of the Hamiltonian turns out to be more efficient to explain certain processes and phenomena in quantum mechanics. For example, complex potentials (optical models) have been used to account for the presence of resonance phenomena in nuclear, atomic or molecular scattering as well as in chemical reactions [26], to avoid boundary effects in wave-packet propagation methods (absorbing boundaries) [27], as the basis to build up semiclassical coherent-state propagation schemes [28, 29] or to analyze and compute complex eigenvalues [30–35]—see also the discussion in Sect. 6.2.4 in connection to the complexification of Bohmian mechanics. Recent interest has been generated in studying the classical dynamics and properties associated with complex Hamiltonians [36–41].

In brief, there are basically four routes (apart from their corresponding combinations) to reach a complex dynamics [41]:

1. Whenever absorbing boundary conditions are assumed, a complex form of the potential function is commonly applied, namely an *optical potential* (in analogy to the complex refraction index from optics [24]). This was first considered, for example, in problems involving nuclear scattering [42, 43]. In these cases, the potential reads as $V(q) = V_r(q) + i V_i(q)$, where V_r and V_i are real functions that rule, respectively, the system dynamics and its attenuation or the damping effects undergone by it. Since V_i acts as an *absorber*, this kind of potentials are typically employed in wave-packet propagation methods to avoid non-physical reflections at the boundaries of the corresponding numerical grids [27].

2. In the semiclassical coherent-state propagator method [28, 29], a complex formulation arises immediately when the change is usually considered within the coherent-state formulation of quantum mechanics [44] and identified with a change of classical variables. That is, the coherent state

$$|z\rangle = e^{-|z|^2/2} e^{z\hat{a}^\dagger} |0\rangle \tag{1.25}$$

is generated by the (quantum) Hamitonian $\hat{H} = \hat{p}/2m + m\omega^2\hat{q}/2$, with $\hat{a}^\dagger = (m\omega\hat{q} + i\hat{p})/\sqrt{2m\omega\hbar}$, $|0\rangle$ being the ground state of the harmonic oscillator and

$$z = \frac{1}{\sqrt{2m\omega\hbar}} (m\omega q + ip) . \tag{1.26}$$

Taking into account this latter relation, the classical dynamics described in terms of the two real variables (q, p) can be replaced by the complex variables (z, z^*), which gives rise to a complex (classical) dynamics. Actually, the semiclassical method leads to a dynamics where both q and p become complexified themselves. This allows us to define two new complex variables,

$$u = \frac{1}{\sqrt{2m\omega\hbar}}\,(m\omega q + ip)\,, \quad v = \frac{1}{\sqrt{2m\omega\hbar}}\,(m\omega q - ip)\,, \tag{1.27}$$

such that $u \neq v^*$, because q and p are complex. These new variables satisfy the (complex) Hamiltonian equations

$$\dot{u} = \frac{1}{i\hbar}\frac{\partial H}{\partial v}\,, \quad \dot{v} = -\frac{1}{i\hbar}\frac{\partial H}{\partial u}\,. \tag{1.28}$$

3. The relations described by (1.27) can be considered within another alternative complexification scheme, where the variables q and p defining the classical Hamiltonian are directly made complex by analytical continuation. That is, one assumes either a direct transformation to an expanded complex space,

$$q = q_1 + iq_2, \quad p = p_1 + ip_2, \tag{1.29}$$

or a sort of rotation in the complex phase space

$$q = q_1 + ip_2, \quad p = p_1 + iq_2. \tag{1.30}$$

In both cases, q and p are complex quantities, while q_1, q_2, p_1 and p_2 are real. Usually the type of transformation (1.29) is considered in problems involving the calculation of eigenstates of the Hamiltonian [30–35], while (1.30) is related to (or arises from) formulations based on the use of coherent states [28, 29, 38–41], as seen in (2).

4. Finally, a fourth type of complexifying a classical Hamiltonian is by just considering a complex parameter, such as the mass, angle, frequency or time (in the latter case there is a *Wick rotation*).

Although one can directly tackle the study of the classical dynamics associated with the complex Hamiltonians arising from these types of schemes, usually all of them come from some requirement (either theoretical or numerical) in quantum mechanics. Note that complex numbers are regarded as a signature proper of this theory (as well as any other wave theory).

In order to illustrate how the classical equations of motion are determined when dealing with complex variables, let us consider, for example, the change of variable (1.29). Consider also that the complexified Hamiltonian can be expressed as $H(q, p) = H_1(q_1, p_1, q_2, p_2) + i H_2(q_1, p_1, q_2, p_2)$ and is such that, before such an operation, Hamilton's equations hold, i.e.,

$$\dot{q} = \frac{\partial H}{\partial p}\,, \quad \dot{p} = -\frac{\partial H}{\partial q}\,. \tag{1.31}$$

Substituting H by its complex expression into (1.31) and then taking into account $\partial/\partial q = (\partial/\partial q_1 - i\partial/\partial q_2)/2$ and $\partial/\partial p = (\partial/\partial p_1 - i\partial/\partial p_2)/2$, one finds

$$\dot{q}_1 = \frac{1}{2}\frac{\partial H_1}{\partial p_1} + \frac{1}{2}\frac{\partial H_2}{\partial p_2}\,, \tag{1.32a}$$

$$\dot{q}_2 = \frac{1}{2}\frac{\partial H_2}{\partial p_1} - \frac{1}{2}\frac{\partial H_1}{\partial p_2}, \tag{1.32b}$$

$$\dot{p}_1 = -\frac{1}{2}\frac{\partial H_1}{\partial q_1} + \frac{1}{2}\frac{\partial H_2}{\partial q_2}, \tag{1.32c}$$

$$\dot{p}_2 = -\frac{1}{2}\frac{\partial H_2}{\partial q_1} + \frac{1}{2}\frac{\partial H_1}{\partial q_2}. \tag{1.32d}$$

If H is assumed to be an analytical function, it will then satisfy the Cauchy–Riemann equations, which in compact form reads as

$$\frac{\partial f}{\partial z^*} = 0, \tag{1.33}$$

where f is a complex analytical function depending on the also complex variable z. Formally, this means that f will only depend on z, but not on its conjugate complex, z^*. By applying (1.33) to our case, one finds the Cauchy–Riemann relations

$$\frac{\partial H}{\partial q^*} = 0 \quad \Rightarrow \quad \frac{\partial H_1}{\partial q_1} = \frac{\partial H_2}{\partial q_2}, \quad \frac{\partial H_2}{\partial q_1} = -\frac{\partial H_1}{\partial q_2}, \tag{1.34a}$$

$$\frac{\partial H}{\partial p^*} = 0 \quad \Rightarrow \quad \frac{\partial H_1}{\partial p_1} = \frac{\partial H_2}{\partial p_2}, \quad \frac{\partial H_2}{\partial p_1} = -\frac{\partial H_1}{\partial p_2}. \tag{1.34b}$$

After substitution of these relations into the equations of motion (1.32),

$$\dot{q}_1 = \frac{\partial H_1}{\partial p_1}, \quad \dot{p}_1 = -\frac{\partial H_1}{\partial q_1}, \tag{1.35a}$$

$$\dot{q}_2 = -\frac{\partial H_1}{\partial p_2}, \quad \dot{p}_2 = \frac{\partial H_1}{\partial q_2}. \tag{1.35b}$$

Notice here that, due to the analyticity of H, the time-evolution of these variables can be obtained by only taking into account the real part of H. According to (1.35), (q_1, p_1) and (q_2, p_2) form two pairs of conjugate variables with the formal structure of their equations of motion following that of Hamilton's equations (although the sign is changed in the second pair). Similarly, if (1.30) was considered instead of (1.29), then

$$\dot{q}_1 = \frac{\partial H_1}{\partial p_1}, \quad \dot{p}_1 = -\frac{\partial H_1}{\partial q_1}, \tag{1.36a}$$

$$\dot{q}_2 = \frac{\partial H_1}{\partial p_2}, \quad \dot{p}_2 = -\frac{\partial H_1}{\partial q_2}, \tag{1.36b}$$

with the Cauchy–Riemann equations being

$$\frac{\partial H_1}{\partial q_1} = \frac{\partial H_2}{\partial p_2}, \quad \frac{\partial H_2}{\partial q_1} = -\frac{\partial H_1}{\partial p_2}, \qquad (1.37a)$$

$$\frac{\partial H_1}{\partial p_1} = \frac{\partial H_2}{\partial q_2}, \quad \frac{\partial H_2}{\partial p_1} = -\frac{\partial H_1}{\partial q_2}, \qquad (1.37b)$$

As can be noticed, the sign of the Hamiltonian relations for (q_2, p_2) is now the appropriate one.

1.3 From Regular to Chaotic Dynamics

The concept of N-dimensional system is indistinctly used to refer either to a system described by N degrees of freedom or to a set of N independent (interacting or not) systems. In the first meaning one considers a single object described by many internal degrees of freedom (e.g., a molecular system), while in the latter each degree of freedom (or group of them) is related to an independent system (e.g., an interaction among several atomic or molecular systems). In either case, many times one is only interested in a subset of the N degrees of freedom, which are considered as the system or subsystem of interest, and the remaining ones, which constitute the *environment* or *bath*. This is, generally speaking, the starting point of the theory of open classical and quantum systems (see Chaps. 2 and 5).

In this regard, consider a Hamiltonian function describing the dynamics of an N-dimensional system,

$$H = \sum_{i=1}^{N} \frac{p_i^2}{2m_i} + V(\{q_i\}_{i=1}^{N}) + \sum_{j=1}^{K} V_{\text{ext}}^{(j)}(\{q_i\}_{i=1}^{N}), \qquad (1.38)$$

where V is the interaction potential coupling the different degrees of freedom (usually called the internal potential) and V_{ext} a certain external potential function acting on them. The evolution of the N degrees of freedom is given, in general, by a set of $2N$ nonlinear coupled differential equations, as seen above. Splitting the degrees of freedom into system and environment, one notes that the dynamical evolution of the former undergoes fluctuations or deviations with respect to its isolated dynamics due to the presence of the environment. Actually, under certain environmental conditions the system dynamics acquires *stochastic* features (see Chaps. 2 and 5). This type of stochasticity, which manifests as a seemingly erratic behavior in the individual evolution of each degree of freedom (or subsystem), arises from the intrinsic *chaotic* dynamics of a relatively complex associated Hamiltonian function [45]. However, this is different from the stochasticity produced by external noise sources, where the total dynamics is not conservative but *dissipative* (see Chaps. 2 and 5).

1.3.1 Deterministic Chaos and Intrinsic Stochasticity

Chaotic dynamics are primarily characterized by a high sensitivity to initial conditions. In general, they appear in dynamical systems described by nonlinear differential
equations. More specifically, according to the *Poincaré–Bendixon theorem* [46],
a system will display a chaotic dynamics if it is described by at least a set of
three autonomous (time-independent) coupled equations or two if the system is non-
autonomous (the third equation is supplied by the time-dependence). Hence, in the
case of an autonomous Hamiltonian system, it is necessary, at least, two degrees
of freedom with a nonseparable potential function. Usually, the sensitivity to initial
conditions manifests as an exponential growth of the distance between neighboring
trajectories, which is measured quantitatively by means of the so-called *Lyapunov
exponent*.

The phase space of a chaotic Hamiltonian system consists of regions of partial
integrability, characterized by the existence of tori and dynamical stability, with
interspersed regions of chaotic behavior [45]. This behavior has its origin in the
dynamical instability caused by the so-called homoclinic and heteroclinic inter-
sections of manifolds of unstable periodic motions [47–50]. The degree of chaos
(or irregularity) of a system can be varied by changing one parameter of the Hamiltonian, for example, some coefficient in the potential function or the total energy in the
case of a conservative system. This gives rise to regimes ranging from total integrability (no dynamical instability is present) to total hyperbolicity (all periodic motions
are unstable). The latter implies *ergodicity*, a situation that can be characterized as
an intrinsic random behavior.[2] The consequences of this intrinsic randomness are far
reaching. For example, in general, detailed descriptions of the system evolution will
not be practical and a statistical approach will be required. In other words, the evolution and relaxation towards equilibrium of certain average quantities will result more
(physically) meaningful than the individual behavior of a trajectory corresponding
to a given set of initial conditions. In this sense, chaos provides a natural justifica-
tion for the introduction of statistical ensembles. Furthermore, since chaos already
appears in two degrees of freedom Hamiltonian systems, statistical mechanics can
be justified even for small classical systems; the presence of many particles is not
a basic requirement for the foundation of statistical mechanics, in particular for the
existence of transport phenomena [51].

At a more quantitative level, system dynamics can be studied in terms of a set of
first-order differential equations with the form of flow equations [51],

$$\dot{\mathbf{x}} = \mathbf{F}(\mathbf{x}), \tag{1.39}$$

where $\mathbf{x} \equiv (q_1, q_2, \ldots, q_N, p_1, p_2, \ldots, p_N)$ is a point representing the state of the
system on its phase space and $\mathbf{F}$ specifies the flow equations. Equivalently, the system
trajectories (in phase space) can be described by a mapping transformation,

---

[2] The term *intrinsic* is used because this behavior arises from a set of deterministic equations of
motion with no need to introduce any external "environment" or fluctuations

$$\mathbf{x} = \mathbf{\Phi}^t(\mathbf{x}_0), \qquad\qquad (1.40)$$

where $\mathbf{\Phi}^t$ is the evolution rule, called the *flow*, which tells us where the initial points in phase space $\mathbf{x}_0$ have moved to at a time t. Usually, the mapping transformation $\mathbf{\Phi}^t$ is a nonlinear function of the initial conditions and time. The evolution of a volume of phase space is controlled by the Jacobian determinant of the transformation (1.40). Formally solving (1.39) or (1.40), this determinant will read as

$$|\det \partial_{\mathbf{x}} \mathbf{\Phi}^t| = e^{\int_0^t \nabla \cdot \mathbf{F} d\tau}, \qquad\qquad (1.41)$$

where $\partial_{\mathbf{x}}$ stands for the partial derivative with respect to $\mathbf{x}$. For Hamiltonian systems the phase-space volume is preserved. Accordingly, $\nabla \cdot \mathbf{F} = 0$ and the Jacobian (1.41) becomes equal to unity. Depending on how $\mathbf{x}$ is affected by the mapping transformation $\mathbf{\Phi}^t$, trajectories can be classified as:

- *Stationary*: if $\mathbf{\Phi}^t(\mathbf{x}) = \mathbf{x}$ for all t.
- *Periodic*: if $\mathbf{\Phi}^t(\mathbf{x}) = \mathbf{\Phi}^{t+T}(\mathbf{x})$ for a given minimum (finite) period T.
- *Aperiodic*: if $\mathbf{\Phi}^t(\mathbf{x}) \neq \mathbf{\Phi}^\tau(\mathbf{x})$ for all $t \neq \tau$.

Stationary points are usually equilibrium points of the potential function. Periodic trajectories can be stable or unstable, which means that nearby aperiodic trajectories will display quasiperiodic and chaotic behavior, respectively.

   A key element in the analysis of the degree of irregularity of a complex dynamical system is the structure of its phase space. This provides us with a qualitative view of the main features given a particular value of the Hamiltonian parameters (typically the total energy), such as equilibrium points and periodic motions. These elements are signatures of the presence of stability regions and chaos or the domain of initial conditions that will lead to chaotic and quasiperiodic behaviors as time evolves—for other more quantitative analyses of the degree of chaos, the calculation of global indicators, such as Lyapunov exponents or entropic measures, is required [51]. In the particular case of two degree-of-freedom conservative systems, the study of the phase space consists in analyzing the associated *Poincaré surface of section* (PSOS), which is a projection (mapping) of the total phase space on a two-dimensional subspace. More specifically, since the energy conserves, one of the variables is kept fixed at a constant value, say the coordinate of the complementary subspace. Each time the momentum of the complementary subspace satisfies a certain condition, the trajectory is recorded and plotted in the subspace considered (i.e., the corresponding coordinate and associated conjugate momentum). A full picture of the PSOS (which is a "reduced" view of the higher-dimensional phase space) is then obtained by sampling the whole phase space with different (properly chosen) initial conditions. In terms of flows, the Poincaré map can be defined as

$$\bar{\mathbf{x}}_{n+1} = \boldsymbol{\Phi}(\bar{\mathbf{x}}_n), \tag{1.42}$$

where $\bar{\mathbf{x}}$ represents the dynamical variables intrinsic to the PSOS. Hence periodic motions are seen as fixed points of the Poincaré map $\bar{\mathbf{x}}_n = \boldsymbol{\Phi}(\bar{\mathbf{x}}_n)$. Quasiperiodic trajectories will give rise to regular islands and chaotic trajectories to randomly distributed points on the PSOS.

The local stability of periodic motions is an important issue. To characterize the stability of an orbit or trajectory, namely the reference trajectory, given by (1.40), the evolution of trajectories starting at conditions which slightly deviate from it in small amounts $\delta\mathbf{x}$ are studied. Substituting the new trajectory $\mathbf{x}' = \mathbf{x} + \delta\mathbf{x}$ into (1.39) and then expanding to linear order in $\delta\mathbf{x}$ yields

$$\delta\dot{\mathbf{x}} = \partial_{\mathbf{x}}\mathbf{F}(\mathbf{x})\delta\mathbf{x}. \tag{1.43}$$

This is a system of linear equations with solutions

$$\delta\mathbf{x}_t = \partial_{\mathbf{x}_0}\boldsymbol{\Phi}^t(\mathbf{x}_0)\delta\mathbf{x}_0 = \mathbf{M}(\mathbf{x}_0, t)\delta\mathbf{x}_0, \tag{1.44}$$

where $\mathbf{M}(\mathbf{x}_0, t)$ is the *fundamental matrix*, which obeys the evolution equation

$$\dot{\mathbf{M}}(\mathbf{x}_0, t) = \partial_{\mathbf{x}_0}\mathbf{F}[\boldsymbol{\Phi}^t(\mathbf{x}_0)]\mathbf{M}(\mathbf{x}_0, t)\delta\mathbf{x}_0. \tag{1.45}$$

Within this context, this matrix is known as the *stability matrix*. If the trajectory $\mathbf{x}$ is periodic with period T, $\mathbf{M}(x_0, T)$ is also known as the *monodromy matrix*. The eigenvalues (λ_i) and eigenvectors of this matrix determine the local behavior of neighboring trajectories, since they describe the deformation of a neighborhood $\delta\mathbf{x}$ for a finite time t. Thus, nearby trajectories separate exponentially along unstable directions (given by the eigenvectors associated with the λ_i, with $|\lambda_i| > 1$,) approach each other along stable directions ($|\lambda_i| < 1$) or maintain their distance along marginal directions ($|\lambda_i| = 1$). Due to the symplectic structure of Hamilton's equations of motion (see Sect. 1.2.1), in Hamiltonian systems real eigenvalues come in pairs $(\lambda, 1/\lambda)$, where one corresponds to the unstable direction and the other to the stable one, while complex eigenvalues may appear in pairs (with $|\lambda| = 1$) or in quartets (i.e., $\lambda, 1/\lambda, \lambda^*, 1/\lambda^*$, but only for systems with more than two degrees of freedom). For example, in two-dimensional Hamiltonian systems, the Poincaré map reduces the monodromy matrix to a 2×2-matrix, which is characterized by a couple of eigenvalues. If both are real, the orbit is unstable; if they are conjugate complex, the orbit is stable (periodic).

The previous analysis refers to trajectory properties for a given value of the Hamiltonian parameters. However, one is often interested in obtaining information for a particular range of values of such parameters, which implies a parametric analysis of evolution of the phase-space structure. The Kolmogorov–Arnold–Moser (KAM) theorem [23, 47] gives a detailed account on the destruction of individual tori in phase space under perturbations. However, in order to obtain a global picture of the phase-space structure at any relevant energy, a good strategy is to follow the evolution with energy of the principal periodic orbits.[3] A suitable starting point consists in defining the main families of periodic orbits according to Weinstein's theorem [52], which in the vicinity of an equilibrium point of the potential guarantees the existence of as many periodic orbits as system degrees of freedom. There are several numerical techniques to locate periodic orbits, even when they are highly unstable [53, 54] (for example, those based on the *shooting* methods employed to solve ordinary differential equations with two-point boundary conditions [55]). Once a periodic orbit is located, the parameter is slightly changed in order to determine the orbit for the new value.

Let us express the eigenvalues of the monodromy matrix as $\lambda = \exp(\alpha T)$. Then, given a two-dimensional Hamiltonian system and a periodic orbit of period 1 (i.e., $T = 1$) on the Poincaré map, the monodromy matrix becomes

$$\mathbf{M}_1 = \begin{pmatrix} e^{\alpha_1} & 0 \\ 0 & e^{-\alpha_1} \end{pmatrix}. \tag{1.46}$$

The stability of the orbit can be inferred directly from the trace of this matrix as follows. If the eigenvalues are complex, $\mathrm{Tr}(\mathbf{M}_1) = 2\cos\sigma_1 \, (\alpha_1 = i\sigma_1)$; if they are real, $\mathrm{Tr}(\mathbf{M}_1) = 2\cosh\alpha_1$. Therefore, the orbit will be:

- *Stable*: if $|\mathrm{Tr}(\mathbf{M}_1)| \leq 2$,
- *Unstable*: if $|\mathrm{Tr}(\mathbf{M}_1)| > 2$.

This criterion is also valid if the periodic orbit is a fixed point of period n on the Poincaré map, although replacing $\mathbf{M}_1$ by $\mathbf{M}_n$. Then, making use of the property $\mathbf{M}_n = \mathbf{M}_1^n$, for stable fixed points one finds that the trace after n iterations of the map will be

$$\mathrm{Tr}(\mathbf{M}_n) = 2\cos(n\sigma_1) = 2\cos\left[n(\cos)^{-1}[\mathrm{Tr}(\mathbf{M}_1/2)]\right]. \tag{1.47}$$

The same result holds for unstable orbits, but replacing the cosines by hyperbolic cosines.

In order to understand now how periodic orbits appear, in general, notice that, as can be shown [56], when the stability matrix has an eigenvalue $\lambda = \pm 1$, (1.45) has a periodic solution. Thus, a periodic orbit of period n on the Poincaré map can only appear (or disappear, if it already existed) whenever

$$\mathrm{Tr}\,(\mathbf{M}_n) = 2. \tag{1.48}$$

This is called a *bifurcation*. Accordingly, from (1.47), fixed points of higher periods n can be obtained from the period-1 fixed point whenever the relation

$$\mathrm{Tr}\,(\mathbf{M}_1) = 2\cos(2\pi m/n) \tag{1.49}$$

---

[3] By "principal" orbits it is meant the simplest ones (i.e., those with smaller periods and the simplest topology in general), since periodic orbits of higher periods usually originate from them.

is satisfied, where m is an integer such that the cosine is modulo π. Unstable periodic orbits do not bifurcate (i.e., do not give rise to new ones), but may change their stability. In two-dimensional Hamiltonian systems there are only five types of bifurcations [57–59]. For example, the period-doubling bifurcation ($n = 2$, $m = 1$, $\mathrm{Tr}(\mathbf{M}_1) = -2$) changes the stability of the period-1 motion. In such a case, locating the most important periodic orbits of period 1 and following the evolution of the monodromy matrix with energy will allow us to obtain valuable information about how the phase space structure changes through this type of bifurcation.

In order to illustrate the theory here presented, in Fig. 1.1 several PSOSs are displayed, showing the gradual transition from regular to chaotic dynamics as a function of the total energy in the case of the well-known Hénon–Heyles Hamiltonian system [60],

$$ H = \frac{p_x^2}{2m_x} + \frac{p_y^2}{2m_y} + \frac{1}{2}m_x\omega_x^2 x^2 + \frac{1}{2}m_y\omega_x^2 y^2 + \lambda\left(x^2 y - \frac{y^3}{3}\right). \qquad (1.50) $$

The PSOSs displayed on the left-hand side panels are obtained by recording the (y, p_y) points of a set of trajectories each time they cross the plane $x = 0$ with $p_x > 0$; similarly, on the right hand side, the PSOSs correspond to the crossings (x, p_x) when $y = 0$ and $p_y > 0$. At low energies, the different crosses with the PSOS give rise to nearly circular distributions of points. If trajectories are periodic, they will cross the corresponding PSOS a finite number of times and, therefore, only a finite number of points associated with it will be observed. On the contrary, quasiperiodic trajectories densely cover phase space regions and, therefore, give rise to close orbits in the PSOS as $t \to \infty$—although the closer they are to a periodic trajectory, the slower they cover the phase space orbit. At higher energies, according to the KAM theorem, the tori "break down" and orbits appear as sets of points scattered throughout the PSOS, which enclose chains of stability region or islands. Actually, in the lower panels (for energies close to the onset of dissociation), chains of islands can be seen.

Chaos plays a fundamental role in areas such as non-equilibrium statistical mechanics, where it provides a connection between the irreversible phenomenological macroscopic equations and the reversible Hamiltonian equations [61]. Moreover, deterministic chaos can also induce transport mechanisms not considered in conventional statistical mechanics. For example, the possibility of anomalous transport, i.e., mean square displacements (MSDs) growing faster or slower than linear in time, thus implying a violation of Einstein's diffusion law.

One of the relevant questions when talking about transport properties is that of the decay of correlation functions with time. Also here dynamical instabilities rule the intrinsic relaxation times of correlation functions, allowing one to obtain transport coefficients from them. Time correlation functions are important from an experimental viewpoint, since the spectra measured by means of the different spectroscopic techniques are related to the power spectra of well-defined dynamical variables (see Appendix B). In general, spectral functions contain information on the system frequencies. The corresponding spectrum can be discrete or continuum. Systems characterized by a discrete spectrum of real frequencies present almost-periodic

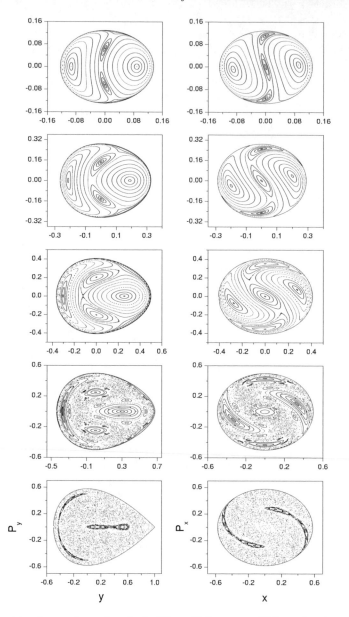

Fig. 1.1 Transition from order to chaos in the Hénon–Heiles system [60] as a function of the total energy E. From top to bottom, $E = 1/120$, $E = 1/24$, $E = 1/12$, $E = 1/8$ and $E = 1/6$, the latter being the *onset of dissociation*. The parameters considered are $m_x = m_y = \omega_x = \omega_y = \lambda = 1$

oscillations, while continuum spectra indicate systematic decays [51]. When spectral functions are allowed to be analytically continued towards complex frequencies, one talks about resonances. The real part of these frequencies thus gives information about oscillation periods and its imaginary part about relaxation rates. This relaxation process can be no longer exponential. More formally, these resonances are defined as the poles of spectral functions and characterize transient behaviors in the time evolution. In nonlinear dynamics, these classical resonances are called Pollicott–Ruelle resoances [51].

1.3.2 Random Walks and Lévy Flights

Chaotic dynamics induce some intrinsic randomness in the system. This randomness is generated by the deterministic equation of motion, with no need to introduce any external fluctuation. In this case, it makes sense to seek for a statistical description of the dynamics. The study of the evolution of trajectory ensembles is then carried out by means of the probability conservation principle, which takes the form of a continuity equation.

If the phase space is sufficiently chaotic, one expects typical diffusive behaviors: at very long times, chaotic trajectories may behave similarly to random walks due to their residence times around stability islands.To illustrate this behavior, consider, for example, the adsorbate diffusion on metal surfaces [62]. Here, adsorbates undergo short jumps or flights between different cells of the substrate combined with short-time intracell residence or localization. Average jump lengths between sites and intracell mean waiting times are then finite. Thus, in agreement with Einstein's relation,

$$\langle |\mathbf{x}(t) - \mathbf{x}(0)|^2 \rangle \sim Dt, \tag{1.51}$$

this gives rise to MSDs that depend linearly with time. Physically, this means that it is not necessary to describe in a very detailed way the deterministic evolution of the adsorbates, but the effect induced by the surface on their positions (or velocities) allows us to accounted for them as Gaussian stochastic variables (see Appendix B). As a consequence, particles exhibit a *Brownian motion*.

Transport processes for which the associated MSDs violate Einstein's relation (1.51)—processes where particles do not undergo a Brownian motion—are generically called *anomalous diffusion processes*. In such cases, (1.51) is replaced by

$$\langle |\mathbf{x}(t) - \mathbf{x}(0)|^2 \rangle \sim D_\alpha t^\alpha. \tag{1.52}$$

If $\alpha < 1$, the diffusion process is slower than ordinary Brownian motion and the corresponding regime is called *subdiffusion*. On the contrary, for $\alpha > 1$, the diffusion process is faster than a Brownian motion and the regime is known as *enhanced diffusion* or *superdiffusion*. Since Richardson's work on turbulence in 1926 [63], anomalous diffusion and transport have been described in many different statistical frameworks. Simple, non-Brownian random-walk models accounted well for the first observations of anomalous diffusion. These models also provide an intuitive physical picture of such processes: superdiffusion is originated by anomalously long jumps of a random walker, namely a *Lévy walk*, while subdiffusion is associated with unusually long waiting times between successive walks. Subdiffusive processes are usually modelled by a continuous-time random walk (CTRW) with a fractal distribution of waiting times and, therefore, are also called "fractal time" processes. Nevertheless, both anomalous regimes can be described within the CTRW formalism, the existence of Lévy or stable probability distribution functions being central to the explanation of general anomalous diffusion processes (see Appendix B). Additional statistical frameworks employed to model anomalous transport include descriptions based on the Langevin and the Fokker–Planck equations (see Sect. 2.3) via fractional derivatives, generalized thermostatistics or combined approaches.

As it was stated above, chaotic dynamics can mimic the behavior displayed by stochastic systems with no need to introduce any external noise source. It has been widely shown that an unbound deterministic dynamical system fulfilling the ergodic property can be described by a diffusion equation, thus exhibiting a normal diffusive behavior. However, dynamical systems with mixed phase spaces (i.e., with coexistence of regular and chaotic regions) are more interesting: inside a chaotic region the ergodic property is expected to be approximately valid, although the existence of stability islands may subtly change the statistical properties of the system. In particular, the island structure can also induce anomalous transport under certain circumstances. For example, it has been shown [62] that the existence of Lévy distributions of jump lengths is crucial to explain anomalous diffusion in two-dimensional Hamiltonian systems.

The concept of *Lévy flight* [64] is currently well established and widely used in the physics of Lévy walks and Lévy statistics. It is used to indicate a random walk in a continuous N-dimensional space displaying a stable or Lévy distribution of jump lengths and a finite average time between jumps. More specifically, consider a random walk described by the jump probability distribution function

$$P(x) = \frac{\lambda - 1}{2\lambda} \sum_{j=0}^{\infty} \lambda^{-j} \left[\delta(x - b^j) + \delta(x + b^j) \right], \qquad (1.53)$$

where $b > \lambda > 1$. According to (1.53), the probability to find a flight or jump of length b^j (backwards or forward) is λ^{-j}, which decreases rapidly with the length of the jump. That is, on average there will be λ jumps of length b^j before observing a jump of length b^{j+1}, this giving rise to a *self-similar* or *fractal* pattern of walks [65]. Due to the finiteness of the mean waiting times as well as the randomness in their distribution, Lévy processes are Markovian in nature. A general feature of Lévy distributions is that the corresponding MSD diverges, which arises from the fact that long jumps are considered to be instantaneous. Obviously, in the diffusion of a massive particle through space, the velocity cannot be infinite. These nonphysical flights are then replaced by Lévy walks, where one takes into account the time needed to complete each jump of the random walk. Consequently, even when the average jump distance is infinite, the MSD after a time t will follow an algebraic dependence on time. In Fig. 1.2, it is observed a Lévy walk typical of Na-atom diffusion on a corrugated Cu(001) surface [62].

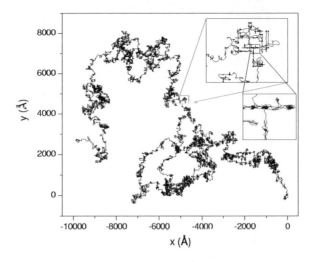

Fig. 1.2 In surface diffusion processes, Brownian-like motions arisedue to thermal vibrations. In the figure, trajectory pursued by a Na atom on a Cu(001) surface at $T = 300$ K. Due to the presence of the surface, the motion is of the type of *Lévy walk*, although at times much larger than the inverse of the surface friction constant, it becomes Brownian-like [62]. In the progressive enlargements, the *clustering* structure typical of this type of motion is observed, which is due to the combination of residence periods within a surface potential well and flights of arbitrary length between cells. Here, the substrate friction constant is $\gamma = 0.5\,\mathrm{ps}^{-1}$ and the evolution is up to $t = 20,000\,\mathrm{ps}$

1.4 Classical Particle Ensembles

1.4.1 Scattering Singularities

In many physical applications one is interested in the behavior of an ensemble or beam of particles after a scattering process. A detailed analysis of scattering singularities provides us with a very rich information of the phenomenon under study as well as the nature of the interactions involved. Originally, due to its simplicity, this kind of analysis was carried out for central force fields. The measurable quantities are mainly derived from a fundamental function, the so-called *scattering cross section*. This gives the scattering probability of an initial particle beam through a transversal section of the beam or, in other words, the number of particles scattered into a solid angle per unit time divided by the incident flux. Usually, this function depends on the final dispersion or scattering angle and the impact parameter, which is defined as the perpendicular distance between the center of force and the beam incidence velocity. It is well-known from standard scattering theory [10] that the cross section displays two types of singularities or zeros in its denominator, which give rise to the rainbow and glory effects. The *rainbow effect* is related to the extrema of the classical deflection function (defined as the final scattering angle versus impact parameter), while the *glory effect* appears when forward or backward scattering takes place.

The rainbow effect is closely related to caustics or accumulation of trajectories at rainbow angles. In optics, the same behavior is found, but in terms of light rays. An appropriate analysis of caustics should be carried out by appealing to the catastrophe theory [66, 67]. Furthermore, *orbiting singularities* can also be observed if some scattering particle is temporarily captured by the central force field.

If a periodic force field is considered (for example, the scattering of a particle beam by a corrugated periodic surface), apart from the rainbow and glory effects, a new singularity has been characterized: the so-called *skipping singularity* [67]. The classical image of this effect is that of skipping stones on the surface of a river. In order to make a stone to skip it is necessary to throw it with the correct incident angle to reach certain impact points on the water surface. Interestingly enough, this singularity has been shown to be responsible for the onset of chaos in this type of multiple scattering since memory of initial conditions is completely lost. In particular, when the rainbow angle reaches $\pi/2$ degrees (motion parallel to the surface), chaotic dynamics start developing.

1.4.2 Liouvillian Dynamics

Classical statistical mechanics was originally formulated by Boltzmann, Maxwell and Gibbs about 200 years ago. It deals with both equilibrium and non-equilibrium systems of distinguishable interacting particles [68–70]. In this theory, where motion is deterministic though unpredictable, natural or spontaneous fluctuations are analyzed. These fluctuations are defined as any deviation undergone by a dynamical variable with respect to its mean value and constitute a fingerprint of complexity or, in other words, a signature of the lack of knowledge or ignorance on the corresponding physical system. This implies that the evaluation of such fluctuations at two different times will render information about the relaxation processes involved and, therefore, leads us to a better understanding of the physical system.

In the limit of high temperature and low density, the behavior of any physical system consisting of N microscopic particles (or, in general, degrees of freedom; see discussion in Sect. 1.3) becomes classical. This can be expressed in a more formal way as follows [68]. If $\Gamma = \hbar\sqrt{2\pi\beta/m}$ is the particle mean thermal de Broglie wavelength and $n = N/V$ is the density of the N particles confined in a volume V, the so-called *degeneracy discriminant* $n\Gamma^3$ turns out to be a very convenient parameter to classify the physical systems as classical or quantum. When $n\Gamma^3 \ll 1$ or goes to zero, the system exhibits a classical behavior; when the degeneracy discriminant approaches unity, the system of indistinguishable particles exhibits quantum or strongly degenerate behavior. Furthermore, the smaller the particle mass, the larger the quantum effects.

Consider now a classical system containing N interacting particles. In statistical mechanics, the positions of these particles are usually specified by $3N$ coordinates, $q_1, \ldots, q_{3N}$, with their corresponding conjugate momenta being $p_1, \ldots, p_{3N}$. According to Gibbs [71], one can construct an associated Euclidean space of $6N$

dimensions, the so-called *phase space*. The dynamical state of the system at a time t is thus given by a point, namely the *phase point*, on the phase space. The trajectory described by the phase point is ruled by $6N$ Hamilton equations of motion and can be denoted generically as $\{q(t), p(t)\}$ (see Sect. 1.2.2), being a mapping between the initial state $\{q(0), p(0)\}$ and the final state $\{q(t), p(t)\}$.

In general, there is a large number of phase points compatible with the information available about the system (total energy, volume, etc.) and, according to Gibbs, the set of all such points constitutes an ensemble of systems. Since the number of systems in a given ensemble is very dense, it is possible to define a density of phase points or distribution function, $f(q, p, t)$. This function represents the fraction of phase points contained within the volume $dqdp$ at a given time t, and satisfies the normalization condition

$$\int f(q, p, t)dqdp = 1. \tag{1.54}$$

By applying the time-derivative operator, d/dt, to (1.54),

$$\frac{df}{dt} = 0, \tag{1.55}$$

which is an alternative way to enunciate the principle of conservation of phase density for Hamiltonian systems. Taking into account Hamilton's equations of motion, it is straightforward to derive from this latter equation the most fundamental equation of classical statistical mechanics, namely the *Liouville equation*,

$$\frac{\partial f}{\partial t} + \{f, H\} = 0. \tag{1.56}$$

This equation can also be written in the more standard form as

$$\frac{\partial f}{\partial t} = -i\mathcal{L}f, \tag{1.57}$$

where $\mathcal{L}f = i\{H, f\}$ is the *Liouville operator*, defined in terms of the Poisson bracket (1.23). The form of the Liouville equation arises from the fact that for Hamiltonian systems the phase-space volume conserves, i.e., $\nabla \cdot \mathbf{u} = 0$, where $\mathbf{u} \equiv (\dot{q}_1, \dot{q}_2, \cdots, \dot{q}_{3N}, \dot{p}_1, \dot{p}_2, \cdots, \dot{p}_{3N})$ is the generalized velocity in phase space. In other words, the phase-space flow is *incompressible*. However, in general, the conservation equation of phase points is described by a *generalized continuity equation*,

$$\frac{\partial f}{\partial t} + \nabla \cdot (\mathbf{u} f) = 0. \tag{1.58}$$

This equation can be recast as a generalized Liouville equation,

$$\frac{\partial f}{\partial t} = \mathcal{L}f, \tag{1.59}$$

where $\mathcal{L}f \equiv -\nabla \cdot (\mathbf{u}f)$. For Hamiltonian systems, (1.59) reduces to the usual operator that appears in (1.57) in terms of the Poisson bracket. Equation (1.55) is equivalent to

$$f(q, p, t)dqdp = f(q_0, p_0, t_0)dq_0dp_0, \tag{1.60}$$

where $q_0 = q(t_0)$ and $p_0 = p(t_0)$. This is an alternative way to see that both the density and the number of phase points in a given volume element remain the same at any time and, therefore, the cloud of phase points behaves as an incompressible fluid—the Jacobian between the final and initial points is unitary. Actually, the point (q, p) can be considered as a function or mapping of the initial point (see Sect. 1.3.1), i.e.,

$$q = q(q_0, p_0, t), \quad p = p(q_0, p_0, t). \tag{1.61}$$

Given a Hamiltonian, the corresponding trajectories can never cross in phase space due to the uniqueness of the classical equations of motion, as seen in Sect. 1.2.1. This classical *non-crossing rule* is comparable to the one arising in Bohmian mechanics (see Sect. 6.2.1), although the latter is more restrictive in the sense that it directly applies on the configuration space defined by q.

The Liouville equation can be seen as the infinitesimal generator of the group of time translations induced in the phase space of the probability densities. In this way, this equation is brought up into the form of a Schrödinger-like equation, equivalent to the $6N$ Hamilton equations of motion associated with the N-body system. The formal solution of this equation is, therefore,

$$f(q, p, t) = e^{-i\mathcal{L}(t-t_0)} f(q_0, p_0, t_0), \tag{1.62}$$

where the operator $e^{-i\mathcal{L}t}$ propagates f ahead a distance $t - t_0$ in time. Since (1.57) is an eigenvalue equation, the solution (1.62) can also be expressed [72–75] as

$$f(q, p, t) = \sum_{\lambda',\alpha'} c_{\lambda',\alpha'} f^{\lambda',\alpha'}(q, p)e^{-i\lambda't}, \tag{1.63}$$

where λ' is an eigenvalue, α' its degeneracy, $f^{\lambda',\alpha'}$ the corresponding eigendistribution and

$$c_{\lambda',\alpha'} = \int f_0(q_0, p_0)[f^{\lambda',\alpha'}(q_0, p_0)]^* dq_0 dp_0 \tag{1.64}$$

its coefficient, and $f_0(q_0, p_0)$ any classical initial distribution function. The formal comparison between the solutions of the quantum and classical Liouville equations may lead to think that the superposition principle also holds classically. Some superpositions of eigendistributions for different eigenvalues of the Liouville operator describing the classical harmonic oscillator can be found in [72]. In those cases where one is only interested in the distribution of some of the degrees of freedom, reduced distribution functions can be easily obtained by integrating over all those phase coordinates which are not of interest. In this regard, the radial distribution functions from the theory of liquids constitute a good example.

Once distribution functions and their respective equations of motion in phase space are obtained, the next step consists of defining the ensemble average of a dynamical variable $A(q, p, t)$ to be

$$\langle A(t) \rangle = \int A(q, p, t) f(q, p, t) dq dp. \tag{1.65}$$

Usually, the thermodynamic properties of a system are expressed as ensemble averages of certain functions of the coordinates and momenta of the constituent particles. At equilibrium, such averages become independent of time, i.e., $\langle A(t) \rangle = \langle A \rangle$. The dynamical variables of interest are often functions of the coordinates and momenta of just a few particles. This leads to the use of reduced distribution functions, with the corresponding equations of motion commonly given in terms of a hierarchy known as the Bogoliubov–Born–Green–Kirkwood–Yvon hierarchy [76–80].

Equilibrium distribution functions are usually defined in terms of the macroscopic parameters chosen to characterize the ensemble. For a microcanonical ensemble (fixed number of particles, volume and total energy E_0),

$$f_0(q, p) = C\delta(H(q, p) - E_0), \tag{1.66}$$

where δ is the Dirac δ-function and C is a normalization constant. The constraint of energy conservation allows us to define time-averages over the dynamical history of the system. Microcanonical ensembles and time averages are identical if the system is *ergodic*. This means that after a large lapse of time the system has visited the whole phase region compatible with (1.66)—time-averages can be evaluated in molecular dynamics simulations, for example. In a similar vein, canonical ensembles are characterized by the same values of number of particles, volume and temperature. The requirement of thermal equilibrium allows us to define energy averages as

$$\langle E \rangle = \frac{1}{Z} \int H e^{-\beta H} dq dp, \tag{1.67}$$

where the classical, canonical partition function can be expressed as

$$Z = \frac{1}{N!h^{6N}} \int e^{-\beta H} dq dp, \qquad (1.68)$$

with h being Planck's constant and $\beta = 1/k_B T$ (k_B is the Boltzmann constant); the factor h^{6N} in (1.68) ensures dimensionless quantities and goes over correctly to the corresponding quantum statistical values.

To deal with nonequilibrium properties, in the 1950s Green and Kubo developed the time-correlation function (TCF) formalism [81]. By definition, the classical TCF of two dynamical variables A and B is

$$C(t) = \langle A(0)B(t)\rangle = \int A(q, p, 0) B(q, p, t) f(q, p) dq dp, \qquad (1.69)$$

where $f(q, p)$ is the equilibrium phase space distribution function (for vectorial variables, a scalar product is used). If $A = B$, $C(t)$ is the so-called autocorrelation function. The time Fourier transform of $C(t)$ gives a generalized susceptibility, where the thermal transport coefficients correspond to its zero–frequency limit. TCFs may depend on space and, in general, describe the thermal fluctuations occurring spontaneously in systems at equilibrium as well as the response of a system to a weak, external perturbation (*linear response theory*). Unlike partition functions, many TCFs can be measured directly [82]. Alternatively, $C(t)$ can also be defined in terms of fluctuations by replacing A and B by $\delta A(0) = A(0) - \langle A \rangle$ and $\delta B(t) = B(t) - \langle B \rangle$. Not far from equilibrium, relaxation processes are governed by a principle first enunciated by Onsager in 1930, the so-called *regression hypothesis*. This hypothesis states that the relaxation of macroscopic non-equilibrium disturbances is governed by the same laws as the regression of spontaneous microscopic fluctuations in an equilibrium system. This is a consequence of the well-known *fluctuation–dissipation theorem* that Callen and Welton proved in 1951. The linear response theory is precisely based on this theorem.

The linear response theory describes the changes that a small external field induces on the macroscopic properties of a system at equilibrium. Consider such an external field is turned on at some initial time t_0 and it is treated as a perturbation. The total Hamiltonian describing this effect is expressed as a sum of two terms, $H = H_0 + H_1$, where the zeroth–order Hamiltonian, H_0, describes the system equilibrium state and the interaction Hamiltonian, H_1, introduces a time-dependence. The phase–space density can then be expressed as a power series expansion of the perturbation,

$$f = f_0 + f_1 + \cdots . \qquad (1.70)$$

Substituting this expression into the Liouville equation and separating terms at different orders, the first two orders render

$$\frac{\partial f_0}{\partial t} + \{f_0, H_0\} = 0, \qquad (1.71a)$$

$$\frac{\partial f_1}{\partial t} + \mathcal{L} f_1 = -\{f_0, H_1\}, \tag{1.71b}$$

whenever the zeroth–order equation can be solved or f_0 is known (typically f_0 is the canonical distribution function). Equation (1.71b) can be formally integrated in terms of the zeroth–order Liouville operator, $\mathcal{L}_0(\cdot) = i\{H_0, \cdot\}$, to yield

$$f_1(t) = e^{(t-t_0)\mathcal{L}_0} f_1(t_0) - \int_{t_0}^{t} e^{-(t-t')\mathcal{L}_0} \{f_0, H_1(t')\} dt', \tag{1.72}$$

where it is usually assumed $f_1(t_0) = 0$. This equation can be considered a precursor of the generalized master equation that appears in quantum mechanics (see Chap. 5).

The Boltzmann equation and one of its main direct consequences, namely the so-called *H-theorem* (and *irreversibility*), are far from the scope of this monograph, so they will not be treated here. However, the interested reader can consult some of the references given throughout this Section, where these issues are discussed in more detail.

1.4.3 Classical Statistical Mechanics as a Field Theory

As will be seen in Chap. 6, Bohmian mechanics [83–85] is based on the possibility to reformulate standard quantum mechanics in terms of a quantum Hamilton–Jacobi equation, which gives rise to a hydrodynamic description of quantum systems [86]. In classical mechanics there is also the possibility to proceed in the opposite way, i.e., to describe the evolution of an ensemble of identical, non-interacting classical particles with mass m under the influence of an external potential V by means of a "classical" Schrödinger equation. Note that, given an external potential V and a total energy E, there is an infinite number of associated classical trajectories—as many as initial conditions (q_0, p_0) consistent with the condition $H(q_0, p_0) = E$. Therefore, as seen in Sect. 1.3, a single trajectory is meaningless; valuable information about the system can only be extracted when they are considered statistically, carrying out an appropriate sampling over initial conditions.

Consider that such an ensemble is described by a certain distribution function $f(q, p, t) \equiv \rho_{cl}(q, p, t)$, while the individual evolution of each particle is ruled by (1.12a). The individual and ensemble dynamics are therefore determined by (1.15) and (1.58), respectively, where $\mathbf{u} = \nabla S/m$. It can be noticed that the system dynamics is independent of the particular choice of the initial ensemble distribution, ρ_0. However, the subsequent evolution of ρ_{cl} will depend on the system (particle) dynamics through ∇S in (1.58). Within this formulation, classical statistical mechanics can then be interpreted as a field theory. Actually, if $\mathbf{u}$ and ρ_{cl} are assigned to the velocity field and density of an incompressible and non-rotational fluid respectively, the formalism can be tackled from a hydrodynamic viewpoint. As can be noticed, by applying the ∇-operator to (1.15) and then using (1.12a), an Euler equation is reached where the velocity field does not depend on the fluid density,

$$\frac{\partial \mathbf{u}}{\partial t} + (\mathbf{u} \cdot \nabla)\mathbf{u} = -\frac{\nabla V}{m}. \tag{1.73}$$

The classical Schrödinger equation can be now derived having in mind that the problem just described can also be formulated considering an associated *Lagrangian density* [85]. As is well known in theoretical mechanics when dealing with fields or waves [8, 10], a Lagrangian L can be understood as the space integral of a certain Lagrangian density $\mathcal{L}$. The corresponding Euler–Lagrange equations, which are more general than (1.3), are

$$\frac{\partial \mathcal{L}}{\partial \varphi_i} - \sum_{k=1}^{N} \frac{\partial}{\partial \tau_k} \frac{\partial \mathcal{L}}{\partial (\partial \varphi_i / \partial \tau_k)} = 0, \tag{1.74}$$

where φ_i represents a field variable depending on N independent parameters τ_k. Consider that the τ_k, with $k = 1, 2, 3$, represent the three space coordinates, $(q_1, q_2, q_3,)$ while $\tau_4 = t$. The only two fields of interest are ρ_{cl} and S. Observe that, unlike the previous Lagrangian description, the solutions obtained from (1.74) are fields associated with the motion of a set of classical particles, but not the trajectories of these particles themselves. This is more apparent when the classical Lagrangian density is considered [85],

$$\mathcal{L} = -\left[\frac{\partial S}{\partial t} + \frac{(\nabla S)^2}{2m} + V\right]\rho_{cl} = -\frac{dS}{dt}\rho_{cl} + \left[\frac{(\nabla S)^2}{2m} - V\right]\rho_{cl}. \tag{1.75}$$

Although the equations of motion for individual particles cannot be directly derived from this Lagrangian density, the information to obtain them is already implicitly given. However, note that if the fields are evaluated along a classical trajectory, the right–hand side of either equality vanishes and $\mathcal{L} = 0$. As seen in Sect. 1.2.2, the term between brackets in the first equality corresponds to the Hamilton–Jacobi equation (1.15). In the second equality, on the other hand, the particle Lagrangian (1.2), with $\dot{q} = \nabla S/m$, is being both added and substracted. This extremal condition appears when applying the Euler–Lagrange equation (1.74) to (1.75) with respect to the field S. If, on the contrary, it is applied with respect to ρ_{cl}, the continuity equation (1.58) is obtained, which here describes a swarm of single, non-interacting particles—each one evolving according to (1.15)—and can be recast as

$$\frac{\partial \rho_{cl}}{\partial t} + \nabla \cdot \mathbf{J}_{cl} = 0. \tag{1.76}$$

Here, $\mathbf{J}_{cl} \equiv \rho_{cl} \nabla S$ is the current density associated with the flow of the density distribution function and, therefore, it will provide information about the flow of classical particles (distributed at any time according to ρ_{cl}). Obviously, since the Hamilton–Jacobi formalism is being considered, ρ_{cl} is not described in terms of the phase–space variables (q, p), but in terms of q and ∇S, this being the reason why (1.76) does not depend directly on p, as the Liouville equation (1.58).

In this regard, note that the evolution of trajectories does not depend on the evolution of the density distribution function, which might be inferred from the fact that (1.75) is separable (factorizable) into two parts, one depending on S and the other one on ρ_{cl}. Therefore, here ρ_{cl} only specifies the way how the initial conditions for a set of individual trajectories are chosen (sampled). In this sense, the relationship between ρ_{cl} and the trajectories is statistical rather than dynamical, differently from what can be found in Bohmian mechanics (see Sect. 6.2.1).

Further proceeding within this description and expressing the information encoded by S (dynamics) and ρ_{cl} (statistics) in a more compact form by defining a "classical" wave function,

$$\Psi_{cl} = \rho_{cl}^{1/2} e^{iS/\hbar}. \tag{1.77}$$

If this form and its conjugate complex, Ψ_{cl}^*, are substituted into (1.75), the Lagrangian density (1.75) becomes

$$\mathcal{L} = i\hbar \left(\Psi_{cl}^* \frac{\partial \Psi_{cl}}{\partial t} - \Psi_{cl} \frac{\partial \Psi_{cl}^*}{\partial t} \right) + \frac{\hbar^2}{8m} \left(\frac{\nabla \Psi_{cl}}{\Psi_{cl}} - \frac{\nabla \Psi_{cl}^*}{\Psi_{cl}^*} \right)^2 |\Psi_{cl}|^2 - V |\Psi_{cl}|^2. \tag{1.78}$$

Applying now variations to (1.78) with respect to Ψ_{cl}^* gives us the Euler–Lagrange equation

$$i\hbar \frac{\partial \Psi_{cl}}{\partial t} = -\frac{\hbar^2}{2m} \nabla^2 \Psi_{cl} + V \Psi_{cl} + \frac{\hbar^2}{2m} \frac{\nabla^2 |\Psi_{cl}|}{|\Psi_{cl}|} \Psi_{cl}, \tag{1.79}$$

which is the classical analogous of the Schrödinger equation [87–90]. Operating with respect to Ψ_{cl}, the (classical) conjugate complex Schrödinger equation is similarly obtained. Unlike its quantum counterpart, the solutions of (1.79) do not satisfy the superposition principle, since the last term is nonlinear in Ψ_{cl}—actually, it has some similarity to the quantum potential (see Sect. 6.2.1)—, which is necessary in order to suppress quantum-like effects in classical mechanics.

1.4.4 Superposition of Classical Distribution Densities

Due to the nonlinearity of the classical Schrödinger equation, the superposition principle does not hold for classical wave functions, but for density distribution functions, since they satisfy the Liouville equation, which is linear (see Sect. 1.4.2). Nonetheless, in order to compare later on with quantum and Bohmian mechanics, consider the problem of an ensemble of identical, non-interacting free classical particles with mass m in one dimension. Moreover, it will also be assumed that their velocity is the same (v) and constant. This is equivalent to say that their energy is the same, since $V = 0$. Essentially, this ensemble is described by a *microcanonical* distribution function. Taking this into account, (1.76) can be expressed in one dimension as

$$\frac{\partial \rho_{cl}}{\partial t} = -v \frac{\partial \rho_{cl}}{\partial x}. \tag{1.80}$$

This is the Euler equation corresponding to an incompressible flow. In order to solve (1.80), it is possible to switch from ρ_{cl} to its Fourier transform or spectral decomposition,

$$\rho_{cl}(x, t) = \frac{1}{\sqrt{2\pi}} \int \tilde{\rho}_{cl}(\alpha, t) e^{-i\alpha x} d\alpha. \tag{1.81}$$

Substituting (1.81) into (1.80) yields

$$\frac{\partial \tilde{\rho}_{cl}}{\partial t} = iv\alpha \tilde{\rho}_{cl}, \tag{1.82}$$

which has the solution

$$\tilde{\rho}_{cl}(\alpha, t) = \tilde{\rho}_{cl}(\alpha, 0) e^{i\alpha vt}. \tag{1.83}$$

Introducing this value in the integrand of (1.81) and rearranging terms, it is found that

$$\rho_{cl}(x, t) = \rho_{cl}(x - vt). \tag{1.84}$$

Physically, this result means that the density distribution function describing an incompressible, free fluid does not spread, but preserves its initial shape along time: it is *dispersionless*. This is a direct (rather obvious) consequence of the fact that all particles have the same energy.

With respect to the dynamics, taking advantage of the fact that v is a constant, (1.15) can be straightforwardly integrated by parts. This yields

$$S(x, t) = S'(x) - Et. \tag{1.85}$$

Proceeding similarly, but starting from (1.12a),

$$S(x, t) = S''(t) + mvx. \tag{1.86}$$

Finally, putting together the results (1.85) and (1.86) gives rise to

$$S(x, t) = S_0 + mv(x - vt/2) = S_0 + p(x - pt/2m), \tag{1.87}$$

where $p = mv$ is the particle momentum. If this expression is substituted into (1.12a) and then the corresponding equation is integrated,

$$x(t) = x_0 + vt. \tag{1.88}$$

A swarm of particles moving as (1.88), with randomly chosen initial positions (according to $\rho_{cl}(x, 0)$) and the same velocity v, will evolve in time according to

Fig. 1.3 Classical trajectories distributed according to (1.89). Trajectories associated with each density distribution function are plotted with different colors to distinguish them. As can be seen, they can pass through the same point at the same time because classical dynamics does not exclude this possibility in configuration space, but only in phase space

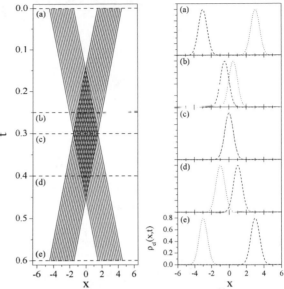

(1.84). If ρ_{cl} is given by two separate Gaussian distributions with width σ, since the superposition principle holds for them,

$$\rho_{cl}(x, t) = \rho_{cl,1}e^{-(x-\bar{x}_1-vt)^2/2\sigma^2} + \rho_{cl,2}e^{-(x-\bar{x}_2+vt)^2/2\sigma^2}. \tag{1.89}$$

Nevertheless, ensembles of trajectories associated with each distribution will evolve independently and, therefore, there is no relevant effect among them (see Fig. 1.3), contrary to what happens in Bohmian mechanics (see the corresponding applications in Volume 2).

1.5 Classical Mechanics of Continuum Media

1.5.1 Coupled Motion

As will be seen in Chap. 2 (and also in Chap. 5) within the framework of the theory of open systems, there are different ways to tackle the problem of the coupling between a system and its surrounding. This coupling usually leads to very important effects on the system, such as energy and coherence losses. For example, it is well known that the action of an external friction on an oscillator gives rise to its gradual energy loss, while the application of driving forces produces the opposite effect. Both friction and driving forces are a simplified or coarse-grained way to describe a mean effect of the surrounding on the system, where there is no *feedback* from the latter on the former. However, it is interesting to understand the basic mechanisms behind this back and forth energy transfer between the two subsystems, namely the system and its environment, for it allows us to better understand more general descriptions coming from the theory of open systems.

Thus, consider a conservative system described by N degrees of freedom, each one associated with a generalized coordinate $q_k (k = 1, 2, \ldots, N)$. The system (stable) equilibrium configuration is specified by $q_k = q_{k0}, \dot{q}_k = 0$ and $\ddot{q}_k = 0$. Since the Lagrange equations are satisfied, it can be shown [22] that the kinetic energy is a homogeneous quadratic function of the generalized velocities,

$$T = \frac{1}{2} \sum_{j,k} m_{jk} \dot{q}_j \dot{q}_k, \tag{1.90}$$

and the potential energy can be expressed in terms of small displacements with respect to the equilibrium position,

$$U = \frac{1}{2} \sum_{j,k} A_{jk} q_j q_k, \tag{1.91}$$

where $U(\{q_{k0}\}) = 0$ (without loss of generality) and

$$A_{jk} \equiv \left. \frac{\partial^2 U}{\partial q_j \partial q_k} \right|_0, \tag{1.92}$$

with $A_{jk} = A_{kj}$. Although the A_{jk} are just numbers, in general the m_{jk} may depend on the coordinates [22]. Hence, in order to express (1.90) at the same order of approximation as (1.91), the m_{jk} is expanded around the equilibrium positions, then keeping only the zeroth-order, $m_{jk}(q_{l0})$ (from now on m_{jk} will denote this value). With this, the system of N coupled Euler–Lagrange equations to be solved reads as

$$\sum_j (A_{jk} q_j + m_{jk} \ddot{q}_j) = 0, \tag{1.93}$$

which, by means of solutions of the type $q_j(t) = a_j e^{i(\omega t - \delta)}$, can also be expressed as

$$\sum_j (A_{jk} - \omega^2 m_{jk}) a_j = 0. \tag{1.94}$$

The determinant

$$|A_{jk} - \omega^2 m_{jk}| = 0 \tag{1.95}$$

ensures the existence of a solution for the system (1.94). The associated equation is called the *characteristic* or *secular equation* and its solutions, the eigenfrequencies ω_r, describe the oscillation *normal modes*. Any solution $q_j(t)$ can then be expressed as a linear combination of the corresponding associated eigenvectors $\mathbf{a}_r$, i.e.,

$$q_j(t) = \sum_r a_{jr} e^{i(\omega_r t - \delta_r)}, \tag{1.96}$$

where the eigenvectors satisfy the normalization relation,

$$\sum_{j,k} m_{kj} a_{jr} a_{ks} = \delta_{rs}. \tag{1.97}$$

Alternatively, the q_j can also be described in terms of a set of new coordinates, namely the normal modes, as

$$q_j(t) = \sum_r a_{jr} \eta_r(t), \tag{1.98}$$

where each $\eta_r(t) \equiv \beta_r e^{i\omega_r t}$ oscillates at only one frequency (ω_r) and satisfies a simpler equation,

$$\ddot{\eta}_r + \omega_r^2 \eta_r = 0. \tag{1.99}$$

The normal modes introduce an important simplification of the system Hamiltonian, which reads as

$$H = \sum_{j,k} \left(\frac{1}{2} m_{jk} p_j p_k + \frac{1}{2} A_{jk} q_j q_k \right) = \sum_r \left(\frac{1}{2} \bar{p}_r^2 + \frac{1}{2} \omega_r^2 \eta_r^2 \right), \tag{1.100}$$

i.e., it is given in terms of N *independent* oscillators, with $p_j = \dot{q}_j$ and $\bar{p}_r = \dot{\eta}_r$.

To illustrate this description, consider two oscillating particles of equal mass, m, both subject to a restoring linear force with constant κ. These particles are also coupled through another linear force with constant κ_{12}. If x_1 and x_2 denote the coordinates describing the position of these particles, the total kinetic and potential energies are

$$T = \frac{1}{2} m \dot{x}_1^2 + \frac{1}{2} m \dot{x}_2^2, \tag{1.101a}$$

$$U = \frac{1}{2} \kappa x_1^2 + \frac{1}{2} \kappa x_2^2 + \frac{1}{2} \kappa_{12} (x_1 - x_2)^2, \tag{1.101b}$$

respectively. The corresponding secular equation is

$$\begin{vmatrix} \kappa + \kappa_{12} - m\omega^2 & -\kappa_{12} \\ -\kappa_{12} & \kappa + \kappa_{12} - m\omega^2 \end{vmatrix} = 0, \tag{1.102}$$

with eigenfrequencies given by

$$\omega_- = \sqrt{\frac{\kappa}{m}}, \quad \omega_+ = \sqrt{\frac{\kappa + 2\kappa_{12}}{m}}. \tag{1.103}$$

The displacements x_1 and x_2 can be therefore expressed as a linear combination of two motions, with frequencies ω_- and ω_+. The motion with frequency ω_- is called the symmetric mode, $\eta_- = x_1 + x_2$, because both particles oscillate in the same direction (*in phase*). On the other hand, ω_+ gives rise to the so-called antisymmetric mode, $\eta_+ = x_1 - x_2$, because of their oscillation in opposite directions (*out of phase*). In terms of these modes, the total Hamiltonian reads as

$$
\begin{aligned}
H &= \frac{p_1^2}{2m} + \frac{1}{2}(\kappa + \kappa_{12})x_1^2 + \frac{p_2^2}{2m} + \frac{1}{2}(\kappa + \kappa_{12})x_2^2 - \kappa_{12}x_1 x_2 \\
&= \frac{\bar{p}_+^2}{2} + \frac{1}{2}\omega_+^2 \eta_+^2 + \frac{\bar{p}_-^2}{2} + \frac{1}{2}\omega_-^2 \eta_-^2 .
\end{aligned}
\tag{1.104}
$$

That is, one passes from two coupled oscillators ($H_1 + H_2 + V_{12}$) to two uncoupled *effective* oscillators ($H_+ + H_-,$) although the phase coherence is preserved implicitly: the coherence remains present in the collective (in phase or out of phase) motion. This is precisely one of the very first steps considered in many analytical approaches implemented to study open quantum systems [91–94] (see Chap. 5).

An interesting case is that of the *weak coupling limit*, $\kappa_{12} \ll \kappa$. In this case, $\omega_\pm \approx \omega_0 (1 \pm \epsilon)$, where $\omega_0 = \sqrt{(\kappa + \kappa_{12})/m}$ is the natural frequency of either oscillator when the other is held fixed and $\epsilon = \kappa_{12}/2\kappa \ll 1$ is a small perturbation introduced by the coupling (the presence of the other oscillator). If $x_1(0) = x_{10}$, $x_2(0) = 0$ and $\dot{x}_1(0) = \dot{x}_2(0) = 0$,

$$
x_1(t) = \frac{x_{10}}{2}(\cos \omega_+ t + \cos \omega_- t) \approx x_{10} \cos \epsilon \omega_0 t \cos \omega_0 t,
\tag{1.105a}
$$

$$
x_2(t) = \frac{x_{10}}{2}(\sin \omega_+ t + \sin \omega_- t) \approx x_{10} \sin \epsilon \omega_0 t \sin \omega_0 t.
\tag{1.105b}
$$

As it can be seen in Fig. 1.4, this means that, as time goes on, the amplitude of x_1 starts to decrease slowly with frequency $\epsilon \omega_0$ due to an energy transfer to the second oscillator. Consequently, the amplitude of the latter starts to increase progressively until, at $t = \pi/2\epsilon \omega_0$, all the energy has been transferred. In the case considered in the figure, this means that $x_1 = 0$ and $x_2 = x_{10}$. Then, the process reverts—energy is transferred back again to the first oscillator. This energy transfer continues indefinitely due to the conservation of the total energy, giving rise to the so-called phenomenon of *beats*.

1.5.2 Superposition of Classical Waves

The present understanding of (quantum) wave–packet propagation (and the dynamical interference displayed when they "collide") is pretty much based on how this process is commonly thought in classical wave mechanics. It is therefore worth revising the case of two (classical) pulses traveling with opposite velocities along a

Fig. 1.4 Beats produced in the oscillations displayed by two coupled, identical particles, where $\epsilon = 1/17$

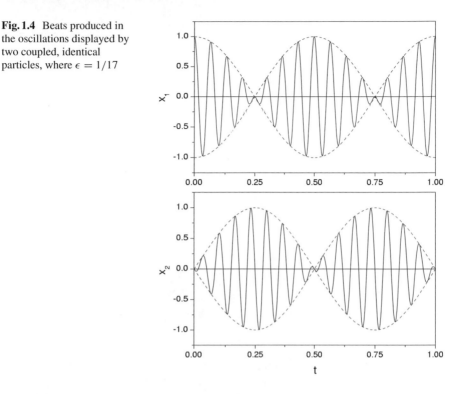

homogeneous string.[4] In this case, according to classical wave mechanics, given an initial perturbation $\xi(x, 0)$, its evolution at any subsequent time is determined by the wave equation

$$\frac{\partial^2 \xi}{\partial t^2} = v^2 \frac{\partial^2 \xi}{\partial x^2},\tag{1.106}$$

where the propagation (or group) velocity v depends on the physical properties of the string. As is well-known, one of the general solutions to this equation is

$$\xi(x, t) = \xi(x - vt),\tag{1.107}$$

a pulse propagating at a constant velocity v along the string and keeping its initial shape $\xi(x, 0)$ at any time. As in Sect. 1.4.4, this result can be easily derived by solving (1.106) by means of the Fourier method. That is, expressing $\xi(x, t)$ as

$$\xi(x, t) = \frac{1}{\sqrt{2\pi}} \int \tilde{\xi}(\alpha, t) e^{-i\alpha x} \, d\alpha\tag{1.108}$$

and introducing this expression into (1.106), the latter becomes

---

4 This example has been chosen here for its simplicity and without loss of generality. Thus, the results could also be readily extended to nonhomogeneous, three-dimensional media.

$$\frac{\partial^2 \tilde{\xi}}{\partial t^2} = -(v\alpha)^2 \tilde{\xi},$$ (1.109)

whose solution is

$$\tilde{\xi}(\alpha, t) = \tilde{\xi}(\alpha, 0)e^{i\alpha vt}.$$ (1.110)

Substituting this expression into (1.108) and taking into account that

$$\xi(x, 0) = \frac{1}{\sqrt{2\pi}} \int \tilde{\xi}(\alpha, 0)e^{-i\alpha x} dx,$$ (1.111)

(1.107) is finally reached,

$$\xi(x, t) = \frac{1}{\sqrt{2\pi}} \int \tilde{\xi}(\alpha, 0)e^{-i\alpha(x-vt)} d\alpha = \xi(x - vt).$$ (1.112)

Consider that the perturbation or pulse has also initially a Gaussian shape with width σ,

$$\xi(x, 0) = \xi_0 e^{-x^2/2\sigma^2}.$$ (1.113)

According to (1.107), its propagation along the string is described by

$$\xi(x, t) = \xi_0 e^{-(x-vt)^2/2\sigma^2},$$ (1.114)

i.e., the Gaussian propagates with positive velocity along the string. If instead of a pulse there are two of them moving with identical velocities in modulus, but opposite directions (see Fig. 1.5), because the superposition principle holds for (1.106), the total perturbation felt by the string will be the sum of the two pulses,

$$\xi(x, t) = \xi_1 e^{-(x-\bar{x}_1-vt)^2/2\sigma^2} + \xi_2 e^{-(x-\bar{x}_2+vt)^2/2\sigma^2},$$ (1.115)

with $\bar{x}_1 < 0$ and $\bar{x}_2 > 0$. As seen in Fig. 1.5, after some time (panels (b) and (c)), both pulses overlap, the total amplitude being the sum of both ξ_1 and ξ_2. However, during their overlapping, the pulses proceed with their respective motions, but without "interacting"—otherwise the superposition principle would not hold. After ξ reaches its maximum amplitude (see panel (c)), according to the standard interpretation for this process—commonly taught in classical mechanics courses—, each pulse goes on unaffected with its initial velocity (see panel (d)). This interpretation follows from a literal reading of (1.115): since the propagation of each pulse can be obtained from the wave equation independently, they also keep moving with their respective initial velocities all the way (during and beyond the overlapping). Now, if one considers a view based on the flow of the energy associated with each pulse, their motions have to be explained in a very different way: during the overlapping the energy flux reverts, this leading to a reflection of the pulses (see panel (e)). No further details will be exposed here about this interesting non-crossing property, but it will appear later on in Chap. 6 as well as in Volume 2 in connection with both Bohmian mechanics and standard electromagnetism. In the case of Bohmian mechanics, it will be related to the probability density flow, while in electromagnetism it arises in a similar context as here, i.e., linked to the (electromagnetic) energy flow.

Fig. 1.5 From **a** to **d**, snapshots illustrating the time-evolution of a classical wave or perturbation consisting of two counter-propagating Gaussian pulses according to the standard vision. **e** Final outcome if one considers the interpretation based on the non-crossing of the energy fluxes associated with each pulse

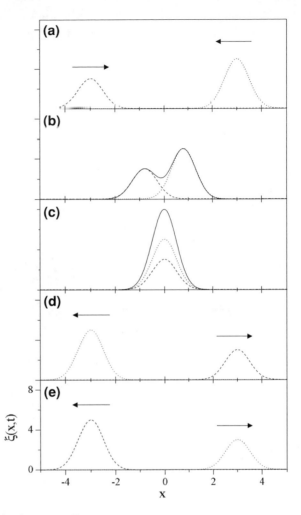

1.5.3 The Hydrodynamic Approach

When dealing with the dynamics of many microscopic particles, both length and time scales need to be simultaneously considered [70]. Typically, wavelengths are compared with the mean free path, l_m, and times with the mean collision time, τ_c. Hence the dispersion plane (frequency versus wave number) can be divided into three regions, which correspond to three different regimes:

- The *hydrodynamics regime*, where $kl_m \ll 1$ or $\omega\tau_c \ll 1$.
- The *kinetic* or *molecular hydrodynamics regime* [82], where $kl_m \sim 1$ or $\omega\tau_c \sim 1$.
- The *free–particle regime*, where $kl_m \gg 1$ or $\omega\tau_c \gg 1$.

In the first regime, the dynamical variables are described by macroscopic equations of motion, while in the second one microscopic equations of motion should be applied. In order to establish a bridge between both regimes it is necessary first to specify the relationship between the two types of dynamical variables. To this end, a "coarse-graining" procedure is carried out. For example, the difference between the microscopic particle density and the (macroscopic) hydrodynamic local density relies on the fact that the latter arises from averaging the first over both a volume and a time interval. The volume is assumed to be macroscopically small, but still sufficiently large to ensure that the relative fluctuation of the particle number is negligible; the time interval is short on a macroscopic scale, but long compared to τ_c. Once this smoothing step is complete, the Onsager regression hypothesis arises—which can be justified on the basis of the linear response theory by means of the fluctuation–dissipation theorem—, thus connecting linearly both regimes.

Hydrodynamics is a part of continuum mechanics dealing with the flow of fluids [69, 70]. One of the most fundamental equations in hydrodynamics is the so-called *mass continuity equation*, which expresses the conservation law for the fluid mass in 3D and reads as

$$\frac{\partial \rho_{fl}(\mathbf{r}, t)}{\partial t} + \nabla \cdot \mathbf{J}_{fl}(\mathbf{r}, t) = 0, \tag{1.116}$$

where ρ_{fl} is the mass density and

$$\mathbf{J}_{fl}(\mathbf{r}, t) = \rho_{fl}(\mathbf{r}, t)\mathbf{v}(\mathbf{r}, t) \tag{1.117}$$

is the mass current density, with $\mathbf{v}$ being the velocity of an element of fluid. The continuity equation (1.116) can also be recast as

$$\frac{d\rho_{fl}(\mathbf{r}, t)}{dt} + \rho_{fl}(\mathbf{r}, t)\nabla \cdot \mathbf{v}(\mathbf{r}, t) = 0, \tag{1.118}$$

with

$$\frac{d}{dt} \equiv \frac{\partial}{\partial t} + \mathbf{v} \cdot \nabla \tag{1.119}$$

being the so-called hydrodynamic or Lagrangian derivative. This operator represents the change in time of a given macroscopic function along the system flow or streamlines. Analogously to the mass density, a conservation law for the fluid momentum can also be enunciated,

$$\rho_{fl}(\mathbf{r}, t)\frac{d\mathbf{v}(\mathbf{r}, t)}{dt} + \nabla \cdot \sigma(\mathbf{r}, t) = 0, \tag{1.120}$$

where the σ function here represents the stress tensor or momentum current. This law can be considered a Newton second law of continuum mechanics assuming that no external force per unit mass is present. Furthermore, a conservation law for the fluid energy can also be specified,

$$\rho_{fl}(\mathbf{r}, t)\frac{dE(\mathbf{r}, t)}{dt} + \nabla \cdot \mathbf{Q}(\mathbf{r}, t) = 0, \tag{1.121}$$

where $E(\mathbf{r}, t)$ is the total energy per unit mass and $\mathbf{Q}$ the energy current or heat flux vector.

In some special cases, the fundamental equations of continuum mechanics can be reduced to some simple forms. For example, if Fick's law of diffusion,

$$\rho_{fl}(\mathbf{r}, t)\mathbf{v}(\mathbf{r}, t) = -D\nabla\rho_{fl}(\mathbf{r}, t), \tag{1.122}$$

where D is the diffusion constant, is substituted into (1.116),

$$\frac{\partial\rho_{fl}(\mathbf{r}, t)}{\partial t} = D\nabla^2\rho_{fl}(\mathbf{r}, t). \tag{1.123}$$

This is the so-called *diffusion equation*. Usually, it is assumed that the heat flux vector is given by Fourier's law,

$$\mathbf{Q}(\mathbf{r}, t) = -\lambda\nabla T(\mathbf{r}, t), \tag{1.124}$$

with λ being the thermal conductivity and T the temperature. Then, substituting (1.124) into (1.122) yields

$$\rho_{fl}(\mathbf{r}, t)C_v\frac{\partial T(\mathbf{r}, t)}{\partial t} = \lambda\nabla^2 T(\mathbf{r}, t), \tag{1.125}$$

where C_v is the heat capacity. This equation is the so-called *heat conduction equation*, which has the same formal solution as the diffusion equation.

The simplest hydrodynamic equation is the *Euler equation*,

$$\left(\frac{\partial}{\partial t} + \mathbf{v}\cdot\nabla\right)\mathbf{v} + \frac{\nabla p(\mathbf{r}, t)}{\rho_{fl}(\mathbf{r}, t)} = 0, \tag{1.126}$$

where p is the pressure. This equation describes the evolution of a non-viscous flow and is obtained by setting the shear (η) and bulk (κ) viscosity coefficients equal to zero. On the contrary, for a viscous flow the hydrodynamics is ruled by the Navier–Stokes equation,

$$\rho_{fl}\left(\frac{\partial}{\partial t} + \mathbf{v}\cdot\nabla\right)\mathbf{v} = \eta\nabla^2\mathbf{v} + \left(\frac{\eta}{3} + \kappa\right)\nabla(\nabla\cdot\mathbf{v}) - \nabla p(\mathbf{r}, t). \tag{1.127}$$

As it will be seen in Chap. 6, these two equations also play a fundamental role in the understanding of the dynamics of quantum systems when they are described or interpreted as quantum fluids [86].

If fluctuations of the mass density and temperature are small, the main conservation laws can be linearized. Then, the decay of long wavelength fluctuations in terms of the well-known collective modes of hydrodynamics can easily be deduced.

References

1. Copernicus, N.: De Revolutionibus Orbium Coelestium (1543). Translated into English by Wallis, C.G.: On the Revolutions of the Heavenly Spheres (St John's College Bookstore, Annapolis, MD, 1939, republished by Prometheus Books, New York, 1995)
2. Kepler, J.: Astronomia Nova (1609). Translated into English by Donahue, W.H.: New Astronomy (Cambridge University Press, Cambridge, 1992)
3. Galilei, G.: Dialogo sopra i due massimi sistemi del mondo tolemaico e copernicano (1632). Translated into English by Drake, S.: Dialogue Concerning the Two Chief World Systems, Ptolemaic and Copernican (University of California Press, Los Angeles, CA, 1981)
4. Descartes, R.: Principia Philosophiae (1644). Translated into English by Miller, V.R., Miller, R.P.: Principles of Philosophy (Kluwer Academic, Amsterdam, 1991)
5. Newton, I.: Philosophiae Naturalis Principia Mathematica. Royal Society, London (1687). Published in English by Cohen, I.B., Whitman, A., Newton, I.: The Principia: Mathematical Principles of Natural Philosophy. A New Translation (University of California Press, Los Angeles, CA, 1999)
6. Bell, J.S.: On the Einstein–Podolsky–Rosen paradox. Physics 1, 195–200 (1964). Reprinted in Bell, J.S.: Speakable and Unspeakable in Quantum Mechanics (Cambridge University Press, Cambridge, 1987)
7. Aspect, A., Grangier, P., Roger, G.: Experimental tests of realistic theories via Bell's theorem. Phys.Rev. Lett. 47, 460–463 (1981)
8. Margenau, H., Murphy, G.M.: The Mathematics of Physics and Chemistry, 2nd edn. D.van Nostrand Company, New York (1956)
9. Landau, L.D., Lifschitz, E.M.: Mechanics. Pergamon Press, Oxford (1960)
10. Goldstein, H.: Classical Mechanics. Addison-Wesley Publising Company, Reading (1980)
11. Landau, L.D., Lifschitz, E.M.: The Classical Theory of Fields, 4th edn. Butterworth–Heinemann, London (1975)
12. Parnovsky, A.S.: Some generalisations of Brachistochrone problem. Acta Phys. Pol. A 93, S55–S64 (1998)
13. Erlichson, H.: Johann Bernoulli's Brachistochrone solution using Fermat's principle of least time. Eur. J. Phys. 20, 299–304 (1999)
14. Tannery, P., Henry, C. (eds.): Œuvres de Fermat, pp. 354–457. Gauthier–Villars, Paris (1894)
15. Hero of Alexandria, Catoptrica (circa 60 AC)
16. Ibn al-Haytham "Alhazen", Book of Optics (1021)
17. de Maupertuis P.L.M.: Accord de différentes lois de la nature qui avoient jusqu'ici paru incompatibles. Histoire de l'Académie Royale des Sciences et des Belles Lettres, pp. 417–426 (1744)
18. de Maupertuis P.L.M.: Les lois de mouvement et du repos, déduites d'un principle de métaphysique. Histoire de l'Académie Royale des Sciences et des Belles Lettres, pp. 267–294 (1746)
19. Hamilton, W.R.: On a general method in dynamics; by which the study of the motions of all free systems of attracting or repelling points is reduced to the search and differentiation of one central relation, or characteristic function. Phil. Trans. R. Soc. Lond. 124, 247–308 (1834)
20. Hamilton, W.R.: Second essay on a general method in dynamics. Phil. Trans. R. Soc. Lond. 125, 95–144 (1835)
21. de Laplace, P.S.: Theorie Analytique des Probabilités. In: Œuvres Complètes de Laplace, vol. VII. Gauthier–Villars, Paris (1820)

22. Thornton, S.T., Marion, J.B.: Classical Dynamics of Particles and Systems, 5th edn. Thomson, Belmont (2004)
23. Arnold, V.I.: Mathematical Methods of Classical Mechanics. Springer, New York (1989)
24. Born, M., Wolf, E.: Principles of Optics. Pergamon Press, Oxford (1980)
25. Elmore, W.C., Heald, M.A.: Physics of Waves. Dover Publications, New York (1985)
26. Moiseyev, N.: Quantum theory of resonances: calculating energies, widths and cross-sections by complex scaling. Phys. Rep. 302, 211–293 (1998)
27. Kosloff, R., Kosloff, D.: Absorbing boundaries for wave propagation problems. J. Comp. Phys. 63, 363–376 (1986)
28. Xavier, A.L., de Aguiar, M.A.M.: Complex trajectories in the quartic oscillator and its semiclassical coherent-state propagator. Ann. Phys. (N.Y.) 252, 458–478 (1996)
29. Xavier, A.L., de Aguiar, M.A.M.: Phase-space approach to the tunnel effect: a new semiclassical transversal time. Phys. Rev. Lett. 79, 3323–3326 (1997)
30. Leacock, R.A., Padgett, M.J.: Hamilton–Jacobi theory and the quantum action variable. Phys. Rev. Lett. 50, 3–6 (1983)
31. Leacock, R.A., Padgett, M.J.: Hamilton–Jacobi/action–angle quantum mechanics. Phys. Rev. D 28, 2491–2502 (1983)
32. Bender, C.M., Boettcher, S.: Real spectra in non-Hermitian Hamiltonians having $\mathcal{PT}$ symmetry. Phys. Rev. Lett. 80, 5243–5246 (1998)
33. Bender, C.M., Brody, D.C., Jones, H.F.: Complex extension of quantum mechanics. Phys. Rev. Lett. 89, 270401(1–4) (2002)
34. Bender, C.M., Boettcher, S., Meisinger, P.N.: quantum mechanics. J. Math. Phys. 40, 2201–2229 (1999)
35. Degiovanni, L., Rastelli, G.: Complex variables for separation of the Hamilton–Jacobi equation on real pseudo-Riemannian manifolds. J. Math. Phys. 48, 073519(1–23) (2007)
36. Bender, C.M., Chen, J.-H., Darg, D.W., Milton, K.A.: Classical trajectories for complex Hamiltonians. J. Phys. A 39, 4219–4238 (2006)
37. Bender C.M., Darg D.W.: Spontaneous breaking of classical $\mathcal{PT}$ symmetry. J. Math. Phys. 48, 042703(1–14) (2007)
38. Kaushal, R.S., Korsch, H.J.: Some remarks on complex Hamiltonian systems. Phys. Lett. A 276, 47–51 (2000)
39. Kaushal, R.S., Singh, S.: Construction of complex invariants for classical dynamical systems. Ann. Phys. 288, 253–276 (2001)
40. Singh, S., Kaushal, R.S.: Complex dynamical invariants for one-dimensional classical systems. Phys. Scr. 67, 181–185 (2003)
41. Kaushal, R.S.: Classical and quantum mechanics of complex Hamiltonian systems: an extended complex phase space approach. Pramana J. Phys. 73, 287–297 (2009)
42. Le Levier, R.E., Saxon, D.S.: An optical model for nucleon-nuclei scattering. Phys. Rev. 87, 40–41 (1952)
43. Feshbach, H., Porter, C.E., Weisskopf, V.F.: Model for nuclear reactions with neutrons. Phys. Rev. 96, 448–464 (1954)
44. Glauber, R.J.: Coherent and incoherent states of the radiation field. Phys. Rev. 131, 2766–2788 (1963)
45. Lichtenberg, A.J., Lieberman, M.A.: Regular and Chaotic Dynamics. Springer, New York (1992)
46. Jordan, D.W., Smith, P.: Nonlinear Ordinary Differential Equations, 3rd edn. Oxford University Press, Oxford (1999)
47. Tabor, M.: Chaos and Integrability in Nonlinear Dynamics. Wiley, New York (1989)
48. Gutzwiller, M.C.: Chaos in Classical and Quantum Mechanics. Springer, Berlin (1990)
49. Ott, E.: Chaos in Dynamical Systems. Cambridge University Press, Cambridge (1993)
50. Wiggins, S.: Introduction to Applied Nonlinear Dynamical Systems and Chaos. Springer, New York (1990)

51. Gaspard, P.: Chaos, Scattering and Statistical Mechanics. Cambridge University Press, Cambridge (1998)
52. Weinstein, A.: Normal modes for nonlinear hamiltonian systems. Invent. Math. **20**, 47–57 (1973)
53. Christiansen, F.: Fixed points, and how to get them classical and quantum chaos. In: Cvitanović, P., Artuso, R., Mainieri, R., Tanner, G., Vattay G. (eds.) The Chaos Book. Niels Bohr Institute, Copenhagen (2001). www.nbi.dk/ChaosBook/
54. Farantos, S.C.: Methods for locating periodic orbits in highly unstable systems. THEOCHEM **341**, 91–100 (1995)
55. Prosmiti, R., Farantos, S.C.: Periodic orbits bifurcation diagrams and the spectroscopy of C_2 H_2 system. J. Chem. Phys. **103**, 3299–3314 (1995)
56. Seydel, R.: Practical Bifurcation and Stability Analysis. Springer, New York (1994)
57. Meyer, K.R.: Generic bifurcation of periodic points. Trans. Amer. Math. Soc. **149**, 95–107 (1970)
58. de Aguiar, M.A.M., Malta, C.P., Baranger, M., Davies, K.T.R.: Bifurcations of periodic trajectories in non-integrable Hamiltonian systems with two degrees of freedom: numerical and analytical results. Ann. Phys. (NY) **180**, 167–205 (1987)
59. Mao, J.-M., Delos, J.B.: Hamiltonian bifurcation theory of closed orbits in the diamagnetic Kepler problem. Phys.Rev. A **45**, 1746–1761 (1992)
60. Hénon, M., Heyles, C.: The applicability of the third integral of motion: some numerical experiments. Astron. J. **69**, 73–79 (1964)
61. Dorfman, J.E.: An Introduction to Chaos in Non-Equilibrium Statistical Mechanics. Cambridge University Press, Cambridge (1999)
62. Vega, J.L., Guantes, R., Miret–Artés, S.: Chaos and transport properties of adatoms on solid surfaces. J. Phys.: Condens. Matter **14**, 6193–6232 (2002)
63. Metzler, R., Klafter, J.: The random walk's guide to anomalous diffusion. a fractional dynamics approach. Phys. Rep. **339**, 1–77 (2000)
64. Klafter, J., Shlesinger, M.F., Zumofen, G.: Beyond Brownian motion. Phys. Today **49**(2), 33–39 (1996)
65. Mandelbrot, B.: The Fractal Geometry of Nature. Freeman, San Francisco (1982)
66. Berry, M.: Waves and Thom's theorem. Adv. Phys. **25**, 1–26 (1976)
67. Guantes, R., Sanz, A.S., Margalef-Roig, J., Miret-Artés, S.: Atom-surface diffraction: a trajectoy description. Surf. Sci. Rep. **53**, 199–330 (2004)
68. Pathria, R.K.: Statistical Mechanics. Pergamon Press, Oxford (1972)
69. McQuarrie, D.A.: Statistical Mechanics. Harper and Row, New York (1976)
70. Hansen, J.P., McDonald, I.R.: Theory of Simple Liquids. Academic Press, New York (1986)
71. Gibbs, J.W.: Elementary Principles in Statistical Mechanics. Scribner's Sons, New York (1902)
72. Jaffé, C., Brumer, P.: Classical Liouville mechanics and intramolecular relaxation dynamics. J. Phys. Chem. **88**, 4829–4839 (1984)
73. Jaffé, C., Brumer, P.: Classical-quantum correspondence in the distribution dynamics of integrable systems. J. Chem. Phys. **82**, 2330–2340 (1985)
74. Wilkie, J., Brumer, P.: Quantum-classical correspondence via Liouville dynamics. I. Integrable systems and the chaotic spectral decomposition. Phys. Rev. A **55**, 27–42 (1997)
75. Wilkie, J., Brumer, P.: Quantum-classical correspondence via Liouville dynamics. II. Correspondence for chaotic Hamiltonian systems. Phys. Rev. A **55**, 43–61 (1997)
76. Bogoliubov, N.N.: Kinetic equations. J. Phys.USSR **10**, 265–274 (1946)
77. Born, M., Green, H.S.: A general kinetic theory of liquids I. The molecular distribution functions. Proc. R. Soc. A **188**, 10–18 (1946)
78. Kirkwood, J.G.: The statistical mechanical theory of transport processes I. General theory. J. Chem. Phys. **14**, 180–201 (1946)
79. Kirkwood, J.G.: The statistical mechanical theory of transport processes II. Transport in gases. J. Chem. Phys. **15**, 72–76 (1947)

80. Yvon, J.: Theorie Statistique des Fluides et l'Equation d'Etat, Actes Scientifique et Industrie, vol. 203. Hermann, Paris (1935)
81. Kubo, R., Toda, M., Hashitsume, N.: Nonequilibrium Statistical Mechanics Statistical Physics II. Springer, Berlin (1985)
82. Boon, J.P., Yip, S.: Molecular Hydrodynamics. Dover, New York (1991)
83. Bohm, D.: A suggested interpretation of the quantum theory in terms of "hidden" variables. I. Phys. Rev. **85**, 166–179 (1952)
84. Bohm, D.: A suggested interpretation of the quantum theory in terms of "hidden" variables. II. Phys. Rev. **85**, 180–193 (1952)
85. Holland, P.R.: The Quantum Theory of Motion. An Account of the de Broglie–Bohm Causal Interpretation of Quantum Mechanics. Cambridge University Press, Cambridge (1993)
86. Madelung, E.: Quantentheorie in hydrodynamischer Form. Z. Phys. **40**, 322–326 (1926)
87. Schiller, R.: Quasi-classical theory of the nonspinning electron. Phys. Rev. **125**, 1100–1108 (1962)
88. Schiller, R.: Quasi-classical transformation theory. Phys. Rev. **125**, 1109–1115 (1962)
89. Rosen, N.: The relation between classical and quantum mechanics. Am. J. Phys. **32**, 597–600 (1964)
90. Rosen, N.: Quantum particles and classical particles. Found. Phys. **16**, 687–700 (1986)
91. Louisell, W.H.: Quantum Statistical Properties of Radiation. Wiley, New York (1990)
92. Weiss, U.: Quantum Dissipative Systems. (World Scientific, Singapore (1999)
93. Breuer, H.P., Petruccione, F.: The Theory of Open Quantum Systems. Oxford University Press, Oxford (2002)
94. Razavy, M.: Classical and Quantum Dissipative Systems. Imperial College Press, London (2005)

Chapter 2
Dynamics of Open Classical Systems

2.1 Introduction

Traditionally, in classical mechanics conservative forces derivable from interaction potentials—and therefore dependent on particle positions—have received much more attention than dissipative or damping ones. The latter have been postulated to follow a power law of the velocity, to be dependent on accelerations or even to be nonlocal in space and time. However, the corresponding nonconservative systems, i.e., systems that lose energy as motion takes place, are ubiquitous in Nature. Therefore, despite of dealing with them is not exempt from difficulties, they are becoming more and more attractive from a theoretical viewpoint in recent years. Nowadays classical dissipation constitutes a very active field of research. For example, in a relatively recent monograph, Razavy [1] surveys the very extensive literature on the subject, paying special attention to the quantization of simple, solvable classical systems.

Open classical systems are usually defined as those where the system of interest is surrounded by an *environment*. When the environment is constituted by many degrees of freedom, characterized by a certain temperature (a measure of its internal energy), it is called a *bath* (*heat bath*) or *reservoir*. In this sense, open systems can also be defined as systems exchanging or *dissipating* energy with another one. The dissipative forces leading to such energy transfers can be derived from a conservative many-body problem. In this case, the full conservative system is typically split up into two interacting parts or subsystems: the subsystem of interest or dissipating system and the heat bath. Because the bath is usually an extended system with many degrees of freedom, according to its definition, energy will not flow equally in both directions. Eventually the system relaxes, losing its energy as time goes on. Hence, dissipation is seen as an *irreversible process*. This fact is in apparent contradiction with the time-symmetry exhibited by the equations of motion of classical mechanics. To understand it, one has to consider that the concept of *irreversibility* is related to the so-called *Poincaré recurrence time*, which is extremely large for an extended system and, therefore, any process will appear to be irreversible. Of course, relaxation processes may also display recurrences if energy flows in both directions, which

A. S. Sanz and S. Miret-Artés, *A Trajectory Description of Quantum Processes.*
I. Fundamentals, Lecture Notes in Physics 850, DOI: 10.1007/978-3-642-18092-7_2,
© Springer-Verlag Berlin Heidelberg 2012

usually happens when the environment has a few degrees of freedom. If bath noise or fluctuations influence importantly the dynamics of the dissipative system, then stochasticity will play a fundamental role. In this Chapter, only white noise will be considered, for which the noise autocorrelation function is governed by a Dirac δ-function. This is also called a *Markovian regimen*. Detailed discussions of the action of colored noise on dynamical systems can be found in [2], for example.

In this context, space–time correlation functions (as defined in Sect. 1.4.2) play a fundamental role since they are measurable quantities. These functions provide a rather complete information about the decay of spontaneous thermal fluctuations through thermodynamic averages of the product of two dynamical variables. Different theoretical approaches have been developed for their calculation at finite wavenumbers and frequencies. These approaches range from continuum description (very low wavenumbers and frequencies) in terms of hydrodynamic equations to molecular dynamics simulations; in between, molecular hydrodynamic descriptions are usually preferred. Irreversible time-dependent nonequilibrium properties are very often analyzed within the so-called *linear response theory*, where systems are supposed to be close to equilibrium. The cornerstone of this theory is the *fluctuation–dissipation theorem* [3] and one of its important consequences: Onsager's regression hypothesis. As has already been mentioned in Chap. 1, this hypothesis states that the relaxation of macroscopic nonequilibrium perturbations is also governed by the regression law of spontaneous microscopic fluctuations in systems at equilibrium. A direct evaluation of such correlation functions is a very difficult task when dealing with many-body system. Thus, the most general formalism starts with the Liouville equation for a dynamical variable—in general, depending on all the system coordinates and momenta. Then, by means of the projection-operator technique, one reaches a generalized Langevin equation for such a dynamical variable. This equation is given in terms of a random force and its autocorrelation or *memory* function (sometimes it is also called delayed function). A similar equation can also be obtained for its normalized space–time autocorrelation function without the random force term, known as the *equation of the memory function*. Modelling memory functions is a very standard procedure to obtain correlation functions. Nonetheless, there are other alternative, well-known approaches to calculate correlation functions, such as the kinetic theory based on the linearization of the Boltzmann equation (in phase space), the so-called mode-coupling theory, or the short-time (sum rules) and long-time behavior (transport coefficients). A detailed presentation and discussion of these interesting and important topics, which can be found in the more specialized literature [4–6], are out of the scope of this monograph.

In dissipative dynamics, there are several oscillator models considered paradigmatic, which can describe phenomenologically many elementary classical and quantum processes. This is the case, for example, of the damped and driven harmonic oscillator, typically considered in applications involving linear damping. This oscillator obeys a differential equation of motion in configuration space given by

$$\ddot{q} + \gamma \dot{q} + \omega_0^2 q = F(t), \qquad (2.1)$$

where ω_0 is the natural or harmonic frequency of the oscillator, γ is the friction parameter and $F(t)$ is the force acting on it. Analogously, there is the so-called parametric driven harmonic oscillator, which arises originally from the study of electric circuits. This oscillator is described by the differential equation

$$\ddot{q} + \gamma(t)\dot{q} + \omega_0^2(t)q - 0, \tag{2.2}$$

where γ and ω_0 depend only on time and not on the state of the oscillator. In general, the dependence on time is assumed to be periodic with the same period. Moreover, an extra driving force, $F(t)$, can also be added. By assuming Ohmic damping, or constant friction, and substituting the position q by $qe^{-\gamma t/2}$ into (2.2), the damping can be formally removed. This yields the equation of motion of an undamped oscillator evolving under the action of a modified potential,

$$\ddot{q} + \left[\omega_0^2(t) - \gamma^2/4\right]q = 0. \tag{2.3}$$

For periodic forces or functions, Floquet's theorem can be applied to this second-order differential equation to find the corresponding periodic solutions (for certain cases, some of them become unstable). Mathieu's oscillator is a special case of this oscillator, where $\omega_0^2(t) = \omega_0^2 + \epsilon \cos(\Omega t)$. This oscillator has been used to interpret several experiments. For example, parametric resonances take place when the external excitation frequency of a given parameter is equal to twice the oscillator natural frequency.

Among the different nonlinear differential equations describing the motion of a classical system, the so-called van der Pol and Duffing equations play a special role. The van der Pol equation is a second order differential equation original from self-sustained electric circuits that displays nonlinear damping. In one-dimension, its general form reads as

$$\ddot{y} + \gamma(y^2 - \alpha)\dot{y} + y = F(t), \tag{2.4}$$

with γ and α real. In a similar vein, the Duffing equation describes the damping motion of an oscillator subject to the influence of a nonharmonic force (Hooke's law is not obeyed). This equation, given by

$$\ddot{y} + \gamma\dot{y} + \alpha y + \beta y^3 = A\cos(\omega t + \phi), \tag{2.5}$$

with γ, α, and β real, leads to chaotic dynamics. These important topics, nonlinear damping [1] or chaotic dissipative motion [7] are also out of the scope of this monograph.

The literature about the four oscillator models mentioned above and their applications is very extensive. The reader interested in a more detailed analysis is addressed to any standard textbook dealing with linear and nonlinear dynamical systems. This Chapter focuses on a general and simple introduction to dissipative and stochastic dynamics in classical mechanics, with the purpose of supplying the means for a better understanding of the dynamics of open quantum systems later on in Chap. 5.

2.2 Dissipative Dynamics

Forces are usually assumed as conservative and derivable from interaction potentials. As it was shown in Chap. 1, the canonical formulations of classical mechanics for conservative systems are essentially the Lagrangian, Hamiltonian and Hamilton–Jacobi formulations. These formalisms have been developed and extended to find quantum analogues of conservative systems. However, for nonconservative systems this extension is much more troublesome. One of the most important issues in analytical dynamics is the so-called *inverse problem*, i.e., the problem of determining the Lagrangian and Hamiltonian functions from the equations of motion (see Sect. 1.2.1). This problem has been widely studied due to the non-uniqueness implicit in those formulations. The conditions for the existence of a Lagrangian function are called the *Helmholtz conditions*, which provide a way to construct Lagrangians. However, not all of them are acceptable because of the violation of some physical requirement. Moreover, when the number of degrees of freedom is equal or greater than two, not always a Lagrangian can be found for a given classical system. The connection between invariance or symmetry properties and conserved quantities is given by *Nöther's theorem* [8]. This theorem allows to determine constants of motion, if they exist, from the equations of motion in those cases where the problem cannot be formulated in terms of the variational principle. This issue becomes critical when dealing with dissipation, for dissipative systems cannot always be described by means of a Lagrangian or a Hamiltonian. Hence finding an appropriate quantum description for these systems is not exempt from difficulties and controversy.

Even when not all forces acting on a system can be derived from a potential function, the Lagrange equations (1.3) can still be written [8] as

$$\frac{\partial L}{\partial q_i} - \frac{d}{dt}\frac{\partial L}{\partial \dot{q}_i} = Q_i, \quad i = 1, 2, \ldots, N, \tag{2.6}$$

where the Lagrangian L contains the potential of the conservative forces and Q_i represents the remaining forces. The simplest way to include dissipation within the Lagrangian formulation is by adding the so-called *dissipation function* to the Lagrange equations,

$$Q_i = -\frac{\partial \mathcal{F}}{\partial \dot{q}_i}. \tag{2.7}$$

This idea, due to Rayleigh, is based on considering that the gradient of Q_i with respect to the velocity just gives the dissipative force. In order to include nonlinear damping forces, this Rayleigh function has been generalized by Lur'e [1]. Dissipative dynamics can be considered by following different routes, which will be briefly analyzed below.

2.2.1 Effective Hamiltonians. The Caldirola–Kanai Model

One of the simpler ways to incorporate environment effects is by considering time-dependent Lagrangian and/or Hamiltonian functions, thus avoiding to deal explicitly with the environment degrees of freedom. This approach allows to preserve the canonical formalism, which can be a good starting point to find out the quantum analogue of the corresponding dissipative dynamics.

The paradigm of the dissipative dynamics is the damped harmonic oscillator model, where the damping force is linear with the velocity. In a one-dimensional configuration space, its equation of motion is (for $m = 1$)

$$\ddot{q} + \gamma \dot{q} + \omega_0^2 q = 0, \qquad (2.8)$$

where γ is the damping constant or friction. Physically, this equation describes a classical dissipative system losing energy at a constant rate γ as time increases. The Hamiltonian model associated with this simple system is the so-called *Caldirola–Kanai (CK) Hamiltonian* [1, 9, 10],

$$H_{CK} = \frac{p^2}{2m} e^{-\gamma t} + V(q) e^{\gamma t}, \qquad (2.9)$$

which was initially considered for a particle with time-dependent mass. In this Hamiltonian,

$$V(q) = \frac{1}{2} m \omega_0^2 q^2 \qquad (2.10)$$

is the potential for a harmonic oscillator with frequency ω_0 and mass m. As shown below, this Hamiltonian has also been considered extensively for damping motion. The corresponding Lagrangian is

$$L_{CK} = \left(\frac{1}{2} m \dot{q}^2 - \frac{1}{2} m \omega_0^2 q^2 \right) e^{\gamma t}. \qquad (2.11)$$

As mentioned above, different Hamiltonians and Lagrangians from those given respectively by (2.9) and (2.11) may also lead to the same equation of motion (2.8) in configuration space. Actually, the correct equation of motion in phase space cannot be obtained from the CK Hamiltonian.

From (2.9), the associated CK Hamilton–Jacobi equation now reads as

$$\frac{\partial S}{\partial t} + \frac{1}{2m} \left(\frac{\partial S}{\partial q} \right)^2 e^{-\gamma t} + \frac{1}{2} m \omega_0^2 q^2 e^{\gamma t} = 0. \qquad (2.12)$$

Following the usual procedure of separation of variables (see Sect. 1.2.2), S can be expressed as

$$S(q; t) = W(q) - \alpha t. \tag{2.13}$$

After some algebraic manipulations, one obtains the solution of (2.8),

$$q(t) = \sqrt{\frac{2\alpha}{m\omega^2}} e^{-\gamma t/2} \sin[\omega(t + \beta)], \tag{2.14}$$

where $\beta = \partial S/\partial \alpha$ and the oscillation frequency is

$$\omega = \sqrt{\omega_0^2 - \frac{\gamma^2}{4}}. \tag{2.15}$$

In the case of a charged particle moving in an external electromagnetic field and subject to a conservative force, the Lorentz force can be added to the equation of motion. Alternatively, using the principle of minimal coupling [1], the canonical momentum can be replaced by a momentum involving the vector potential—adding to the corresponding Hamiltonian a term with the scalar potential. For conservative systems, since H represents the total energy of the particle both formulations lead to the same result. However, for dissipative systems, such as the problem of radiation damping, the minimum coupling scenario does not apply, for it does not lead to the correct equation of motion. On the other hand, the classical equation of motion for a harmonically bound electron coupled to an electromagnetic field gives rise to the classical theory of line widths. As mentioned above, other very well-known models can also be found in the literature, such as the driven damped oscillator, Raleigh's oscillator or the variable mass oscillator.

The previous examples are all formulated in real space. However, the discussion can be extended to a complex coordinate formulation. In this sense, the so-called Dekker Hamiltonian [11] plays a special role. From the damped harmonic oscillator, complex coordinates are introduced according to the change of variable

$$\xi = \frac{1}{\sqrt{\omega}} \left[p + \left(\frac{\gamma}{2} - i\omega \right) q \right]. \tag{2.16}$$

Then, given the Lagrangian

$$L = \frac{i}{2} \left[\xi^* \dot{\xi} - \xi \dot{\xi}^* \right] - \left(\omega - \frac{i\gamma}{2} \right) \xi^* \xi, \tag{2.17}$$

the complex Hamiltonian that arises from it reads as

$$H = - \left(i\omega + \frac{\gamma}{2} \right) \pi \xi, \tag{2.18}$$

where $\pi = \partial L/\partial \dot{\xi}$. In terms of the physical (real-valued) variables, this Hamiltonian can also be expressed as

$$H = \frac{1}{2}p^2 + \frac{\gamma}{2}(pq + qp) + \frac{1}{2}\left(\omega^2 + \frac{\gamma^2}{4}\right)q^2 - i\frac{\gamma}{2}. \qquad (2.19)$$

2.2.2 Lagrangians for Dissipative Systems and Diffusion Equations

An alternative way to tackle the problem of the Lagrangian formulation for dissipative systems is as follows. Consider the equation of motion (2.8) for the damped harmonic oscillator. As mentioned above, in principle this equation cannot be derived from any Lagrangian, since there is no stationary solution. In order to find out a suitable Lagrangian, one can assume that the energy lost by the system goes into another system, namely a *mirror–image system*, which absorbs it [12]. That is, if the energy of the oscillator described by (2.8) is lost at a rate γ, it will be gained at the same rate (with negative friction, $-\gamma$) by the mirror–image system, here denoted by $\bar{q}$. This implies a zero total energy balance and, more importantly, that stationary (extremal) solutions for the larger system can be found. Thus, consider the Lagrangian describing these coupled systems is

$$L = m\dot{q}\dot{\bar{q}} - \frac{1}{2}m\gamma(\bar{q}\dot{q} - \dot{\bar{q}}q) - m\omega_0^2 q\bar{q}, \qquad (2.20)$$

where m is the system mass. Applying variations [13] with respect to $\bar{q}$ and q, one obtains (2.8) as well as its homologous for $\bar{q}$,

$$\ddot{\bar{q}} - \gamma\dot{\bar{q}} + \omega_0^2\bar{q} = 0. \qquad (2.21)$$

By further proceeding, it is possible to extract the Hamiltonian equations of motion. Thus, applying the expression corresponding to the calculation of generalized momenta from Lagrangian mechanics (see Chap. 1),

$$p = \frac{\partial L}{\partial \dot{q}} = m\dot{\bar{q}} - \frac{1}{2}m\gamma\bar{q} = me^{-\gamma t/2}\frac{d}{dt}\left(e^{-\gamma t/2}\bar{q}\right), \qquad (2.22a)$$

$$\bar{p} = \frac{\partial L}{\partial \dot{\bar{q}}} = m\dot{q} - \frac{1}{2}m\gamma q = me^{-\gamma t/2}\frac{d}{dt}\left(e^{-\gamma t/2}q\right). \qquad (2.22b)$$

Taking into account the functional form displayed by the last equalities in each equation, a new set of generalized coordinates and momenta can be defined,

$$Q \equiv e^{-\gamma t/2}q, \quad P \equiv e^{-\gamma t/2}p, \quad \bar{Q} \equiv e^{-\gamma t/2}\bar{q}, \quad \bar{P} = e^{-\gamma t/2}\bar{p}, \qquad (2.23)$$

such that (2.22) become

$$P = m\dot{\bar{Q}}, \quad \bar{P} = m\dot{Q}. \tag{2.24}$$

These relations somehow show the energetic balance in the full system described by the Lagrangian (2.20), where the energy loss due to dissipation in the system of interest is balanced with an energy increase, at the same rate, in its image. The associated Hamiltonian is then given by

$$H = p\dot{q} + \bar{p}\dot{\bar{q}} - L = \frac{p\bar{p}}{m} + \frac{1}{2}\gamma(\bar{q}\bar{p} - qp) + m\omega^2 q\bar{q}, \tag{2.25}$$

where ω is given by (2.15). With this change of variables, the Hamiltonian equations of motion now read as

$$\dot{q} = \frac{\bar{p}}{m} - \frac{1}{2}\gamma q, \tag{2.26a}$$

$$\dot{\bar{p}} = -m\omega^2 q - \frac{1}{2}\gamma\bar{p}, \tag{2.26b}$$

$$\dot{\bar{q}} = \frac{p}{m} + \frac{1}{2}\gamma\bar{q}, \tag{2.26c}$$

$$\dot{p} = -m\omega^2\bar{q} + \frac{1}{2}\gamma p. \tag{2.26d}$$

In these equations, the intertwining between coordinates is very apparent—in the pairs $(q, \bar{p})$ and $(\bar{q}, p)$. This intertwining eventually leads to the system energy dissipation and its absorption by the image system. This can also be noticed from the eigenvalues of the matrix associated with the system of equations (2.26), when the latter as expressed in symplectic notation. These eigenvalues are $\lambda_-^{\pm} = -\gamma/2 \pm \omega$ and $\lambda_+^{\pm} = \gamma/2 \pm \omega$, where $\lambda_-^{\pm}$ describes the system damping and $\lambda_+^{\pm}$ the image-system energy absorption (at the same rate that the system losses it).

The mirror–image method thus allows to apply the variational techniques to dissipative problems, since a Lagrangian density can be defined for them. In other words, whenever one deals with dissipative problems with a gradual energy loss at a constant rate and there is no knowledge on the bath dynamics (neither it is necessary), this technique can be used to derive the corresponding equations of motion. This is the case, for example, of the *heat equation*, which describes the time-evolution of temperature (heat distribution) in a certain space region. In this case, one can construct a Lagrangian density for this *diffusion equation* [12], which reads as

$$\mathcal{L} = -\frac{1}{2}\left(\bar{\psi}\frac{\partial\psi}{\partial t} - \psi\frac{\partial\bar{\psi}}{\partial t}\right) - D\nabla\psi \cdot \nabla\bar{\psi}, \tag{2.27}$$

where ψ represents the density of the diffusing heat ($\bar{\psi}$ represents the mirror–image of ψ) and D is the so-called *diffusion constant* or *diffusion coefficient*. Proceeding as before, one finds the Euler–Lagrange equations

$$D\nabla^2\psi = \frac{\partial\psi}{\partial t}, \tag{2.28a}$$

$$D\nabla^2\bar{\psi} = -\frac{\partial\bar{\psi}}{\partial t}, \tag{2.28b}$$

where (2.28a) is the heat equation and (2.28b) is an equation describing the absorption of the heat flux leaving the system. Equation (2.28a) can also be regarded as the diffusion equation with constant D for a swarm of identical, noninteracting particles. In this case, $\psi \equiv P$ is the probability density function describing the position of one of such particles, which move pursuing random trajectories. This type of motion is called *Brownian motion*. If all particles start at $t_0 = 0$ and $\mathbf{r}_0 = \mathbf{0}$ (with $q \equiv \mathbf{r} = (x, y, z)$), the subsequent time-evolution of the ensemble will be described (see Sect. 2.3.2) by

$$P(\mathbf{r}, t) = \left(\frac{1}{4\pi Dt}\right)^{3/2} e^{-\mathbf{r}^2/4Dt}. \tag{2.29}$$

2.2.3 The Many-Body Problem

Dissipative forces can also be derivable from conservative many-body problems. In general, any system of N interacting particles can be split up into two interacting parts or subsystems, S_1 and S_2. This splitting is introduced on purpose to analyze the time-evolution of one of these subsystems, say S_1, while the other one (S_2, in this case) is regarded to play the role of an environment. If both subsystems have a few degrees of freedom, the energy exchange goes in both directions, from S_1 to S_2 and vice versa, as seen in Sect. 1.5.1. However, if one of them has a very large number of degrees of freedom, say S_2, and its dynamics becomes rather complex, the energy will only flow in one direction, from S_1 to S_2. The dynamics of S_1 then becomes dissipative, and the corresponding force is determined by the nature of the coupling with the extended system S_2. For example, well-known models exhibiting this type of dynamics are [1]:

- The Schrödinger chain, formed by an infinite number of mass points coupled by elastic springs. Here the decay law of any of its constituents is non-exponential.
- The Rubin model, where a massive particle is coupled to a semi-infinite chain of oscillators.
- The dynamics of a nonuniform chain (different masses and elastic spring couplings), where the decay law for a given particle is exponential.

A special case arises when a collection of harmonic oscillators is analyzed, all of them linearly coupled to a given system, e.g., a particle or a harmonic oscillator. For example, the van Kampen model describes an electron harmonically coupled to an electromagnetic field expressed in terms of confined waves in a large but finite sphere. Another example is Sollfrey's model, which describes an oscillator coupled to a string of finite or infinite length. Among these models, the most celebrated one is the Ullersma Hamiltonian model [14–17],

$$H = \frac{p^2}{2M} + V(q) + \sum_{i=1}^{N} \left(\frac{p_i^2}{2m_i} + \frac{1}{2} m_i \omega_i^2 q_i^2 + \kappa_i q_i q \right), \qquad (2.30)$$

where κ_i stands for the system-environment coupling coefficients. Usually, it is assumed that the masses of all oscillators are equal. At $t = 0$, all of them are at rest at their equilibrium positions. The formal solution for each q_i is

$$q_i(t) = -\frac{\kappa_i}{\omega_i} \int_0^t \sin[\omega_i (t - t')] q(t') dt', \qquad (2.31)$$

while the equation of motion for q reads as

$$\ddot{q} + \frac{\partial V}{\partial q} + \int_0^t K(t - t') q(t') dt' = 0. \qquad (2.32)$$

In the latter equation, the kernel $K(t - t')$ has the general form

$$K(t - t') = \sum_{i=1}^{N} \frac{\kappa_i^2}{\omega_i} \sin[\omega_i (t - t')], \qquad (2.33)$$

although other different forms can also be envisaged, for example, $K(t - t') = 2\gamma \delta(t - t')$, which is widely used. Thus, by extending to infinity the upper limit of the integral in (2.32) and assuming that $V(q)$ is harmonic, this equation reduces to the damped harmonic oscillator equation of motion (2.8). In this particular case, the Hamiltonian (2.30) becomes a quadratic function of the coordinates and momenta. As is well known, this type of Hamiltonian can be diagonalized exactly by means of a canonical transformation. The resulting Hamiltonian is also harmonic and consists of $N + 1$ independent oscillators with renormalized frequencies or normal mode frequencies [1].

Another interesting case arises when $V(q)$ describes a potential barrier, so that only tunneling allows a particle to pass through. In classical mechanics the momentum becomes imaginary and therefore an imaginary time formulation for the particle motion can be used. The corresponding Lagrangian (Euclidean Lagrangian) is expressed as

$$L = \frac{\dot{q}^2}{2} + V(q) + \sum_{j=1}^{N} \left(\frac{\dot{q}_j^2}{2} + \frac{1}{2} \omega_j^2 q_j^2 + \kappa_j q_j q \right), \qquad (2.34)$$

where the time derivatives are taken with respect to the imaginary time $\tau = it$ (*Wick rotation*). Now, the classical dynamics occurs in the inverted potential, with the equation of motion for q being

$$\ddot{q} - \frac{\partial V}{\partial q} + \int_{-\infty}^{+\infty} K(\tau - \tau')q(\tau')d\tau' = 0, \qquad (2.35)$$

where the explicit form of the kernel is

$$K(\tau - \tau') = -\sum_{j-1}^{N} \frac{\kappa_j^2}{2\omega_j} e^{-\omega_j|\tau - \tau'|}. \qquad (2.36)$$

2.3 Stochastic Dynamics

As mentioned at the beginning of Sect. 1.3, when dealing with many-body systems described by Hamiltonian functions like (1.38), dynamics may exhibit stochastic features. This is a "coarse-grained" effect arising when one only focuses on the dynamics of the system of interest, neglecting details about the environment dynamics. At present there are high performance numerical techniques (e.g., the so-called *Molecular Dynamics* methods [18]), which carry out sophisticated simulations of many degree-of-freedom classical systems. Relatively large sets of Hamiltonian or Newtonian coupled differential equations can be solved provided there is a complete information of the initial conditions for all the degrees of freedom involved. To some extent, these simulations mimic the own experiment (of course, at the level of accuracy of the model employed). Indeed, in those cases where no experiment is available, they play the role of an experiment itself. This is a very important advantage, although there are also some disadvantages. For example, among the main disadvantages, one finds that in these approaches some physical insight is unavoidably lost, for statistical methods have to be eventually considered in order to understand the underlying physics—the study of isolated trajectories in systems described by a large number of degrees of freedom is useless. This flaw can be surmounted through the use of some theoretical model devised within the framework of the theory of open classical systems. From this viewpoint, N-body problems can be replaced by simpler single-body ones, where an effective (phenomenological) interaction between the system of interest and the environment is assumed. In general, the effective interaction is introduced by means of a noise or fluctuating force coming from the bath and whose intensity is accounted for by a friction coefficient, and typically linear with temperature. The case where the friction coefficient is constant in space and time, as in (2.8), is called *Ohmic friction*. The drastic reduction of dimensionality of the original problem arising when stochastic models are assumed is very advantageous computationally, since the specific dynamics of the environment—with dimensions typically much larger than those associated with the subsystem of interest—is neglected. Furthermore, more importantly, this allows us to apply analytical statistical treatments to study the subsystem of interest, so that its dissipation mechanisms can be better characterized and understood.

The stochastization or randomization of a general physical process thus consists of carrying out a sort of coarse-graining in space and time [3]. The degree of "crudeness" required by a stochastization is directly related to the level of accuracy required by the spatial and temporal measurements of dynamical variables. A stochastic physical process is called *Markovian* if its time evolution is determined by the present and not its past (see Appendix B), losing very quickly any memory of its past. As a consequence, Markovian laws of motion are first-order differential equations with respect to time. Delayed effects and nonlocal properties are therefore not taken into account. The paradigm of stochastic processes is the Brownian motion, i.e., the seemingly random movement of particles suspended in a fluid, but also, in a more modern conception, the mathematical model used to describe similar random motions in other systems [19–21]. The random-walk problem is often considered as a model for such a motion. Brownian motion is not only Markovian, but also Gaussian, since the central limit theorem applies for sufficiently long times, at least longer than the system correlation time, so that the system has lost memory of its initial conditions. If the number of particles is not too large and the particularities of the interactions among them can be ignored, Brownian particles are governed by the standard diffusion equation. The mean time between collisions of Brownian particles and their surrounding is of the same order of magnitude or even slightly shorter than the average period of the environment fluctuating force.

There are mainly three ways to introduce stochasticity. First, phenomenologically, describing Brownian-like motions by means of the standard Langevin equation, where the system-environment interaction is governed by two parameters: temperature and friction [6, 22]. Second, starting from the Liouville equation, which is satisfied by any dynamical variable. Within this approach, Fokker–Planck-type equations can be easily reached. Actually, projection-operator techniques are very often used to obtain a generalized Langevin equation [4], where its kernel or memory function also fulfills a given integro-differential equation written in terms of its corresponding time-correlation function [5]. And third, as shown in Sect. 1.2.2, following the Ullersma model [14–17] or the so-called Caldeira–Leggett Hamiltonian model [23], the equations of motion can be expressed in terms of a generalized Langevin equation whenever the oscillators are not assumed to be at rest at $t_0 = 0$. The trajectories issued from solving such equations are called (classical) stochastic trajectories. Notice that this stochasticity is due to an external noise source, quite different from the inherent or intrinsic stochasticity related to chaotic dynamics (see Sect. 1.3).

A central issue which is not going to be treated here is the role played by external noise in nonequilibrium phase transitions, also called *noise-induced transitions* [24].

2.3.1 Brownian Motion and the Langevin Equation

Stochastic dynamics deals with random or stochastic variables and stochastic processes (see Appendix B), Brownian motion being a paradigm of this type of dynamics. This singular motion was formerly described by Ingen-Housz [25] in 1785 as an irregular motion of coal dust on a surface of alcohol—similar conclusions were drawn by Bywater [26] in 1819—and later on by Brown [27, 28] in 1827 when studying pollen particles suspended on water. Some of the mathematics behind the

Brownian motion are already incipient in Thiele's works on the least-square method in the 1880s [29]. However, it was not until 1880 when the first stochastic model to describe the stock option market as a Brownian motion was proposed by Bachelier [30]. Then, shortly after, independent physical solutions to the problem of Brownian motion were given by Einstein [31, 32], in 1905, and Smoluchowski [33], in 1906, who used this type of motion as an indirect proof of the existence of atoms and molecules. According to Einstein, the MSD of a Brownian particle is proportional to the first power of time—a result reminiscent of the random-walk problem—, this being the main feature defining Brownian motion.

For simplicity, a one-dimensional description of Brownian motion is going to be considered, since the essential physics is well contained in this simple case. This motion, for example, takes place when a particle is adsorbed on a flat surface. Due to the fact the particle–surface interaction is zero, no direction is privileged. Furthermore, this dynamics will be the starting point to discuss simple physical processes in terms of quantum stochastic trajectories in Volume 2. The equation of motion describing a Brownian particle of mass m embedded in a fluid, proposed by Langevin [34] in 1908, is given by

$$m\dot{v} = -m\gamma v + mR_G(t), \tag{2.37}$$

or, in the form of a stochastic differential equation, as

$$mdv = -m\gamma vdt + mdW(t), \tag{2.38}$$

where $dW(t) = R_G(t)dt$ is a *Wiener process* [35, 36] (see Appendix B). The right-hand side of this equation can be split up into two contributions:

1. A deterministic part, characterized by the friction force $-m\gamma v$, with γ being the friction coefficient depending on the fluid viscosity.
2. A random part, governed by the random force $mR_G(t)$ or *Gaussian white noise*.

Since the random force is described by a Wiener process, it satisfies the two conditions of a typical *Gaussian white noise*:

1. The stochastic process $R_G(t)$ is Gaussian with zero mean, i.e.,

$$\langle R_G(t) \rangle = 0.$$

2. The force–force time-correlation function is infinitely short, i.e.,

$$m^2 \langle R_G(0)R_G(\tau) \rangle = A\delta(\tau),$$

with A being a constant giving the strength of the coupling between particle and environment and determined by the energy equipartition theorem.

Fig. 2.1 As an example of Brownian motion driven by a Gaussian white noise, as in Fig. 1.2, the classical stochastic trajectory pursued by a Na atom is also displayed here at $T = 300$ K, though on a flat surface ($V = 0$). The friction constant is $\gamma = 0.5\,\text{ps}^{-1}$ and the evolution is up to $t = 20,000\,\text{ps}$

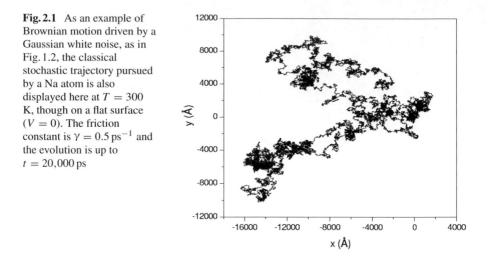

The validity of this model relies on the fact that the Brownian particle is much heavier than the environmental ones. This implies that the kicks received by the particle, although relatively weak, they are very effective when considered in a very large number—the central limit theorem holds and, therefore, the noise becomes Gaussian. Actually, these kicks can be seen as coming from thermal fluctuations of the surroundings. Remember that the detailed time-evolution of the environment degrees of freedom is not taken into account because their correlations decay faster than those of the particle (Markovian approximation), as expressed by the property 2. Thus, they are accounted for by assuming the presence of fluctuations that perturb the free evolution of the particle. In Fig. 2.1 one classical stochastic trajectory driven by a Gaussian white noise is plotted, which simulates a realization of a two-dimensional Brownian motion. In order to obtain information about the diffusion process, a swarm of these trajectories (i.e., a sampling over many Brownian realizations) should be considered.

The relationship between the friction in the Langevin equation and the fluctuations of the random force is given by the *fluctuation–dissipation theorem* [3], which reads as

$$\gamma(\omega) = \frac{m}{2k_B T} \int_{-\infty}^{\infty} \langle \delta R_G(0)\, \delta R_G(\tau) \rangle\, e^{-i\omega\tau} d\tau, \qquad (2.39)$$

where

$$\delta R_G(t) \equiv R_G(t) - \langle R_G(t) \rangle \qquad (2.40)$$

is the fluctuation due to the random noise function $R_G(t)$ and k_B is the Boltzmann constant. Whenever properties 1 and 2 for a Gaussian white noise apply, the friction coefficient becomes independent of the frequency

$$\gamma(\omega) = \frac{A}{2mk_BT},$$ (2.41)

with T being the heat bath temperature. Thus, the frequency spectrum of the friction force is flat or *white*, in the sense that all frequencies contribute equally to it, in analogy to white light, i.e.,

$$\gamma(\omega) \equiv \gamma.$$ (2.42)

The strength of the coupling between the Brownian particle and the environment will be

$$A = 2m\gamma k_BT,$$ (2.43)

and the Gaussian white noise correlation function

$$\mathcal{G}_G(\tau) \equiv \langle \delta R_G(0)\, \delta R_G(\tau) \rangle = \frac{2\gamma k_BT}{m}\, \delta(\tau).$$ (2.44)

Physically, this dynamics implies that, at thermal equilibrium, the equipartition theorem holds.

In general, there exists an interaction between the particle and the surface (for example, if the surface is corrugated). Thus, (2.37) can be rewritten as

$$\ddot{q}(t) = -\gamma\,\dot{q}(t')dt' + F(q(t)) + \delta R_G(t),$$ (2.45)

where q represents the particle position and $F = -\nabla V$ is the deterministic force per mass unit derived from the interaction or external potential, V. The solution of this equation can be readily obtained by formal integration, to yield

$$v(t) = v_0 e^{-\gamma t} + \int_0^t e^{-\gamma(t-t')} F(x(t'))dt' + \int_0^t e^{-\gamma(t-t')}\delta R_G(t')dt',$$ (2.46a)

$$q(t) = q_0 + \frac{v_0}{\gamma}\left(1 - e^{-\gamma t}\right) + \frac{1}{\gamma}\int_0^t \left[1 - e^{-\gamma(t-t')}\right]F(q(t'))dt'$$

$$+ \frac{1}{\gamma}\int_0^t \left[1 - e^{-\gamma(t-t')}\right]\delta R_G(t')dt',$$ (2.46b)

where $v_0 = v(0)$ and $q_0 = q(0)$. As can be seen, for $\delta R_G = 0$, (2.46) are the formal solutions of purely deterministic equations of motion. Therefore, without loss of generality, they can be expressed as

$$v(t) = v_d(t) + v_s(t),$$ (2.47a)
$$q(t) = q_d(t) + q_s(t),$$ (2.47b)

where d refers to the deterministic terms of the solutions and s to those associated with the stochastic force. Nevertheless, note that when $\delta R_G(t) \neq 0$ the deterministic part will also present some stochastic features due to the evaluation of $F(q)$ along the trajectory $q(t)$, which is a stochastic process.

Taking advantage of properties 1 and 2 for a Gaussian white noise, the main average quantities can be written as (see Appendix B)

$$\langle v(t) \rangle = \bar{v}_d(t), \tag{2.48a}$$

$$\langle v^2(t) \rangle = \bar{v}_d^2(t) + \langle v_s^2(t) \rangle, \tag{2.48b}$$

$$\langle q(t) \rangle = \bar{q}_d(t), \tag{2.48c}$$

$$\langle q^2(t) \rangle = \bar{q}_d^2(t) + \langle q_s^2(t) \rangle, \tag{2.48d}$$

where the "barred" magnitudes indicate the respective averages of the deterministic part of the solution and

$$\langle v_s^2(t) \rangle = e^{-2\gamma t} \int_0^t dt' e^{2\gamma t'} \int_{-t'}^{t-t'} e^{\gamma \tau} \mathcal{G}_G(\tau) \, d\tau, \tag{2.49a}$$

$$\langle q_s^2(t) \rangle = \frac{1}{\gamma^2} \int_0^t dt' \left[1 - e^{-\gamma(t-t')} \right] \int_{-t'}^{t-t'} \left[1 - e^{-\gamma(t-t'-\tau)} \right] \mathcal{G}_G(\tau) \, d\tau. \tag{2.49b}$$

The final form of these expressions thus reads as

$$\langle v_s^2(t) \rangle = \frac{1}{\alpha^2} \left(1 - e^{-2\gamma t} \right), \tag{2.50a}$$

$$\langle q_s^2(t) \rangle = \frac{1}{\gamma^2 \alpha^2} \left[2\gamma t + 1 - \left(2 - e^{-\gamma t} \right)^2 \right], \tag{2.50b}$$

with $\alpha = \sqrt{m/k_B T}$. For example, if $V = 0$, the system is initially thermalized (i.e., it follows a Maxwell–Boltzmann velocity distribution) and has a uniform probability distribution in positions around $q = 0$, then $\bar{v}_0 = 0$, $\bar{v}_0^2 = k_B T/m$, and $\bar{q}_0 = 0$. This leads to

$$\langle v(t) \rangle = 0, \tag{2.51a}$$

$$\langle v^2(t) \rangle = \frac{k_B T}{m}, \tag{2.51b}$$

$$\langle q(t) \rangle = 0, \tag{2.51c}$$

$$\langle q^2(t) \rangle = \bar{q}_0^2 + \frac{k_B T}{m\gamma^2} \left[2\gamma t + 1 - \left(2 - e^{-\gamma t} \right)^2 \right], \tag{2.51d}$$

as it was found by Wiener in his description of Brownian motion [35].

From (2.51), two dynamical regimes can be clearly distinguished depending on the value of γt. For $\gamma t \ll 1$, collision events are rare and the particle shows an almost free motion with relatively long mean free paths. This is the *ballistic* or *free-diffusion regime*, characterized by the MSD

$$\langle q^2(t) \rangle \sim \frac{k_B T}{m} t^2. \tag{2.52}$$

On the other hand, for $\gamma t \gg 1$, there is no free diffusion, since the effects of the stochastic force (collisions) are dominant. This is the *diffusive regime*, where the MSD is linear with time,

$$\langle q^2(t) \rangle \simeq \frac{2k_B T}{m\gamma} t = 2Dt. \tag{2.53}$$

This is the so-called *Einstein's law* for diffusion. As can be inferred from (2.53), by lowering the friction γ acting on the particle, a faster diffusion takes place (the diffusion coefficient D increases). Transport processes characterized by a MSD violating Einstein's law are generically called *anomalous transport processes* [37] (see Sect. 1.3.2).

2.3.2 Brownian Motion and the Liouville Equation

As mentioned above, one can also seek for a statistical description of the dynamics and study the evolution of (statistical) ensembles of stochastic trajectories. In Sect. 1.4.1, it was already briefly discussed how classical dynamics generate probability densities in phase space. This settled down the basis to define statistical ensembles. The time-evolution of these probability densities is governed by the Liouville equation. Similarly, the time-evolution of any general dynamical variable A, which is a function of the phase–space point at any time, is given by an analogous equation,

$$\frac{\partial A}{\partial t} = -\mathcal{L}A. \tag{2.54}$$

In order to describe experimental macroscopic quantities, a coarse graining in time or time-average of the dynamical variable of interest should be carried out. As seen in Sect. 1.3.1, if the equations of motion are fully deterministic (as it happens, for example, in Molecular Dynamics simulations), the cause leading to make time-averages equivalent to phase–space averages (ergodic hypothesis) and then justifying the calculation of thermodynamic properties, is the intrinsic dynamical instability or chaos. However, when talking about transport properties, the key element is the decay of correlation functions with time, where the Green–Kubo relations constitutes the ordinary link between Liouvillian dynamics and transport coefficients. Although ergodicity is an important condition, it is not very useful, since one cannot define a timescale based only on this property. Again, the intrinsic relaxation times of correlation functions are ruled by the dynamical instability, which also allows to obtain transport coefficients from it. Time-correlation functions are also of experimental relevance, since the spectra measured by various spectroscopic techniques are the power spectra of well-defined dynamical variables. As mentioned in Chap. 1, resonant behavior can be extracted from the complex frequency spectrum of such spectral functions.

The evolution of a Brownian particle can be described (in a Cartesian three-dimensional configuration space) by a dynamical equation [3]

$$\dot{\mathbf{r}} = \mathbf{v}(t), \tag{2.55}$$

where $\mathbf{v}(t)$ is a velocity associated with the particle displacement $\mathbf{r}(t)$. Now, due to the type of motion, $\mathbf{v}(t)$ is a stochastic process and, therefore, $\mathbf{r}(t)$ will also be a stochastic process, which is obtained from $\mathbf{r}(t)$ (of course, not in the usual way how the concept of time-derivative is understood). Thus, consider the microscopic density distribution function (or, more specifically, the probability density) is $\rho(\mathbf{r}, t)$. This quantity gives the probability that a Brownian particle can be found within a volume element $d\mathbf{r}$ of the configuration space at a time t. By invoking the probability conservation property, it can be shown that $\rho(\mathbf{r}, t)$ obeys a stochastic Liouville equation,

$$\frac{\partial \rho(\mathbf{r}, t)}{\partial t} = -\mathcal{L}\rho(\mathbf{r}, t), \tag{2.56}$$

where the differential operator $\mathcal{L}\rho(\mathbf{r}, t) \equiv -\nabla \cdot \{\mathbf{v}(t)\rho(\mathbf{r}, t)\}$ is itself a stochastic operator because $\mathbf{v}(t)$ is a stochastic process. In the reciprocal $\mathbf{k}$-space, (2.56) becomes

$$\frac{\partial \tilde{\rho}(\mathbf{k}, t)}{\partial t} = -i\mathbf{k} \cdot \mathbf{v}(t)\tilde{\rho}(\mathbf{k}, t), \tag{2.57}$$

where $\tilde{\rho}(\mathbf{k}, t)$ is the Fourier transform of $\rho(\mathbf{r}, t)$. The solution of (2.57) is readily obtained to yield

$$\tilde{\rho}(\mathbf{k}, t) = \tilde{\rho}(\mathbf{k}, 0)e^{-i\int_0^t \mathbf{k} \cdot \mathbf{v}(t')dt'}. \tag{2.58}$$

Since $\mathbf{v}(t)$ is a stochastic process, the probability distribution in Fourier space is given by averaging over all possible paths,

$$I(\mathbf{k}, t) = \langle \tilde{\rho}(\mathbf{k}, t)\tilde{\rho}(\mathbf{k}, 0)\rangle \propto \langle e^{-i\int_0^t \mathbf{k} \cdot \mathbf{v}(t')dt'}\rangle. \tag{2.59}$$

This is the definition of the characteristic function (see Appendix B) for the stochastic variable or stochastic trajectory $\mathbf{v}(t)$, also called *intermediate scattering function* within the context of diffusion processes [38]. In the same context, the time Fourier transform of (2.59) is the so-called *scattering law* or *dynamic structure factor* (except for some normalization factor),

$$S(\omega, t) = \int e^{-i\omega t} I(\mathbf{k}, t)dt, \tag{2.60}$$

which is directly related to the observable in diffusion experiments.

Assuming that the position stochastic variable is Gaussian, (2.59) can be reexpressed as

$$I(\mathbf{k}, t) = e^{-\mathbf{k}^2 \langle v_\mathbf{k}^2 \rangle \int_0^t (t-\tau)\phi(\tau)d\tau}, \tag{2.61}$$

where $\phi(t)$ is the normalized velocity autocorrelation function and $v_\mathbf{k}$ stands for the velocity vector projected along $\mathbf{k}$. According to the discussion above, for the motion of the physical particle to behave as a Gaussian, Markovian process, a coarse graining in time should be imposed by taking the long-time limit. Now, if the correlation time is defined as

$$\tau_c = \int_0^\infty \phi(\tau)d\tau, \tag{2.62}$$

equation (2.61) is approximated by

$$I(\mathbf{k}, t) \sim e^{-\mathbf{k}^2 \langle v_\mathbf{k}^2 \rangle \tau_c t + \delta}, \tag{2.63}$$

with $\delta = \mathbf{k}^2 \int_0^\infty \tau\phi(\tau)d\tau$. The intermediate scattering function (2.63) is the solution of the differential equation

$$\frac{\partial I(\mathbf{k}, t)}{\partial t} = -\mathbf{k}^2 \langle v_\mathbf{k}^2 \rangle \tau_c I(\mathbf{k}, t). \tag{2.64}$$

If this equation is Fourier-transformed back to the configuration space, one obtains the standard diffusion equation,

$$\frac{\partial P(\mathbf{r}, t)}{\partial t} = D\nabla^2 P(\mathbf{r}, t). \tag{2.65}$$

In this equation, $P(\mathbf{r}, t)$ is the normalized autocorrelation function of the microscopic number density $\rho(\mathbf{r}, t)$, with initial condition

$$P(\mathbf{r}, 0) = \delta(\mathbf{r} - \mathbf{r}_0), \tag{2.66}$$

and $D = \tau_c \langle v_\mathbf{k}^2 \rangle$ is the diffusion coefficient. The solution of the diffusion equation (2.65) is given by

$$P(\mathbf{r}, t) = \left(\frac{1}{4\pi Dt}\right)^{3/2} e^{-(\mathbf{r}-\mathbf{r}_0)^2/4Dt}, \tag{2.67}$$

which means that, as time goes on, the probability distribution gradually broadens, leading to an irreversible motion.

So far diffusion has been described in the configuration space. Obviously, it also admits a description in the velocity space or in the phase space. In the velocity space, the diffusion equation is described by the *Fokker–Planck equation* [39–41],

$$\frac{\partial P(\mathbf{v}, t)}{\partial t} = \gamma \nabla_\mathbf{v} \cdot [\mathbf{v} P(\mathbf{v}, t)] + \frac{\gamma k_B T}{m} \nabla_\mathbf{v}^2 P(\mathbf{v}, t), \tag{2.68}$$

with γ and $D = \gamma k_B T/m$ being the drift and diffusion coefficients, respectively. The probability density is again sharply peaked at v_0 and, asymptotically in time, it approaches the stationary Maxwellian distribution at temperature T. Regarding the phase–space description, the corresponding diffusion equation is the so-called *Kramers–Chandrasekhar equation* [6, 42, 43]. Within a more general context, it can be shown [44] that all these diffusion equations can be obtained from the reduced Liouville equation for one particle.

In 1940, Kramers [42] proposed a one-dimensional diffusion model for chemical reactions based on the motion of a Brownian particle under the action of an external potential V. In particular, originally V had the form of an asymmetric double well potential. Within this model, the Langevin equation describing the evolution of the reaction coordinate reads as

$$dq = v\,dt, \tag{2.69a}$$

$$m\,dv = -\left(\frac{dV}{dq} + \gamma m v\right)dt + \sqrt{2m\gamma k_B T}\ dW(t). \tag{2.69b}$$

Kramers was interested in the escape rate of the particle from a well. Two different regimes were thus considered for the rate:

1. *Strong friction*, where the friction coefficient is greater than the barrier frequency and the rate is limited by a spatial diffusion, decreasing as γ^{-1}.
2. *Weak friction*, where, on the contrary, the rate is limited by an energy diffusion process and increases linearly with γ.

These two extreme behaviors imply a maximum in between, namely the *Kramers turnover problem* [45]. For example, in the strong friction regime, after a time of the order of γ^{-1} all inertial effects have died out. This means that the left-hand side of (2.69b) is equal to zero (i.e., $dv = 0$) and therefore (2.69a) can be approximated by

$$dq = -\frac{1}{m\gamma}\frac{dV}{dq}dt + \sqrt{\frac{2k_B T}{m\gamma}}\,dW(t). \tag{2.70}$$

The corresponding Fokker–Planck or Smoluchowski equation can then be expressed as

$$\frac{\partial P(q,t)}{\partial t} = -\frac{\partial}{\partial q}\left[\frac{1}{m\gamma}\frac{dV}{dq}P(q,t)\right] + \frac{1}{2}\frac{\partial^2}{\partial q^2}\left[\frac{2k_B T}{m\gamma}P(q,t)\right]. \tag{2.71}$$

In general, solving numerically the Langevin set of equations (2.69) turns out to be easier at any regime than dealing with partial differential equations, such as the Fokker–Planck equation (2.71). In general, the Fokker–Planck equation can be solved as an eigenvalue problem [21]. A quantum and classical theory of surface diffusion based on Kramers' theory of activated escape over one-dimensional potential barriers was developed by Pollak et al. [46–48] and Mel'nikov [49, 50]. Applications to Na atom diffusion on (corrugated) Cu surfaces can be found in the literature [51–53].

In this case, the stochastic trajectories issued from the numerical resolution of Langevin equations similar to the system constituted by (2.69) were used to build the corresponding intermediate scattering function (2.59) and scattering law (2.60).

2.3.3 The System-plus-Bath Approach

The system-plus-bath approach is perhaps the most successful and useful way to deal with stochastic dynamics, since it starts from a total system (system-plus-bath) which is conservative. In the particular case of open quantum systems, it seems to be the most natural approach. Notice that the passage from the classical system to the quantum-mechanical one, i.e., the *quantization* of the classical system, can be done in a straightforward way, applying different standard methods available in quantum mechanics (e.g., via quantum-classical correspondence).

Within the system-plus-bath approach, the corresponding dynamics is commonly described by a total Hamiltonian which is split up into three different parts,

$$H = H_S + H_B + H_{SB}, \tag{2.72}$$

where H_S is the system Hamiltonian, H_B is the bath Hamiltonian, and H_{SB} is the term describing the system-bath interaction or coupling. As mentioned above, the system usually consists of a few degrees of freedom, while the environment is formed by a huge number of them (even infinity). Moreover, it is reasonable to assume that the coupling between them is a linear function of the bath coordinates. This property of linearity is very convenient, since it is then very easy to eliminate the bath coordinates in an exact way. In this regard, for extensive systems, like a reservoir, it is very common to assume a set of N harmonic oscillators,

$$H_B = \frac{1}{2} \sum_{i=1}^{N} \left(\frac{p_i^2}{m_i} + m_i \omega_i^2 q_i^2 \right), \tag{2.73}$$

where q_i and p_i are the position and momentum of the ith oscillator, and m_i and ω_i its mass and frequency, respectively (this one-dimensional Hamiltonian can be extended straightforwardly to three dimensions). Very often, the dissipation mechanism is independent of the choice of this type of bath.

This kind of approach is widely used to describe stochastic processes where dissipation and damping play a fundamental role. Without loss of generality, consider the system is formed by only one degree of freedom and its Hamiltonian is written as

$$H_S = \frac{P^2}{2M} + V(Q), \tag{2.74}$$

where M is the particle mass and $V(Q)$ is an external potential acting on the particle. The coupling between system and bath is generally expressed as a linear interaction term with the following expression

$$H_{SB} = \sum_{i=1}^{N} \left(\frac{c_i^2}{m_i \omega_i^2} Q^2 - 2 c_i q_i Q \right), \tag{2.75}$$

c_i being the strength of the linear coupling. The classical equations of motion for a global system described by the total Hamiltonian H, which in the field of condensed matter physics is known as the *Caldeira–Leggett model Hamiltonian* [46–48, 54], leads to a generalized Langevin equation for $Q(t)$,

$$M \ddot{Q}(t) + M \int_0^t dt' \gamma(t - t') \dot{Q}(t') + \frac{\partial V(Q)}{\partial Q} = R_G(t). \tag{2.76}$$

Here, the memory kernel or time-dependent friction reads as

$$\gamma(t) = \sum_i \frac{c_i^2}{m_i \omega_i} \cos(\omega_i t) \tag{2.77}$$

and the external force $R_G(t)$ as

$$R_G(t) = -\sum_i c_i \left\{ \left[q_i(0) + \frac{c_i}{m_i \omega_i^2} Q(0) \right] \cos(\omega_i t) + \frac{p_i(0)}{m_i \omega_i} \sin(\omega_i t) \right\}. \tag{2.78}$$

Given a suitably defined thermal distribution of initial conditions, denoted by $(Q(0), P(0))$ and $(q_i(0), p_i(0))$, the external force is Gaussian distributed with zero mean, obeying the classical fluctuation–dissipation theorem. The bath or reservoir at a given temperature T is thus a source of noise displaying memory effects. The friction, in this case, is not a constant, but a time-dependent function. Nonetheless, in many physical situations, the memory kernel is a δ-function of time, which leads to a constant (Ohmic) friction. Then, as mentioned above, the system dynamics becomes Markovian, losing track of its past. As can be noticed, the generalized Langevin equation (2.76) is equivalent to (2.32) when all initial conditions are set to zero (no temperature) and the kernel (2.77) is similar to (2.33). Moreover, (2.76) reduces to a standard Langevin equation in the Markovian approximation,

$$\ddot{Q} + \frac{\partial V}{\partial Q} - \gamma Q = R_G(t), \tag{2.79}$$

with $R_G(t) = 2\gamma \delta(t)$.

Nonlinear functions in (2.75) can also be envisaged [23]. In such a case, the open classical system becomes a state-dependent dissipation process and the random force exhibits multiplicative noise (see Appendix B). This leads to noise-induced transitions. This situation will not be considered in this monograph, although it is worth mentioning that nonlinear environments are the subject of intensive research at present, since many important physical processes are better described in this way.

Very recently, some of the classical scattering singularities (rainbow, glory and skipping effects) mentioned in Chap. 1 have been considered under a stochastic view-point. The corresponding analysis has been carried out for adsorbate diffusion [53] on surfaces as well as atom surface scattering [55]. This analysis could be easily extended to any type of scattering. The main features observed and interpreted in terms of a stochastic analysis lead to broadenings and shiftings as a function of the surface temperature. Friction-induced energy loss spectra have also been predicted in atom-surface scattering [56].

2.4 The Stochastic Hamilton–Jacobi Equation

Even if the overall dynamics observed is conservative, one could consider the possibility to understand this "regularity" as the result of an underlying stochastic dynamics. The theory of open classical systems could be then applied to describe this underlying motion. Specifically, assuming the corresponding motions are Brownian-like, they could be described in terms of an *Itô stochastic differential equation* (see Appendix B). In this case, the associated Itô stochastic equation reads [36, 57, 58] as

$$dr(t) = a_+(r, t)dt + b \, dW(t), \tag{2.80}$$

where a_+ is the *mean forward derivative* of the particle position or displacement, and b accounts for the strength of the stochastic force. The diffusion equation for the corresponding probability distribution is described by the Fokker–Planck equation

$$\frac{\partial P(r, t)}{\partial t} = -\nabla \cdot [a_+(r, t)P(r, t)] + \frac{b^2}{2}\nabla^2 P(r, t). \tag{2.81}$$

Under time-inversion, this Fokker–Planck equation can also be written as

$$\frac{\partial P(r, t)}{\partial t} = -\nabla \cdot [a_-(r, t)P(r, t)] - \frac{b^2}{2}\nabla^2 P(r, t), \tag{2.82}$$

where a_- now denotes the *mean backward derivative* of the particle position. From the definitions for two mean derivatives, the *particle mean derivative* is now defined as

$$v(r, t) = \frac{1}{2}(a_+ + a_-). \tag{2.83}$$

This allows to express the continuity equation for this process as

$$\frac{\partial P(r, t)}{\partial t} + \nabla \cdot [v(r, t)P(r, t)] = 0 \tag{2.84}$$

after adding the Fokker–Planck equations (2.81) and (2.82). On the other hand, after substraction of the two Fokker–Planck equations, an additional vector field can be defined, namely the *osmotic velocity*,

$$\mathbf{u}(\mathbf{r}, t) = \frac{1}{2} (\mathbf{a}_+ - \mathbf{a}_-) = \frac{b^2}{2} \nabla \ln P(\mathbf{r}, t), \tag{2.85}$$

with its time-derivative being

$$\frac{\partial \mathbf{u}}{\partial t} = -\frac{b^2}{2} \nabla(\nabla \cdot \mathbf{v}) - \nabla(\mathbf{v} \cdot \mathbf{u}). \tag{2.86}$$

It can be shown [36] that a mean acceleration in this kind of processes can also be defined as

$$\mathbf{a} = \frac{1}{2} \frac{\partial}{\partial t} (\mathbf{a}_+ + \mathbf{a}_-) + \frac{1}{2} \mathbf{a}_+ \cdot \nabla \mathbf{a}_- + \frac{1}{2} \mathbf{a}_- \cdot \nabla \mathbf{a}_+ - \frac{b^2}{4} \nabla^2 (\mathbf{a}_+ - \mathbf{a}_-). \tag{2.87}$$

Hence, taking into account the definitions given above for the mean velocity (2.83) and the osmotic velocity (2.86), the time-derivative resulting for $\mathbf{v}$ reads as

$$\frac{\partial \mathbf{v}}{\partial t} = \mathbf{a} - \mathbf{v} \cdot \nabla \mathbf{v} + \mathbf{u} \cdot \nabla \mathbf{u} + \frac{b^2}{2} \nabla^2 \mathbf{u}. \tag{2.88}$$

The above elements provide a full general hydrodynamic description of Brownian motion. For an overall conservative dynamics—i.e., as arisen from a conservative Markovian diffusion process—, it is assumed that the Brownian particle is moving in an external potential $V(\mathbf{r})$ and the stochastic mean acceleration is given as $\mathbf{a} = -\nabla V(\mathbf{r})/m$. Then the time-derivative of the mean velocity field (2.88) can be expressed as

$$\frac{\partial \mathbf{v}}{\partial t} = -\frac{1}{m} \nabla V(\mathbf{r}) - \mathbf{v} \cdot \nabla \mathbf{v} + \mathbf{u} \cdot \nabla \mathbf{u} + \frac{b^2}{2} \nabla^2 \mathbf{u}. \tag{2.89}$$

This equation provides now a complete general description of the hydrodynamics of Brownian motion under the action of an external potential V. Actually, one can further proceed. Thus, if the probability density is defined by a scalar field $R(\mathbf{r}, t)$ as

$$P(\mathbf{r}, t) = e^{2R(\mathbf{r}, t)}, \tag{2.90}$$

then the osmotic velocity (2.85) is given by

$$\mathbf{u}(\mathbf{r}, t) = b^2 \nabla R(\mathbf{r}, t). \tag{2.91}$$

Similarly, the velocity field can also be defined in terms of the gradient of another scalar field $S(\mathbf{r}, t)$,

$$\mathbf{v}(\mathbf{r}, t) = \frac{\nabla S(\mathbf{r}, t)}{m}. \tag{2.92}$$

As it can be noticed (see Chap. 1), this equation is analogous to the corresponding one for (classical) conservative systems, which establishes a relationship between the particle velocity and its associated action. Taking this into account, (2.84) and (2.89) can also be rewritten as

$$\frac{\partial R}{\partial t} + \frac{\nabla^2 S}{2m} + \frac{1}{m}\nabla R \cdot \nabla S = 0, \tag{2.93a}$$

$$\frac{\partial S}{\partial t} + \frac{(\nabla S)^2}{2m} + V - \frac{mb^4}{2}[(\nabla R)^2 + \nabla^2 R] = 0. \tag{2.93b}$$

The fields R and S can be determined except for a time-dependent phase. Equations (2.93) can be regarded as the hydrodynamic formulation of Newtonian mechanics and constitute respectively the stochastic mechanic counterpart of the classical continuity and Hamilton–Jacobi equations seen in Chap. 1. Note that (2.93b) reduces to the classical Hamilton–Jacobi equation (1.15) for $b = 0$. Furthermore, it is worth stressing that, within this framework, a conservative diffusion process has been generated. As will be seen in Chap. 6, similar coupled equations are also found in Bohmian mechanics.

References

1. Razavy, M.: Classical and Quantum Dissipative Systems. Imperial College Press, London (2005)
2. Hänggi, P., Jung, P.: Colored noise in dynamical systems. Adv. Chem. Phys. **89**, 239–326 (1995)
3. Kubo, R., Toda, M., Hashitsume, N.: Statistical Physics II. Nonequilibrium Statistical Mechanics. Springer-Verlag, Berlin (1985)
4. Hansen, J.P., McDonald, I.R.: Theory of Simple Liquids. Academic Press, New York (1986)
5. Boon, J.P., Yip, S.: Molecular Hydrodynamics. Dover, New York (1991)
6. McQuarrie, D.A.: Statistical Mechanics. Harper and Row, New York (1976)
7. Lichtenberg, A.J., Lieberman, M.A.: Regular and Chaotic Dynamics, 2nd edn. Springer-Verlag, New York (1992)
8. Goldstein, H.: Classical Mechanics. Addison–Wesley Publising Company, Reading, MA (1980)
9. Caldirola, P.: Forze non conservative nella meccanica quantistica. Nuovo Cimento **18**, 393–400 (1941)
10. Kanai, E.: On the quantization of the dissipative systems. Prog. Theor. Phys. **3**, 440–442 (1948)
11. Dekker, H.: Classical and quantum mechanics of the damped harmonic oscillator. Phys. Rep. **80**, 1–110 (1981)
12. Morse, P.M., Feshbach, H.: Methods of Theoretical Physics. McGraw-Hill, New York (1953)
13. Bateman, H.: On dissipative systems and related variational principles. Phys. Rev. **38**, 815–819 (1931)
14. Ullersma, P.: An exactly solvable model for Brownian motion: I Derivation of the Langevin equation. Physica (Utrecht) **32**, 27–55 (1966)
15. Ullersma, P.: An exactly solvable model for Brownian motion: II Derivation of the Langevin equation. Physica (Utrecht) **32**, 56–73 (1966)

16. Ullersma, P.: An exactly solvable model for Brownian motion: III Motion of a heavy mass in a linear chain. Physica (Utrecht) **32**, 74–89 (1966)
17. Ullersma, P.: An exactly solvable model for Brownian motion: IV Susceptibility and Nyquist's theorem. Physica (Utrecht) **32**, 90–96 (1966)
18. Frenkel, D., Smit, B.: Understanding Molecular Simulation: From Algorithms to Applications. Academic Press, New York (2002)
19. van Kampen, N.G.: Stochastic Processes in Physics and Chemistry. North-Holland, Amsterdam (1981)
20. Gardiner, C.W.: Handbook of Stochastic Methods. Springer-Verlag, Berlin (1983)
21. Risken, R.: The Fokker–Planck Equation. Springer-Verlag, Berlin (1984)
22. Pathria, R.K.: Statistical Mechanics. Pergamon Press, Oxford (1972)
23. Weiss, U.: Quantum Dissipative Systems. World Scientific, Singapore (1999)
24. Horsthemke, W., Lefever, R.: Noise-Induced Transition. Springer-Verlag, Berlin (1984)
25. Ingen-Housz, J.: Vermischte Schriften physisch-medizinischen Inhalts, vol. 2, pp. 123–126. Wappler, Vienna (1784)
26. Bywater, J.: Physiological Fragments: To Which are Added Supplementary Observations, to Shew That Vital Energies are of the Same Nature, and Both Derived From Solar Light, pp. 127–128. R. Hunter, London (1824)
27. Brow, R.: A brief account of microscopical observations made in the months of June, July and August 1827, on the particles contained in the poller of plants; and on the general existence of active molecules in organic and inorganic bodies. Phys. Mag. **4**, 161–173 (1828)
28. Brown, R.: Additional remarks on active molecules. Phys. Mag. **6**, 161–166 (1829)
29. Lauritzen, S.L.: Thiele: Pioneer in Statistics. Oxford University Press, Oxford (2002)
30. Bachelier, L.: Théorie de la Spéculation, Annales Scientifiques de l' École Normale Supérieure, vol. 3, pp. 21–86 (1900). Translated into English by Davis, M. Etheridge, A.: Louis Bachelier's Theory of Speculation: The Origins of Modern Finance. Princeton University Press, Princeton (2006)
31. Einstein, A.: Über die von der molekularkinetischen Theorie der Wärme geforderte Bewegung von in ruhenden Flüssigkeiten suspendierten Teilchen. Ann. Physik **17**, 549–560 (1905)
32. Einstein, A.: Investigations on the Theory of Brownian Movement. Dover, New York (1956)
33. Smoluchowski, M.: Zur kinetischen Theorie der Brownschen Molekularbewegung und der Suspensionen. Ann. Physik **21**, 756–780 (1906)
34. Langevin, P.: Sur la théorie du mouvement brownien. Comptes Rendus **146**, 530–532 (1908)
35. Wiener, N.: The average of an analytical functional and the Brownian movement. Proc. Nat. Acad. Sci. U. S. A. **7**, 294–298 (1921)
36. Wolfgang, P., Baschnagel , J.: Stochastic Processes. From Physics to Finance. Springer-Verlag, Berlin (1999)
37. Metzler, R., Klafter, J.: The random walk's guide to anomalous diffusion: a fractional dynamics approach. Phys. Rep. **339**, 1–77 (2000)
38. Lovesey, S.W.: Theory of Neutron Scattering From Condensed Matter. Clarendon Press, Oxford (1984)
39. Fokker, A.D.: Die mittlere Energie rotierender elektrischer Dipole im Strahlungsfeld. Ann. Physik **43**, 810–820 (1914)
40. Planck, M.: Über einen Satz der statistichen Dynamik und eine Erweiterung in der Quantumtheorie. Sitzb. Akad. Berlin, pp. 324–341 (1917)
41. Kolmogorov, A.N.: Über die analytischen Methoden in der Wahrscheinlichkeitsrechnung. Math. Ann. **104**, 415–458 (1931)
42. Kramers, H.A.: Brownian motion in a filed of force and the diffusion model of chemical reactions. Physica **8**, 284–304 (1940)
43. Chandrasekhar, S.: Stochastic problems in physics and astronomy. Rev. Mod. Phys. **15**, 1–89 (1943)
44. Balazs, N.L.: On the quantum-mechanical Fokker–Planck and Kramers–Chandrasekhar equation. Physica **94A**, 181–191 (1978)

45. Hänggi, P., Talkner, P., Borkovec, M.: Reaction-rate theory: fifty years after Kramers. Rev. Mod. Phys. **62**, 251–341 (1990)
46. Pollak, E.: Theory of activated rate processes: a new derivation of Kramers' expression. J. Chem. Phys. **85**, 865–867 (1986)
47. Pollak, E., Grabert, H., Hänggi, P.: Theory of activated rate processes for arbitrary frequency dependent friction: solution of the turnover problem. J. Chem. Phys. **91**, 4073–4087 (1989)
48. Georgievskii, Y., Pollak, E.: Semiclassical theory of activated diffusion. Phys. Rev. E **49**, 5098–5102 (1994)
49, Mel'nikov, V.I., Meshkov, S.V.: Theory of activated rate processes: exact solution of the Kramers problem. J. Chem.Phys. **85**, 1018–1027 (1986)
50. Mel'nikov, V.I.: The Kramers problem: Fifty years of development. Phys. Rep. **209**, 1–71 (1991)
51. Vega, J.L., Guantes, R., Miret-Artés, S.: Mean first passage time and the Kramers turnover theory in activated atom-surface diffusion. Phys. Chem. Chem. Phys. **4**, 4985–4991 (2002)
52. Guantes, R., Vega, J.L., Miret-Artés, S., Pollak, E.: Kramers' turnover theory for diffusion of Na atoms on a Cu(001) surface measured by He scattering. J. Chem. Phys. **119**, 2780–2791 (2003)
53. Miret-Artés, S., Pollak, E.: The dynamics of activated surface diffusion. J. Phys.: Condens. Matter **17**, S4133–S4150 (2005)
54. Caldeira, A.O., Leggett, A.J.: Quantum tunneling in a dissipative system. Ann. Phys. **149**, 374–456 (1983)
55. Pollak, E., Miret-Artés, S.: Classical theory of atom surface scattering. The rainbow effect. Surf. Sci. Rep. (2012, to appear)
56. Moix, J.M., Pollak, E., Miret-Artés, S.: Friction-induced energy-loss rainbows in atom surface scattering. Phys. Rev. Lett. **104**, 116103(1–4) (2010)
57. Nelson, E.: Derivation of the Schrödinger equation from Newtonian mechanics. Phys. Rev. **150**, 1079–1085 (1966)
58. Bohm, D., Hiley, B.J.: Non-locality and locality in the stochastic interpretation of quantum mechanics. Phys. Rep. **172**, 93–122 (1989)

Chapter 3
Elements of Quantum Mechanics

3.1 Introduction

By the end of the nineteenth century, the available physical theories found a serious drawback: they were not able to provide an explanation to a series of experimental data. Well-known phenomena by the time, such as the black-body radiation or the heat capacity of solids at low temperatures, as well as new surprising ones, such as the X-rays or the radioactivity, required of a new atomic model. The simple mathematical models built up to explain such phenomena gave then rise to two new concepts in physics: the dual nature of electromagnetic radiation and the discreteness of certain physical magnitudes. This conceptual framework constituted the core of the so-called "old" *theory of quanta* [1], proposed by Planck and Einstein, and later on Bohr's atomic model was also incorporated.

The theory of quanta, however, presented a series of practical and conceptual flaws. For example, it could not be applied to systems whose classical dynamics is aperiodic or chaotic, thus giving rise to qualitative and incomplete descriptions of spectral lines. On the other hand, conceptually, it did not explain why electrons move in stationary orbits around the nuclei without emitting electromagnetic radiation, quantization rules were postulated *a priori*, no mechanism was provided to explain emission and absorption processes or the assumption of the duality of radiation seemed to be contradictory—it behaves as a corpuscle in the emission and absorption processes, but as a wave during the transit. However, in spite of Bohr's efforts to keep this theory by postulating the *correspondence principle*, with which certain success was achieved, between 1925 and 1927 a new mathematical model appeared: *quantum mechanics*. This new mechanics made apparent that Hamiltonian mechanics transcends classical mechanics. Not only it constitutes an elegant framework to describe the motion of the objects of the everyday life through the *least action principle*, as seen in Chap. 1, but also the quantum world, as shown by Schrödinger, who derived his renown wave equation by combining the so-called *Hamiltonian analogy* (see Sects. 3.2.1 and 7.2.2) with de Broglie's ideas of associating a wavelength to matter particles.

A. S. Sanz and S. Miret-Artés, *A Trajectory Description of Quantum Processes.*
I. Fundamentals, Lecture Notes in Physics 850, DOI: 10.1007/978-3-642-18092-7_3,
© Springer-Verlag Berlin Heidelberg 2012

Due to the relevance of quantum mechanics in the evolution of modern physics, in particular, and science, in general, there are many interesting books about the historical and conceptual development of this theory [1–4], which can be consulted by the interested reader. Here these matters will not be discussed more than necessary; this chapter mainly focusses on some of the most fundamental questions related to the topic covered by this book. In particular, more detail accounts on interpretational issues can be found in [2, 5, 6].

3.2 Fundamentals of Wave Mechanics

3.2.1 Hamiltonian Analogy and Calculus of Variations

As in classical mechanics, the fundamental equation of quantum mechanics, namely the *Schrödinger equation*, can also be obtained through the calculus of variations, though some simple physical hypotheses are eventually needed. Instead of considering a single independent variable (see Appendix A), consider now the more general case of a function of several independent variables, the variational problem consisting of finding a field function[1] ψ which makes stationary the integral [7]

$$\mathcal{F}[I] = \int_{x_{1,a}}^{x_{1,b}} \cdots \int_{x_{n,a}}^{x_{n,b}} I(x_1, \cdots, x_n, \psi, \psi_{x_1}, \cdots \psi_{x_n}) dx_1 \cdots dx_n, \qquad (3.1)$$

where $\psi_{x_i} \equiv \partial \psi / \partial x_i$. Keeping fixed the initial and final configurations (i.e., $\delta x_1 = \cdots = \delta x_n = 0$), and then assuming the condition

$$\delta \mathcal{F} = \int \cdots \int \delta I \, dx_1 \cdots dx_n = 0 \qquad (3.2)$$

holds, one finds

$$\frac{\partial I}{\partial \psi} - \sum_{i=1}^{n} \frac{\partial}{\partial x_i} \frac{\partial I}{\partial \psi_{x_i}} = 0, \qquad (3.3)$$

which is equivalent (see Appendix A) to (A.5) for the ψ field. Of course, if instead of a single function ψ there is a set of field functions $\psi_1, \psi_2, \cdots, \psi_n$, the analogous to (A.11) is obtained.

If some constraints are also specified, the problem belongs to the so-called *isoperimetric class*, as happens with Schrödinger's equation (and any wave equation, in general). Thus, consider one would like to obtain the real field $\psi(x, y, z)$ such that the square of its gradient has an extremum mean value within a certain space region regardless of the particular boundary for such a region. This means that

---

[1] Unless otherwise stated, the concept of *field* function (or, in brief, a *field*) will be used to denote a function which depends on a set of independent variables.

$$\delta \iiint (\nabla \psi)^2 dx dy dz = \delta \iiint (\psi_x^2 + \psi_y^2 + \psi_z^2) dx dy dz = 0, \qquad (3.4)$$

so, after applying (3.3), the so-called *Laplace equation* is found,

$$\nabla^2 \psi = \left(\frac{\partial^2}{\partial x^2} + \frac{\partial^2}{\partial y^2} + \frac{\partial^2}{\partial z^2} \right) \psi = 0. \qquad (3.5)$$

This equation describes the field ψ (regardless what this field represents physically) in vacuum. Additionally, it is required the integral of the square of ψ acquires some given value, which implies the extra condition $J_1 = \psi^2$ (see Appendix A). According to (A.14),

$$\left(\frac{\partial^2}{\partial x^2} + \frac{\partial^2}{\partial y^2} + \frac{\partial^2}{\partial z^2} - \lambda \right) \psi = (\nabla^2 - \lambda)\psi = 0, \qquad (3.6)$$

which is the *wave equation* describing a single-frequency or *monochromatic* wave in vacuum, i.e., a sinusoidal perturbation of the field ψ in vacuum, such that both ψ^2 and $(\nabla \psi)^2$ have a minimum mean value.

In order to find Schrödinger's equation, let us consider the following reasoning. As seen in Chap. 1, in classical mechanics the dynamics of a system is described by a trajectory, which arises after integrating the corresponding equations of motion. These equations come from considering that the system energy is constant along time. The same prescription can be followed in quantum mechanics, although the system Hamiltonian is given in terms of an operator,

$$\hat{H} = \frac{\hat{p}^2}{2m} + V(\hat{q}) = -\frac{\hbar^2}{2m}\nabla^2 + V(x, y, z), \qquad (3.7)$$

instead of a function of the generalized coordinates and momenta. Here, Cartesian coordinates have been chosen for simplicity (but without loss of generality), where $\hat{p}_j \equiv -i\hbar\partial/\partial q_j$ is the momentum operator, with $j = x, y, z$ and, e.g., $q_x = x$. In order to determine ψ, it is required that the average total energy,

$$\langle \hat{H} \rangle = \iiint \psi^* \hat{H} \psi dx dy dz, \qquad (3.8)$$

becomes a minimum with the additional (normalization) condition

$$\iiint \psi^* \psi dx dy dz = 1. \qquad (3.9)$$

Assuming ψ vanishes sufficiently fast at the boundaries of the integration limit, (3.8) can be expressed [8] as

$$\iiint \left[\frac{\hbar^2}{2m}(\psi_x^* \psi_x + \psi_y^* \psi_y + \psi_z^* \psi_z) + \psi^* V \psi \right] dx dy dz = 0, \qquad (3.10)$$

which is symmetric in ψ and its conjugate complex, ψ^*. Considering $K = I - \lambda J_1$ (see (A.13)) for convenience (this does not affect the final result, as will be seen), and then applying the calculus of variations,

$$\left(-\frac{\hbar^2}{2m}\nabla^2 + V\right)\psi = \hat{H}\psi = \lambda\psi, \tag{3.11a}$$

$$\left(-\frac{\hbar^2}{2m}\nabla^2 + V\right)\psi^* = \hat{H}\psi^* = \lambda\psi^*, \tag{3.11b}$$

the latter being the conjugate complex equation of the former, this implying λ has to be a real quantity.

The functions ψ and ψ^* obtained from (3.11a) and its conjugate complex (3.11b) are extremals that make the integral (3.8) to be stationary. If the stationary value is regarded as the system energy E, then, by multiplying the last term in (3.11a) by ψ^* and then integrating over space (or proceeding similarly with (3.11b), but considering ψ), one finds that $\lambda = E$. Therefore, (3.11a) (or, equivalently, its conjugate complex) can be expressed in the more familiar form

$$\hat{H}\psi = E\psi, \tag{3.12}$$

which is the well-known time-independent *Schrödinger equation*. In general, in most cases of physical interest E is a minimum, this being the basis of the so-called *variational* or *Ritz method* [7, 9], devised to obtain approximate solutions to Schrödinger's equation.

Now a description on how Schrödinger derived (3.12) [8, 10–12] and the wave theory by combining the so-called *Hamiltonian analogy* [13, 14] (see also Chap. 7) with de Broglie's ideas of associating a wavelength to matter particles [15] will be presented. This physical model establishes a direct, formal correspondence between optics and the Hamiltonian description of classical mechanics [16, 17]. Thus, consider a system of mass m with an energy E described by a certain Hamiltonian within the Hamilton–Jacobi prescription, i.e.,

$$H(q_k, \partial S/\partial q_k) = E, \tag{3.13}$$

where q_k denotes the system generalized coordinates, S is the system action and the generalized momenta are given by $p_k = \partial S/\partial q_k$ (see Chap. 1). In Cartesian coordinates,

$$E = \frac{1}{2m}\left(p_x^2 + p_y^2 + p_z^2\right) + V(x, y, z), \tag{3.14}$$

which can be expressed as

$$(\nabla S)^2 = 2m(E - V) \tag{3.15}$$

within the Hamilton–Jacobi scenario. Equation (3.15) is very similar to the *eikonal equation* from geometric optics (see Chap. 4),

$$(\nabla S^{\text{opt}})^2 = n^2, \tag{3.16}$$

which describes a light ray when it travels through a medium with refractive index n, and where S^{opt} is a constant-phase surface (the superscript "*opt*" is used in order to distinguish it from the mechanical action S). If the medium is homogeneous, n is constant and the solution to (3.16) is

$$S^{\text{opt}} = n(\alpha x + \beta y + \gamma z), \tag{3.17}$$

where α, β and γ are the direction cosines in the three directions, x, y and z, respectively, and satisfy the relationship $\alpha^2 + \beta^2 + \gamma^2 = 1$. Geometric optics is the $k \to \infty$ limit (or, equivalently, the $\lambda = 2\pi/k \to 0$ limit) of wave optics, where the basic equation is the so-called *Helmholtz equation*[2] (see Chap. 4),

$$(\nabla^2 + k^2)\Psi = 0. \tag{3.18}$$

The solution to this equation is a plane wave that can be expressed as

$$\Psi(x, y, z) = Ae^{i\mathbf{k}\cdot\mathbf{r}} = e^{ik(\alpha x + \beta y + \gamma z)} = e^{ikS^{\text{opt}}/n}. \tag{3.19}$$

As it can be noticed, substituting the last expression in (3.19) into (3.18) and then assuming the limit $k \to \infty$, the eikonal equation (3.16) is recovered.

The whole point of the Hamiltonian analogy is now to establish the relationship between optical and mechanical systems. Actually, quoting Sommerfeld [17],

> Ray optics is the mechanics of light particles; in optically inhomogeneous media the paths of these particles are by no means straight lines, but are determined by Hamilton's ordinary differential equations or Hamilton's principle which is equivalent to them. From the viewpoint of wave optics, on the other hand, the rays of light are given by the orthogonal trajectories of a system of wave surfaces or wave fronts.

Thus, in analogy to (3.19), the general solution to (3.15) reads as

$$S = \kappa \ln \Psi, \tag{3.20}$$

---

[2] Actually, this point in common between scalar optics and quantum mechanics has allowed that many solutions to problems within the latter were directly adapted from well-known nineteenth century solutions from the former [18].

which leads to the equation

$$(\nabla \Psi)^2 - \frac{2m}{\kappa^2}(E - V)\Psi^2 = 0,$$ (3.21)

where it is apparent the closeness between this expression and the one given by $I - \lambda J_1$ above. In order to find the solution Ψ, variations over the field Ψ are considered, which renders the well-known Schrödinger equation,

$$\nabla^2 \Psi + \frac{2m}{\kappa^2}(E - V)\Psi = 0.$$ (3.22)

Applying this equation to the hydrogen atom and seeking for its solutions, Schrödinger found κ had to be equal to $\hbar$. Thus, Eq. (3.22) is the explicit form of (3.12).

In order to derive now the time-dependent Schrödinger equation, note that its time-independent version relies on the classical assumption that, within the (classical) Hamilton–Jacobi framework, the energy arises from the relation (1.17). Therefore, if the energy does not have a definite value, E can be replaced by (1.17) in the time-independent Schrödinger equation (3.12). Then, replacing S by Ψ,

$$H(q, i\hbar\partial/\partial q)\Psi(q, t) = i\hbar\frac{\partial}{\partial t}\Psi(q, t),$$ (3.23)

where the time-dependent Schrödinger equation is readily recognized. This equation can also be expressed in the more commonly used form

$$-\frac{\hbar^2}{2m}\nabla^2\Psi + V\Psi = i\hbar\frac{\partial\Psi}{\partial t}.$$ (3.24)

Alternatively, (3.24) can also be derived by means of the mirror-image method seen in Sect. 2.2.3, which allows us to establish a closer connection with the derivation leading to (3.12) in terms of the variational principle. Note that assuming that D is an imaginary diffusion constant with value $i\hbar/2m$ and substituting it into (2.28a), this equation becomes the time-dependent Schrödinger equation for a free particle ($V = 0$) and Ψ (in the corresponding complex diffusion equation) becomes complex valued. Since Ψ is a complex quantity, this function and its complex conjugate can be considered as independent variables, with the latter being the mirror-image ($\bar{\Psi} = \Psi^*$) of the former. Hence, the problem of a particle of mass m subject to the action of an external potential V can be formulated in terms of (2.27) as

$$\mathcal{L} = -\frac{\hbar}{2i}\left(\Psi^*\frac{\partial\Psi}{\partial t} - \Psi\frac{\partial\Psi^*}{\partial t}\right) - \frac{\hbar^2}{2m}\nabla\Psi \cdot \nabla\Psi^* - \Psi^*V\Psi.$$ (3.25)

Requiring now the integral in space and time $\mathcal{L}$ becomes an extremum for both Ψ and Ψ^* gives rise to the time-dependent Schrödinger equation (and its complex conjugate).

By further proceeding within this theoretical framework and computing the respective canonical momenta, one finds

$$p_t = \frac{\partial \mathcal{L}}{\partial \dot{\psi}} = -\frac{\hbar}{2i}\psi^*, \qquad \bar{p}_t = \frac{\partial \mathcal{L}}{\partial \dot{\psi}^*} = \frac{\hbar}{2i}\psi, \qquad (3.26a)$$

$$p_q = \frac{\partial \mathcal{L}}{\partial \psi_q} = -\frac{\hbar^2}{2m}\nabla\psi^*, \qquad \bar{p}_q = \frac{\partial \mathcal{L}}{\partial \psi_q^*} = -\frac{\hbar^2}{2m}\nabla\psi, \qquad (3.26b)$$

which correspond to the time and space momentum densities associated with the Lagrangian density $\mathcal{L}$. Taking this into account, a "stress tensor" can be defined as

$$\mathcal{W}_{mn} = \psi_m^* \frac{\partial \mathcal{L}}{\partial \psi_n^*} + \psi_m \frac{\partial \mathcal{L}}{\partial \psi_n} - \delta_{mn}\mathcal{L}, \qquad (3.27)$$

where $\psi_m \equiv \partial\psi/\partial\alpha_m$, with $\alpha_m = (x, y, z, t)$ $(m = 1, 2, 3, 4)$. In this way, if $m = n = 4$, the *Hamiltonian energy density* can be expressed as,

$$\mathcal{W}_{44} \equiv \mathcal{H} = \frac{\hbar^2}{2m}\nabla\psi^* \cdot \nabla\psi + \psi^* V\psi, \qquad (3.28)$$

while the *energy density flow vector* arises from the $\mathcal{W}_{4k}$ components of the stress tensor, i.e.,

$$\mathbf{S} = \mathcal{W}_{41}\mathbf{i} + \mathcal{W}_{42}\mathbf{j} + \mathcal{W}_{43}\mathbf{k} = -\frac{\hbar^2}{2m}\left[\left(\frac{\partial\psi^*}{\partial t}\right)\nabla\psi + \left(\frac{\partial\psi}{\partial t}\right)\nabla\psi^*\right]. \qquad (3.29)$$

These two quantities satisfy the continuity equation for the energy density,

$$\frac{\partial \mathcal{H}}{\partial t} + \nabla \cdot \mathbf{S} = 0. \qquad (3.30)$$

The Hamiltonian energy density and the energy flow vector are the analogous to the electromagnetic energy density and Poynting vector, respectively, as will be seen in Chap. 7, which also satisfy a continuity equation similar to (3.30). On the other hand, from the momentum density vector field,

$$-\mathbf{J} = \mathcal{W}_{14}\mathbf{i} + \mathcal{W}_{24}\mathbf{j} + \mathcal{W}_{34}\mathbf{k} = -\frac{\hbar}{2i}\left[\psi^*\nabla\psi - \psi\nabla\psi^*\right], \qquad (3.31)$$

the *quantum probability current density* can be obtained, which also satisfies a continuity (or conservation) equation similar to (3.30) with respect to the quantum probability density,

$$\frac{\partial \rho}{\partial t} + \nabla \cdot \mathbf{J} = 0. \tag{3.32}$$

with $\rho = |\Psi|^2$.

3.2.2 Waves and Uncertainty

Usually the wave function is expressed in the configuration or coordinate representation, i.e., in the space of all possible configurations a physical system, process or phenomenon may explore. From this wave function, any physical information or observable $\mathcal{A}$ can be extracted by calculating the corresponding *expectation* or *average value*,

$$\mathcal{A} \equiv \langle \hat{A} \rangle = \int \psi^*(\mathbf{r}) \hat{A} \psi(\mathbf{r}) d\mathbf{r}. \tag{3.33}$$

In configuration space, $\psi(\mathbf{r})$ is diagonal, which means that the action of any operator depending only on the coordinates will be equivalent to multiply $\psi(\mathbf{r})$ by a function of the coordinates with the same functional form of such an operator. This is the case, for example, of the potential energy operator, where $\hat{V}(\hat{\mathbf{r}})\psi(\mathbf{r}) = V(\mathbf{r})\psi(\mathbf{r})$. However, the momentum operator, $\hat{\mathbf{p}} = -i\hbar\nabla$, or the kinetic energy one, $\hat{\mathcal{K}} = -(\hbar^2/2m)\nabla^2$, are not diagonal in the configuration space and imply to carry out some operations on ψ (in these cases, to compute $\nabla\psi$ or $\nabla^2\psi$, respectively). Hence, working with the corresponding operator and the wave function in the same space results advantageous.

In those cases where properties associated with the momentum must be computed, it is interesting to express the wave function within the momentum representation. Since position and momentum are canonically conjugate variables, there is a relationship between them, which in quantum mechanics translates into a non-commutativity when operating with them, i.e., it is not the same to operate first with the position and then with the momentum than the other way around. This is formally expressed through the well-known non-vanishing commutation relation,

$$[\hat{x}, \hat{p}] = i\hbar. \tag{3.34}$$

In order to illustrate the passage from one representation to the other, first the eigenfunctions must be computed for the eigenvalue equation

$$\hat{p}\phi(x) = -i\hbar\frac{\partial\phi(x)}{\partial x} = p\phi(x), \tag{3.35}$$

which can be easily shown to be

$$\phi(x) = \frac{1}{\sqrt{2\pi}}e^{ikx} \tag{3.36}$$

after formal integration by parts. The solutions (3.36) are plane waves with momentum p, which are normalized with respect to a cycle of phase, 2π (or to a phase–space cell, $2\pi\hbar$, if the argument of the exponential is expressed as $ipx/\hbar$). Taking this into account, any wave function can be decomposed in configuration space as a sum of plane waves,

$$\psi(x) = \frac{1}{\sqrt{2\pi}} \int \tilde{\psi}(k)e^{ikx}dk = \frac{1}{\sqrt{2\pi\hbar}} \int \tilde{\psi}(p)e^{ipx/\hbar}dp, \qquad (3.37)$$

which is just the inverse Fourier transform of the wave function in momentum space. Obviously, the latter is the Fourier transform of $\psi(x)$,

$$\tilde{\psi}(p) = \frac{1}{\sqrt{2\pi\hbar}} \int \psi(x)e^{-ipx/\hbar}dx, \qquad (3.38)$$

where $e^{-ipx/\hbar}/\sqrt{2\pi\hbar}$ represents, in general, a plane wave eigenfunction of the position operator in the momentum space, $\hat{x} = i\hbar\partial/\partial p$.

The relationship between position and momentum through a Fourier transform is already a "suspicious" indication of the non-commutativity of their corresponding operators. This brings about another very important physical consequence, the so-called *Heisenberg's uncertainty principle* [19], according to which it is not possible to measure with an infinite accuracy the values of both members of a canonically conjugate pair of physical variables. In other words, the relationship between the dispersion of these variables,

$$\Delta x \, \Delta p \geq \frac{\hbar}{2}, \qquad (3.39)$$

where $(\Delta x)^2 = \langle(\hat{x} - \langle\hat{x}\rangle)^2\rangle$ and $(\Delta p)^2 = \langle(\hat{p} - \langle\hat{p}\rangle)^2\rangle$, always holds. However, there is also something suspicious here: as happens with the case of evanescent waves and tunneling, the uncertainty relation (3.39) and, therefore, the uncertainty principle is not a result particular of quantum mechanics, but general of any wave theory, since Δx and Δp only measure the dispersions associated with a wave, without specifying the physics described by such a wave. Actually, the uncertainty relation is valid for any square integrable function and its Fourier transform. As it can be shown [20, 21], if a function $f(x)$ is integrable, square-integrable and normalized (L^2-normalized), it will satisfy *Plancherel's theorem* [22],

$$\int_{-\infty}^{\infty} |f(x)|^2 dx = \int_{-\infty}^{\infty} |\tilde{f}(\xi)|^2 d\xi = 1, \qquad (3.40)$$

where $\tilde{f}(\xi)$ is the Fourier transform of f, and therefore it can be shown that their dispersions around their corresponding mean values, x_0 and ξ_0, obey the relation

$$D_{x_0}(f)D_{\xi_0}(\tilde{f}) \geq \frac{1}{16\pi^2}, \qquad (3.41)$$

where

$$D_{x_0}(f) = \int (x^2 - x_0^2)|f(x)|^2 dx, \qquad x_0 = \int x|f(x)|^2 dx, \qquad (3.42a)$$

$$D_{\xi_0}(\tilde{f}) = \int (\xi^2 - \xi_0^2)|\tilde{f}(\xi)|^2 d\xi, \qquad \xi_0 = \int \xi|\tilde{f}(\xi)|^2 d\xi. \qquad (3.42b)$$

In the problem of the particle in a box, for example, one finds that the eigenstates satisfy the uncertainty relation

$$\Delta x \Delta p = n\pi\hbar > \hbar/2, \qquad (3.43)$$

for all n, while in the case of the linear harmonic oscillator,

$$\Delta x \Delta p = (n + 1/2)\hbar \geq \hbar/2. \qquad (3.44)$$

In this latter case, the equality holds for $n = 0$. This is related to the fact that Gaussian functions, which have an also Gaussian Fourier transform, have minimum spreading [23]. This is the reason why in quantum mechanics Gaussian wave packets are called *minimum uncertainty wave packets*.

In classical mechanics, the uncertainty relation does not hold (except for classical wave mechanics) and, therefore, in principle one could have $\Delta x \Delta p = 0$ if the dispersion in both coordinates and momenta is zero (e.g., when dealing with a point on phase space, since there is no limitation to the accuracy in the measurements). This is the usual argument considered to emphasize the difference between classical mechanics and quantum mechanics. However, it is important to stress that, if ψ is regarded as a quantity providing statistical information, it should not be compared with a single classical phase–space point, but with a classical distribution or a classical wave, which in general do present a dispersion. On the other hand, it is interesting to note that in the same way that the relation $\Delta x \Delta p \to 0$ describes the passage to Newtonian mechanics (statistically speaking), in optics there is a "twin" relation, $\Delta x \Delta k \gg 1$, which describes the passage from wave to geometric optics [17]: whenever the light wavelength is much smaller than the dimensions of the objects it finds along its pathway, the geometric optical description will be valid.

3.2.3 Eigenvalues, Probabilities and Time-Evolution

The spectral lines observed by Fraunhofer about 200 years ago correspond to solutions of the Schrödinger equation derived about 100 years after Fraunhofer's findings. This equation gives us the energy at which spectral lines should be observed, but it also gives us information about how electrons move around nuclei. According to Born's *statistical interpretation* [24–29], the wave amplitude ψ_n provides no information about individual processes or systems, but it is just a statistical quantity which describes the distribution (over the corresponding sampling space) of an ensemble of

identical processes or systems. Actually, such a measurement is given by the prob-
ability density, $\rho_n \equiv |\psi_n|^2$, which is a real quantity. There are, of course, different
interpretations of the wave function, this issue still remaining open [30]. However,
the importance of Born's interpretation relies on the fact that it directly comes from
the empirical evidence (remember that Born's proposal was based on his studies on
spectra obtained from scattering experiments); a single experimental detection or
measurement is meaningless, only a collection of them is of interest. For example, in
the case of the hydrogen atom, ρ_n describes how electrons distribute around nuclei in
a sample or, equivalently, the probability to find the electron at a certain place. When
the state ψ_n is expressed in configuration space, i.e., as a continuous function $\psi_n(r)$
of the radial coordinate r, Bohr radius indicates the most likely position to find the
electron in the case that it is in the lowest or *ground* energy level ψ_0 (i.e., the value
of r at which ρ_0 reaches its maximum value), which corresponds to the electron's
orbit radius within Bohr's former atomic model.

Apart from being at different distances r from the nucleus, the electron can also
be in different *excited* energy levels ψ_n with $n \neq 0$. To account for this fact, the
state of the atom is described by a wave function consisting of a *superposition*
of all possible eigensolutions or eigenstates associated with the time-independent
Schrödinger equation,

$$|\psi\rangle = \sum_n c_n |\psi_n\rangle, \tag{3.45}$$

which has been expressed purposely in Dirac's notation. From (3.45), the *density
operator* (it specifies the *state* of the system) is defined as

$$\hat{\rho} \equiv |\psi\rangle\langle\psi|, \tag{3.46}$$

whose associated density matrix with (m, n) element is given by

$$\rho_{mn} = \langle\psi_m|\hat{\rho}|\psi_n\rangle = \mathrm{Tr}\,[|\psi_n\rangle\langle\psi_m|\hat{\rho}] = c_m^* c_n, \tag{3.47}$$

where $\mathrm{Tr}[\,\cdot\,] \equiv \langle\psi|\cdot|\psi\rangle$ denotes the trace of the operator in between the square
brackets. Thus, when the measurement is carried out, the probability to find the
electron in the state ψ_n (or, equivalently, to observe a transition to this state) is given
by $\rho_{nn} = |c_n|^2$. Because there is a large number of hydrogen atoms when this kind
of experiments are carried out, each one will absorb or emit a photon in a different
transition, which in a photograph plate appears as a series of several well-defined
spectral lines—the problem of how a particular ρ_n is detected constitutes the central
problem of the *theory of measurement* [31], which goes beyond the scope of this
chapter and book (it is briefly discussed in Appendix B). In this sense, the value of
the dynamical variable describing the system or observable is given as in classical
mechanics by an averaging (see (3.33)), the density matrix being obtained from the
also called *statistical operator*. Usually three conditions are required to be satisfied
by the density operator representing the state of a system:

1. It is normalized, i.e., $\text{Tr}(\hat{\rho}) = 1$.
2. It is self-adjoint (Hermitian), i.e., $\hat{\rho}^\dagger = \hat{\rho}$, for the observable to be real.
3. It is nonnegative, i.e., $\langle u|\hat{\rho}|u\rangle \geq 0$, for any arbitrary vector $|u\rangle$ of unit norm which is associated with a certain observable (represented by a projection operator $P_u = |u\rangle\langle u|$).

From these three properties, and taking into account (3.45)–(3.47), one finds the following properties for the matrix elements ρ_{mn}:

1. $\sum_n \rho_{nn} = 1$.
2. $\rho_{nm}^* = \rho_{mn}$ (and, therefore, ρ_{nn} is real).
3. $\rho_{nn} \geq 0$.

Note from these properties that $0 \leq \rho_{nn} \leq 1$. A set of mathematically acceptable state operators, $\{\hat{\rho}^{(i)}\}$, is said to be *convex* if all these operators satisfy the above three properties and therefore a linear (convex) combination $\hat{\rho} = \sum_i a_i \hat{\rho}^{(i)}$ can be formed, such that $0 \leq a_i \leq 1$ and $\sum_i a_i = 1$. *Pure states* constitute a particular class, characterized by their relatively simple properties. For example, their associated density matrix is *idempotent*, i.e., $\hat{\rho}^2 = \hat{\rho}$, and therefore

$$\langle\hat{\rho}\rangle = \text{Tr}[\hat{\rho}] = \text{Tr}[\hat{\rho}^2] = 1. \tag{3.48}$$

On the contrary, statistical mixtures are described by *nonpure* or *mixed states*[3] the corresponding density matrix is no longer idempotent and, therefore, $\text{Tr}[\hat{\rho}^2] \leq \text{Tr}[\hat{\rho}] = 1$. The theory of open quantum systems (see Chap. 5) makes special emphasis on this kind of distinction, for non isolated or open (quantum) systems have to be described by means of mixed states.

   In general, for a given observable $\mathcal{A}$, its expectation value can also be calculated from

$$\mathcal{A} \equiv \langle\hat{A}\rangle = \text{Tr}[\hat{\rho}\hat{A}], \tag{3.49}$$

which is the generalization of (3.33). For time-dependent averages, the temporal evolution of the density operator needs to be known. The corresponding equation of motion is the so-called quantum Liouville equation which in the Schrödinger representation reads as

$$\frac{d\hat{\rho}(t)}{dt} = -\frac{i}{\hbar}[\hat{H}, \hat{\rho}(t)] \equiv \mathcal{L}\hat{\rho}(t), \tag{3.50}$$

where $\mathcal{L}$ is the Liouville operator. This equation is sometimes converted to an integral equation, which is solved in an iterative way. Very often, the time-evolution of the

---

[3]   For a more detailed discussion on the properties (and differences) between pure and nonpure or mixed states, see, for example, [28, 29].

matrix elements of the density operator are necessary for quantum transitions within a perturbation treatment as, for example, the optical Bloch equations in the theory of coherent optical phenomena. In Chap. 5, the Liouville equation is extended to consider open quantum systems. This equation is also necessary when quantum correlation functions have to be evaluated. Similar equations can be derived depending on the representation or picture of quantum mechanics chosen for any dynamical variable. Thus, for example, in the Heisenberg representation, the equation of motion of $\hat{A}$ becomes

$$\frac{d\hat{A}(t)}{dt} = -\frac{i}{\hbar}[\hat{A}(t), \hat{H}], \qquad (3.51)$$

which differs from (3.50) in a negative sign.

According to von Neumann [32], in quantum mechanics there are two processes involved. The first process is causal and reversible, and it carries our system from a certain initial state to another one following Schrödinger's time-dependent equation. The second one, the measurement process mentioned above, which is noncausal and irreversible, accounts for the fact a discrete atomic spectrum is observed. A relatively simple way to tackle the problem of time-dependence is by considering the fact that any wave function can be decomposed as a superposition of eigenstates of the Hamiltonian, i.e., eigenstates in an energy representation. Since the time-dependence of one of such eigenstates is given by a phase depending linearly on time (because of the separation of space and time variables allowed by the time-independent Schrödinger equation), the evolution of the corresponding wave function, say (3.45), can be readily written as

$$|\Psi(t)\rangle = \sum_n c_n e^{-iE_n t/\hbar}|\psi_n\rangle. \qquad (3.52)$$

If the eigenstates are given in the coordinate representation, i.e., $\phi_n(\mathbf{r}) = \langle \mathbf{r}|\psi_n\rangle$, then

$$\Psi(\mathbf{r}, t) = \sum_n c_n \phi_n(\mathbf{r}) e^{-iE_n t/\hbar}, \qquad (3.53)$$

where

$$c_n = \int \phi_n^*(\mathbf{r})\Psi(\mathbf{r}, 0)d\mathbf{r}. \qquad (3.54)$$

Consider, for example, a Gaussian wave packet initially centered around $x = a$ in a linear harmonic oscillator potential $V = m\omega^2 x^2/2$,

$$\Psi(x, 0) = \left(\frac{1}{2\pi\sigma_0^2}\right)^{1/4} e^{-(x-a)^2/4\sigma_0^2}, \qquad (3.55)$$

whose exact time-evolution is described by

$$\Psi(x,t) = \left(\frac{1}{2\pi\sigma_0^2}\right)^{1/4} e^{-(x-a\cos\omega t)^2/4\sigma_0^2 - i\omega t/2 - im\omega(4xa\sin\omega t - a^2\sin 2\omega t)/4\hbar},$$

(3.56)

with $\sigma_0^2 = \hbar/2m\omega$. Taking into account the spectral decomposition corresponding to a harmonic oscillator (see Volume 2), (3.56) can be expressed as

$$\Psi(x,t) = \sum_{n=0}^{\infty} A_n u_n(x) e^{-iE_n t/\hbar} = e^{-i\omega t/2} \sum_{n=0}^{\infty} A_n u_n(x) e^{-in\omega t},$$

(3.57)

where $E_n = (n+1/2)\hbar\omega$, u_n are the corresponding eigenfunctions or eigenstates, and

$$A_n = \int_{-\infty}^{\infty} u_n^*(x) \Psi(x,0) dx = \frac{\alpha^2 e^{-\alpha^2/4}}{\sqrt{2^n n!}},$$

(3.58)

with $\alpha^2 = a^2/2\sigma_0^2$. As it can be noticed, the probability density associated with the wave function (3.56),

$$\rho(x,t) = \sqrt{\frac{1}{2\pi\sigma_0^2}} \, e^{-(x-a\cos\omega t)^2/2\sigma_0^2},$$

(3.59)

moves back and forth in time periodically, but does not change its shape, which remains the same as $\rho(x,0)$ at any time. On the other hand, two limits are interesting when (3.58) is considered. If $a \to 0$, then $n \to 0$ and the wave function becomes the ground state of the linear harmonic oscillator. On the contrary, if a increases and $n \to \infty$, using Stirling's formula in (3.58) and neglecting terms of the order $\ln n$ or lower,

$$\ln A_n \approx n\left(\ln\alpha - \frac{\ln 2}{2}\right) - \frac{n}{2}(\ln n - 1),$$

(3.60)

which has a maximum for $\bar{n} \approx \alpha^2/2$. Substituting this value into E_n, one finds the classical expression for the energy of an oscillator with maximum amplitude a, $E_{\bar{n}} \approx E_{cl} = m\omega^2 a^2/2$.

In the case of scattering problems, on the other hand, it is convenient to express the wave function as a linear combination of *plane waves*, which are eigenstates of the momentum operator, $\hat{\mathbf{p}} = -i\hbar\nabla$. Actually, at asymptotic distances, where $V \approx 0$, this operator and the kinetic energy one, $\hat{\mathcal{K}} = -(\hbar^2/2m)\nabla^2$, commute (i.e., $[\hat{\mathcal{K}}, \hat{\mathbf{p}}] = 0$) and, therefore, both share the same basis of eigenfunctions with eigenvalues $E = p^2/2m = \hbar^2 k^2/2m$—indeed, the momentum eigenfunctions are degenerate in the energy representation, since two eigenvalues p and $-p$, for example, correspond to the same eigenenergy E. Thus, as before, the wave function is expressed as

$$\Psi(x, t) = \int A_k u_k(x) e^{-i E_k t/\hbar} dk, \tag{3.61}$$

where

$$A_k = \int u_k^*(x) \Psi(x, 0) dx \tag{3.62}$$

and $E_k = \hbar^2 k^2/2m$ is the energy eigenvalue associated with the eigenfunctions u_k for the momentum eigenvalues k and $-k$. If some boundary conditions are imposed (e.g., box normalization or Born–von Karman periodic boundary conditions), a quantization of the solutions appears and the basis of momenta (energies) becomes discrete. This implies that (3.61) has to be expressed as a sum, i.e.,

$$\Psi(x, t) = \sum_k A_k u_k(x) e^{-i E_k t/\hbar}. \tag{3.63}$$

As an example, consider the case of the minimum uncertainty wave packet [23], i.e., a wave packet for which the equality holds in the position-momentum uncertainty relation (see Sect. 3.2.2), which can initially be described by

$$\Psi(x, 0) = \left(\frac{1}{2\pi \sigma_0^2}\right)^{1/4} e^{-(x-x_0)^2/4\sigma_0^2 + i p_0 x/\hbar}. \tag{3.64}$$

In free space (but normalizing to a box of length L), the optimum basis set consists of plane waves $u_k(x) = e^{ikx}/\sqrt{L}$, with momentum and energy eigenvalues $k = 2\pi n/L$ and $E_k = \hbar^2 k^2/2m$, respectively. Taking this into account,

$$A_k = \left(\frac{8\pi \sigma_0^2}{L^2}\right)^{1/4} e^{-k^2 \sigma_0^2} \tag{3.65}$$

and, therefore, the time-evolution of (3.64) will be expressed by

$$\Psi(x, t) = \sum_k A_k u_k(x) e^{-i E_k t/\hbar}. \tag{3.66}$$

If $L \to \infty$, the basis set approaches a continuum and then

$$\sum_k \quad \to \quad \frac{L}{2\pi} \int dk. \tag{3.67}$$

Introducing this approximation into (3.66),

$$\Psi(x, t) = \left(\frac{\sigma_0^2}{2\pi^2}\right)^{1/4} \int e^{-k^2 x^2} e^{ikx} e^{-i E_k t/\hbar} dk$$

$$= \left(\frac{1}{2\pi \tilde{\sigma}_t^2}\right)^{1/4} e^{-(x-v_0 t)^2/4\tilde{\sigma}_t \sigma_0 + i p_0 (x - v_0 t)/\hbar + i E t/\hbar}, \tag{3.68}$$

where $v_0 = p_0/m = \langle \hat{p}/m \rangle$, $E = p_0^2/m = \langle \hat{H} \rangle$, $\tilde{\sigma}_t = \sigma_0(1 + i\hbar t/2m\sigma_0^2)$ and

$$\sigma_t = \sigma_0 \sqrt{1 + \left(\frac{\hbar t}{2m\sigma_0^2}\right)^2} \tag{3.69}$$

is the time-dependent spreading of the wave packet. Note that, contrary to the case of the linear harmonic oscillator, here the wave packet spreads as it propagates, since

$$\rho(x, t) = \frac{1}{\sqrt{2\pi\sigma_t^2}} e^{-(x-v_0 t)^2/2\sigma_t^2}. \tag{3.70}$$

By inspecting (3.66) and the first line of (3.68), one finds the right-hand side in both cases can be expressed as

$$\Psi(x, t) = \sum_n \tilde{\Psi}(k, t)e^{ik_n x}, \tag{3.71a}$$

$$\Psi(x, t) = \frac{1}{\sqrt{2\pi}} \int \tilde{\Psi}(k, t)e^{ikx} dk, \tag{3.71b}$$

respectively. That is, the wave function in configuration space can be expressed as the Fourier transform of a certain wave function $\tilde{\Psi}(k, t)$ in momentum space (and vice versa)—obviously, the same holds at $t = 0$ between $\psi(x)$ and $\tilde{\psi}(k)$. This is a very important issue, because it establishes a *correlation* between coordinates and momenta (i.e., in phase space), explaining why there is an uncertainty principle to be satisfied in quantum mechanics (see Sect. 3.2.2) or why phase–space distributions are characterized by regions where negative values are reached, like the Wigner one (see Sect. 3.3.1).

3.2.4 Probability Current Densities and Tunneling

Consider a stream or flux of identical and independent particles traveling through configuration space (for simplicity, in one dimension) with a given energy E. As seen above, this can be represented by a plane wave,

$$\psi(x) = Ae^{ikx}, \tag{3.72}$$

such that $E = \hbar^2 k^2/2m$. Note that $\rho(x)$ is independent of x and, therefore, according to Born's statistical interpretation the probability to find a particle everywhere is the same, which is equivalent to say that statistically the particles distribute equally through the whole configuration space. Since the probability to find the particle in the whole space should be unity, one could assume a large box and normalize with respect to this box, as in the previous Section. However, in the continuum it is also very common to normalize with respect to the incident flux, i.e., $A \sim k^{-1/2}$, which has to do with the quantum probability current density, as seen below.

Because there is a flux of particles, apart from the probability density, one can focus on the number of particles that passes through a certain region or area (and per time unit, if time is involved). One way to determine this quantity is as follows. Multiplying (3.11a) and (3.11b) on the left by ψ^* and ψ, respectively, assuming $\lambda = E$, and then subtracting the resulting equations, after rearranging terms one finds

$$\nabla \cdot \left(\frac{\psi^* \hat{\mathbf{p}} \psi - \psi \hat{\mathbf{p}} \psi^*}{2m} \right) = 0, \tag{3.73}$$

where $\hat{\mathbf{p}} = -i\hbar\nabla$ is the momentum operator. If the wave function is expressed in polar form, $\psi = \rho^{1/2} e^{iS/\hbar}$, (3.73) can be recast as

$$\nabla \cdot \left(\rho \frac{\nabla S}{m} \right) = 0. \tag{3.74}$$

Now, defining the (phase) velocity $\mathbf{v} \equiv \nabla S/m$, the expression between brackets can be identified with a current density, specifically the *quantum probability current density*,

$$\mathbf{J} = \frac{1}{m} \text{Re}[\psi^* \hat{\mathbf{p}} \psi] = \frac{\hbar}{2im} [\psi^* \nabla \psi - \psi \nabla \psi^*] = \mathbf{v}\rho. \tag{3.75}$$

Accordingly, (3.73) establishes that for a given stationary state the net flux of particles is always zero, with the *stationary* flux of such particles being described by $\mathbf{J}$, whose direction at each point of configuration space is given by $\mathbf{v}$; at such a point, the particle density (probability density) is given by ρ. This result is very similar to the one known from electromagnetism (see Chap. 4), where the role of $\mathbf{J}$ is played by the so-called *Poynting vector*. Substituting (3.72) into (3.75), one finds $\mathbf{J} = (\hbar k/m)|A|^2 = v|A|^2$, i.e., (3.72) effectively describes a (stationary) particle flow which moves with velocity $v = \hbar k$. One can therefore consider this flux to be unity (i.e., $\mathbf{J} = 1$), which would imply the normalization condition mentioned above, $A = v^{-1/2} \sim k^{-1/2}$, commonly used in scattering problems.

It is sometimes useful to deal with the so-called flux operator. If $\hat{P} = |\mathbf{r}\rangle\langle\mathbf{r}|$ is the position projection operator, then $\langle\hat{P}\rangle = |\Psi(\mathbf{r})|^2 = \rho(\mathbf{r})$. One then could write a continuity-like equation for $\hat{P}$ from the Heisenberg equation given by (3.51) for $\hat{P}$ as

$$\frac{d\hat{P}}{dt} + \frac{d\hat{\mathbf{J}}}{d\hat{\mathbf{r}}} = 0 \tag{3.76}$$

where the probability flux density operator $\hat{\mathbf{J}}$ is defined according to

$$\hat{\mathbf{J}} = \frac{1}{2m}[\hat{P}\hat{\mathbf{p}} + \hat{\mathbf{p}}\hat{P}] \tag{3.77}$$

being $\langle \hat{\mathbf{J}} \rangle = \mathbf{J}(\mathbf{r})$.

Now, let us consider the effect of placing an obstacle in the way of the wave (3.72), which will lead to the phenomenon of *quantum tunneling*. This effect has been recently treated in two books [33, 34]. The quantum tunnel effect, the possibility for a system to pass from one state A to a state B *through* an energetic barrier (see Volume 2), is considered paradigmatic in quantum mechanics. Nevertheless, this is a general phenomenon appearing whenever a physical system is describable by a wave equation and there is a coupling of "evanescent" waves [35], as also happens in optics [14] (see Sect. 4.4). It was proposed as a physical mechanism in 1928, shortly after the appearance of Schrödinger's equation, to explain the field electron emission [36] and the alpha decay [37, 38] (it is remarkable that these effects were known since the end of the nineteenth century [39, 40], about 30 years before they could be satisfactorily explained). Nevertheless, a year before, in 1927, Friedrich Hund was the first to notice the possibility of tunneling, which he called barrier penetration, in a calculation of the ground state in a double-well potential. The phenomenon arises, for example, in the inversion transition of the ammonia molecule. Nowadays it can be found in a myriad of applications and not only in tunnel microscopy or nuclear physics, the most important possibly being within the fields of semiconductors and superconductors or enzyme chemical reactions (catalysis reactions) because of their direct technological impact on society.

In order to illustrate the physics of the tunnel effect, consider the case of a particle scattered by a square barrier of width a and height V_0 [23], i.e., $V(x) = V_0$ for $0 \leq x \leq a$ and zero everywhere else. Asymptotically, the particle is force-free and, therefore, it can be described by a plane wave like (3.72), i.e.,

$$\psi_{\mathrm{I}}(x) = e^{ikx} + re^{-ikx}, \tag{3.78a}$$

$$\psi_{\mathrm{III}}(x) = te^{ik}, \tag{3.78b}$$

where I and III label the regions $x < 0$ and $x > a$, respectively. Physically, (3.78a) describes an incident or incoming wave with momentum $\hbar k$ and the reflected wave, with opposite momentum, $-\hbar k$; (3.78b) represents the transmitted wave behind the barrier, which evolves with momentum $\hbar k$. Moreover, the number or density of particles is normalized to 1, so that the number of particles reflected and transmitted will be $R = |r|^2$ and $T = |t|^2$, respectively (in this way, $R + T = 1$). Actually, if (3.78a) and (3.78b) are introduced into (3.75),

$$J_{\mathrm{I}}(x) = J_i + J_r = v - vR, \tag{3.79a}$$

$$J_{\mathrm{III}}(x) = J_t = vT, \tag{3.79b}$$

i.e., $J_i = J_t - J_t$.

Now, R and T can be determined as follows. Consider again the continuity of the wave function and its first derivative at the boundaries with the barrier. This implies that within the barrier (i.e., for $0 \leq x \leq a$) solutions should look like

$$\psi_{II}(x) = \alpha e^{ik'x} + \beta e^{-ik'x}, \tag{3.80}$$

such that $\psi_I(0) = \psi_{II}(0)$, $\psi_I'(0) = \psi_{II}'(0)$, $\psi_{II}(a) = \psi_{III}(a)$ and $\psi_{II}'(a) = \psi_{III}'(a)$, and where $k' = \sqrt{2m(E - V_0)/\hbar^2}$. Two cases can then happen. If $E > V_0$, ψ_{II} will consist of a superposition of two plane waves which reflect or transmit above the barrier. In this case,

$$R = \frac{(k^2 - k'^2)^2 \sin^2 k'a}{4k^2k'^2 + (k^2 - k'^2)^2 \sin^2 k'a}, \tag{3.81a}$$

$$T = \frac{4k^2k'^2}{4k^2k'^2 + (k^2 - k'^2)^2 \sin^2 k'a}. \tag{3.81b}$$

On the contrary, if $E < E_0$, k' becomes complex and the arguments of the exponentials in (3.80) become real. In this case, only the solution with $\beta = 0$ has physical meaning, since it implies a gradual (along x) attenuation of the "wave". This kind of damped solutions are called *evanescent waves* [41] because they correspond to waves which are progressively attenuated through a medium and the phenomenon is the well-known *tunnel effect* or *quantum tunneling*. Taking this into account,

$$R = \frac{(k^2 + \kappa^2)^2 \sinh^2 \kappa a}{4k^2\kappa'^2 + (k^2 + \kappa^2)^2 \sinh^2 \kappa a}, \tag{3.82a}$$

$$T = \frac{4k^2\kappa^2}{4k^2\kappa^2 + (k^2 + \kappa^2)^2 \sin^2 \kappa a}, \tag{3.82b}$$

where $\kappa = \sqrt{2m(V_0 - E)/\hbar^2}$. The *evanescent wave coupling* phenomenon is, however, a general property of any wave equation and can, therefore, occur in any context where a wave equation applies (e.g., optics, acoustics, quantum mechanics or waves on strings). These waves appear at the boundary between two media with different wave motion properties, being more intense within one third of a wavelength ($\lambda = 2\pi/k'$) from the surface of formation. In Chap. 4, this issue will be revisited within the context of optics and electromagnetism where, for example, the opacity index of a material or the radiation losses in waveguides can be explained in terms of evanescent waves. It is interesting to note that, due to the leading role of quantum mechanics in modern physics, the well-known concept of evanescent wave coupling is being substituted in optics and electromagnetism by that of tunneling, thus becoming very common in the literature the use of terms such as photon tunneling or acoustic tunneling.

3.3 Ensemble Distributions and the Density Matrix

As shown above, the wave function can be expressed either in the configuration representation or in the momentum one by simply using a Fourier transformation. From them, the density matrix can be obtained in each representation,

$$\hat{\rho}(x, x') \equiv \langle x|\hat{\rho}|x'\rangle \quad \text{or} \quad \tilde{\hat{\rho}}(p, p') \equiv \langle p|\tilde{\hat{\rho}}|p'\rangle, \tag{3.83}$$

respectively. The probability densities in the respective representation follow by only considering their diagonal elements, which is equivalent to tracing $\hat{\rho}$ over the configuration or the momentum variables, respectively, i.e.,

$$\rho(x) \equiv \rho(x, x) = \langle x|\hat{\rho}|x\rangle \quad \text{or} \quad \tilde{\rho}(p) \equiv \tilde{\rho}(p, p) = \langle p|\tilde{\hat{\rho}}|p\rangle. \tag{3.84}$$

Now, the fact that $\rho(x)$ or $\tilde{\rho}(p)$ have an appropriate classical limit does not mean necessarily one has a classical description. Note that in classical mechanics the distribution of positions and momenta is described by a joint probability distribution, $\rho_{cl}(x, p)$ and, therefore, one would desire to deal with similar joint probability distributions in quantum mechanics (mainly when describing open quantum systems, as will be seen in Chap. 5), i.e., a quantum probability distribution in phase space [42]. In other words, joint probability distributions, $\rho_Q(x, p)$, satisfying the properties

$$\int \rho_Q(x, p)dp = \rho(x), \quad \int \rho_Q(x, p)dx = \tilde{\rho}(p), \tag{3.85a}$$

$$\rho_Q(x, p) \geq 0, \qquad \text{for all } (x, p). \tag{3.85b}$$

In principle, one could assume ρ_Q can be expressed uniquely in terms of a pure state, as $\rho_Q(x, p) = \langle \Psi|M(x, p)|\Psi\rangle$, with $M(x, p)$ being a self-adjoint operator. However, such a ρ_Q cannot satisfy both properties (3.85a) and (3.85b) [43]; only mixed states of the form $\hat{\rho} = \sum_i w_i|\psi_i\rangle\langle\psi_i|$ fulfill the requirement that the phase space distribution only depends on the state operator ρ and not on the particular way how it is represented. Thus, when dealing with pure states, one of the two properties has to be "sacrificed" in order to provide a phase–space description of the quantum system. Two possibilities are usually considered in the literature. Either one works with the *Wigner distribution* [44], $\rho_W(x, p)$, which satisfies (3.85a) but not (3.85b), or with the *Husimi distribution* [45], $\rho_H(x, p)$, which works the other way around. Nevertheless, other quantum phase–space distribution functions can also be found in the literature [46].

3.3.1 The Wigner Distribution

For simplicity, consider a single particle in one dimension (though the generalization to many particles in three dimensions is straightforward). The Wigner representation of the density matrix or *Wigner distribution* is defined as

$$\rho_W(x, p) = \frac{1}{2\pi\hbar} \int \langle x - s/2|\hat{\rho}|x + s/2\rangle e^{ips/\hbar} ds. \qquad (3.86)$$

Particularly, for a pure state described by a wave function Ψ, this expression can be expressed as

$$\rho_W(x, p) = \frac{1}{2\pi\hbar} \int \Psi(x - s/2)\Psi^*(x + s/2) e^{ips/\hbar} ds. \qquad (3.87)$$

For example, if the state is given by a Gaussian wave packet,

$$\Psi(x) = \left(\frac{1}{2\pi\sigma_0^2}\right)^{1/4} e^{-(x-x_0)^2/4\sigma_0^2 + ip_0 x/\hbar}, \qquad (3.88)$$

the associated Wigner distribution will also be a Gaussian function, but in phase space, i.e.,

$$\rho_W(x, p) = \frac{1}{\pi\hbar} e^{-(x-x_0)^2/2\sigma_0^2 - 2\sigma_0^2(p-p_0)^2/\hbar^2}. \qquad (3.89)$$

If the system state is described by a superposition of two identical Gaussian wave packets symmetrically centered around $x = 0$ (at $x = \pm x_0$), i.e.,

$$\Psi(x) = \frac{N}{(2\pi\sigma_0^2)^{1/4}} \left[e^{-(x+x_0)^2/4\sigma_0^2} + e^{-(x-x_0)^2/4\sigma_0^2}\right], \qquad (3.90)$$

with $N = 1/\sqrt{2(1 + e^{-x_0^2/2\sigma_0^2})}$, the corresponding Wigner distribution will be

$$\rho_W(x, p) = \frac{N^2}{\pi\hbar} e^{-2\sigma_0^2 p^2/\hbar^2} \left[e^{-(x+x_0)^2/2\sigma_0^2} + e^{-(x-x_0)^2/2\sigma_0^2}\right.$$
$$\left. + 2e^{-x^2/2\sigma_0^2} \cos\left(\frac{2px_0}{\hbar}\right)\right]. \qquad (3.91)$$

This latter case is illustrated in Fig. 3.1(a), where (3.91) displays three peaks, two around $x = \pm x_0$ and a third one in between modulating the oscillatory factor that represents the interference of the two Gaussian wave packets. It is precisely in this oscillatory part where the Wigner distribution reaches both positive and negative

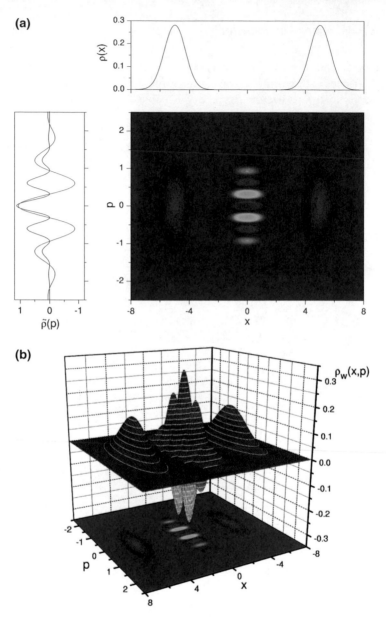

Fig. 3.1 a Wigner distribution function associated with the superposition of two Gaussian wave packets (3.90). In the *contour plot*, the transition from *green* to *red* denotes passing from negative to positive values of this distribution function. In the *upper panel*, the probability density in configuration space, $\rho(x)$; in the left-hand side panel, the probability density in momentum space, $\tilde{\rho}(p)$ (*black line*), and the corresponding wave function, $\tilde{\Psi}(p)$ (*red line*). **b** Surface representation of the Wigner distribution displayed in part (**a**)

values, as can be better appreciated in Fig. 3.1(b)—in general, whenever interference features are present, the Wigner distribution will show negative values. It is also remarkable that this middle, oscillating peak does not disappear as x_0 increases (i.e., when the wave packets are far enough from each other and $N \to 1/\sqrt{2}$), but remains, this being a signature of the *quantum coherence* describing the superposition.

The probability densities in momentum space, $\tilde{\rho}(p)$, associated with (3.88) and (3.90) can be straightforwardly obtained from the corresponding wave functions,

$$\tilde{\Psi}(p) = \left(\frac{2\sigma_0^2}{\pi \hbar^2} \right)^{1/4} e^{-\sigma_0^2(p-p_0)^2/\hbar^2 + i(p-p_0)x_0/\hbar} \tag{3.92}$$

and

$$\tilde{\Psi}(p) = 2N \left(\frac{2\sigma_0^2}{\pi \hbar^2} \right)^{1/4} e^{-\sigma_0^2 p^2/\hbar^2} \cos\left(\frac{px_0}{\hbar} \right), \tag{3.93}$$

respectively. For the particular case of the superposition of two Gaussian wave packets, Fig. 3.1(a) shows that the values of the Wigner distribution along the axes $p = 0$ or $x = 0$ are proportional to $\rho(x)$ (see the upper panel in Fig. 3.1(a)) and $\tilde{\rho}(p)$ (see the left-hand side panel), respectively, in agreement with property (3.85a). It is also observed that the maxima along $x = 0$ of the Wigner distribution correspond to maximum values of $\tilde{\rho}(p)$ or, equivalently, extrema of $\tilde{\Psi}(p)$, while the negative minima are associated with $\tilde{\rho}(p) = 0 (= \tilde{\Psi}(p))$.

In these two examples, the Gaussian wave packet and the coherent superposition fulfill

$$\iint \rho_W(x, p) dx dp = \text{Tr}[\hat{\rho}] = 1, \tag{3.94}$$

which can be easily shown taking into account that (3.86) satisfies (3.85a). Furthermore, ρ_W is always real since $\rho = \rho^\dagger$. However, it is also observed that ρ_W can take both positive and negative values (see Fig. 3.1(b)), thus not satisfying the nonnegativity property required for a state operator (see Sect. 3.2.3); only for pure Gaussian or coherent states ρ_W is always positive. Because of this, Wigner distributions are sometimes referred to as *quasi-probability* distributions. Nevertheless, if the following property

$$0 \le \text{Tr}[\hat{\rho}^{(i)} \hat{\rho}^{(j)}] \le 1 \tag{3.95}$$

for two different density matrices $\hat{\rho}^{(i)}$ and $\hat{\rho}^{(j)}$ is considered together with the definition (3.98) below, one finds

$$0 \le \iint \rho_W^2(x, p) dx dp \le 1. \tag{3.96}$$

Apart from the Wigner distribution, the Wigner representation of a general operator $\hat{A}$ is given by

$$A_W(x, p) = \int \langle x - s/2|\hat{A}|x + s/2\rangle e^{ips/\hbar} ds. \qquad (3.97)$$

In particular, for the potential and kinetic energy operators, $V_W(x, p) = V(x)$ and $K_W(x, p) = K(p) = p^2/2m$, respectively and where x and p are dynamical variables and not operators anymore. Therefore, the Wigner representation of the Hamiltonian operator is just as in classical mechanics, i.e., $H_W(x, p) = p^2/2m + V(x)$. The expectation value of the dynamical variable represented by the operator $\hat{A}$ can also be evaluated within this representation, becoming

$$\bar{A} = \langle \hat{A} \rangle = \iint \rho_W(x, p) A_W(x, p) dx dp \qquad (3.98)$$

when (3.33) has been used. Thus, if $\hat{A} = \hat{H}$ the average value of the system energy is reached.

In spite of the negativeness displayed by the Wigner distribution, it is employed not only in quantum mechanics [42], but also in optics [47], because it allows us to connect in a very straightforward way the classical phase space with a quantum phase space (which has some regions with "negative" probability). Actually, consider the evolution of the Wigner distribution function, given by

$$\frac{\partial \rho_W}{\partial t} = -\frac{p}{m} \frac{\partial \rho_W}{\partial q} + \sum_{n=0}^{\infty} \frac{(\hbar/2i)^{2n}}{(2n+1)!} \frac{\partial^{2n+1} V}{\partial q^{2n+1}} \frac{\partial^{2n+q} \rho_W}{\partial p^{2n+1}} = \{H, \rho_W\}_M, \qquad (3.99)$$

where $\{\cdot, \cdot\}_M$ is the so-called *Moyal bracket* [48, 49]. Given two any general functions A and B of q and p (i.e., in phase space), these brackets are defined as

$$\{A(q, p), B(q, p)\}_M \equiv \frac{2}{\hbar} A(q, p) \sin\left[\frac{\hbar}{2} \left(\overleftarrow{\partial}_q \overrightarrow{\partial}_p - \overleftarrow{\partial}_p \overrightarrow{\partial}_q \right) \right] B(q, p), \qquad (3.100)$$

where the arrow over the partial derivatives indicate the term over which they act (e.g., $A \overleftarrow{\partial}_q B = (\partial A/\partial q) B$). If the third and higher-order derivatives of the potential vanish (i.e., harmonic approximations or lower order ones are considered), the equation of motion (3.99) is exactly the same as its classical counterpart for the classical distribution function,

$$\frac{\partial \rho_{cl}}{\partial t} = -\frac{p}{m} \frac{\partial \rho_{cl}}{\partial q} + \frac{\partial V}{\partial q} \frac{\partial \rho_{cl}}{\partial p} = \{H, \rho_{cl}\}, \qquad (3.101)$$

which has already appeared before in this monograph in different contexts—see, for example, (1.56) in Sect. 1.4.2, where ρ_{cl} was denoted by f. As can be noticed, the passage from (3.101) to (3.99) consists only of substituting the Moyal brackets by the Poisson ones, defined by (1.23), since in the limit $\hbar \to 0$,

$$\{A, B\}_M \approx A \left(\overleftarrow{\partial}_q \overrightarrow{\partial}_p - \overleftarrow{\partial}_p \overrightarrow{\partial}_q \right) B = \{A, B\}. \qquad (3.102)$$

In this way, the Moyal brackets can be seen as a generalization of the standard Poisson ones when the latter are "deformed" by introducing higher-order derivative terms [50, 51]. Thus, whenever this limit satisfies, the Wigner phase–space distribution description of quantum mechanics reduces to the classical one based on Hamilton's equations of motion. This is precisely the basis of numerical approaches (see Volume 2) such as the so-called *Wigner method* [52], where—at a first level of approximation—observables are evaluated from a sampling over classical trajectories distributed in phase space according to a Wigner distribution.

3.3.2 The Husimi Distribution

In the case of the Husimi distribution, the positivity is warranted and, therefore, it acquires a probability interpretation. The idea behind the construction of this type of distribution arises from the way how the probability density arises in configuration space in terms of position eigenvectors, $|x\rangle$. These vectors satisfy both the orthonormality condition, $\langle x|x'\rangle = \delta(x - x')$, and the completeness relation, $\int |x\rangle\langle x|dx = 1$. The system state is then given by $\rho(x) = \langle x|\hat{\rho}|x\rangle$—in the case of a pure state, $\rho(x) = |\Psi(x)|^2$. Now, due to the complementarity of the configuration and momentum spaces, no eigenvectors of both position and momentum exist and, therefore, the same procedure cannot be followed. Nevertheless, one can consider instead the closest functions, namely minimum uncertainty wave packets, denoted by $|q, p\rangle$ and expressed in configuration space as

$$\langle x|q, p\rangle = \left(\frac{1}{2\pi s^2} \right)^{1/4} e^{-(x-q)^2/4s^2 + ipx/\hbar}. \qquad (3.103)$$

These wave packets are not associated with a certain phase space point (q, p), but at least they are localized around it with minimum dispersion in both q ($\Delta q = s$) and p ($\Delta p = \hbar/2s$). The width parameter s defines the basis set $|q, p\rangle$ and, therefore, different basis sets can be obtained by varying it. By construction, these vectors are not orthogonal, but form an over-complete set satisfying the completeness relation

$$\int |q, p\rangle\langle q, p|dqdp = 2\pi\hbar. \qquad (3.104)$$

Nonetheless, note that in the limit $s \to 0$, (3.103) approaches a position eigenvector, while in the limit $s \to \infty$ it approaches a momentum eigenvector. In this sense, as in the case of the Wigner representation, the Husimi representation also constitutes an intermediate step between the position and momentum representations.

With all of this in mind, the Husimi distribution is defined as

$$\rho_H(q, p) = \frac{1}{2\pi\hbar}\langle q, p|\hat{\rho}|q, p\rangle, \qquad (3.105)$$

which in the case of a pure state becomes

$$\rho_H(q, p) = \frac{1}{2\pi\hbar}|\Psi(q, p)|^2 = \frac{1}{2\pi\hbar}\left|\int\langle q, p|x\rangle\Psi(x)dx\right|^2. \qquad (3.106)$$

As can be noted from (3.106), the Husimi distribution is always positive and normalized, for (3.104) ensures $\int \rho_H(q, p)dqdp = 1$. To some extent, the Husimi distribution can be interpreted as a Gaussian smoothing (or coarse-graining) of the Wigner distribution with a "filter" of size larger than $\hbar$—the uncertainty relation for a Husimi distribution is $(\Delta q)_H(\Delta p)_H \geq \hbar$, twice larger than the bound for a usual quantum state, $\Delta q \Delta p \geq \hbar/2$. Actually, Husimi distributions can be expressed as the product of two Wigner distributions,

$$\rho_H(q, p) = \frac{1}{2\pi\hbar}\mathrm{Tr}\big[|q, p\rangle\langle q, p|\hat{\rho}\big] = \iint \rho_{qpW}(q', p')\rho_W(q', p')dq'dp', \qquad (3.107)$$

where ρ_W and ρ_{qpW} are the Wigner distributions associated with the state ρ and the minimum uncertainty wave packet $|q, p\rangle$, respectively. Taking (3.89) into account, the latter reads as

$$\rho_W(q', p') = \frac{1}{\pi\hbar}e^{-(q'-q)^2/2s^2-2s^2(p'-p)^2/\hbar^2}. \qquad (3.108)$$

In spite of the positivity of the Husimi distribution, its momentum and position integrals do not render the position and momentum probability distributions (3.85a), but a Gaussian-averaging version of them,

$$\bar{\rho}_H(q) = \int \rho_H(q, p)dp = \int \sqrt{\frac{1}{2\pi s^2}}\, e^{-(x-q)^2/2s^2}|\Psi(x)|^2dx, \qquad (3.109a)$$

$$\bar{\rho}_H(p) = \int \rho_H(q, p)dq = \int \sqrt{\frac{2s^2}{\pi\hbar^2}}\, e^{-2s^2(p'-p)^2/\hbar^2}|\tilde{\Psi}(p')|^2dp'. \qquad (3.109b)$$

In agreement with the statement above about the limits on s, in the limit $s \to 0$ (3.109a) approaches the position probability density $\rho(q) = |\Psi(q)|^2$, while in the limit $s \to \infty$ (3.109b) will approach the momentum probability density

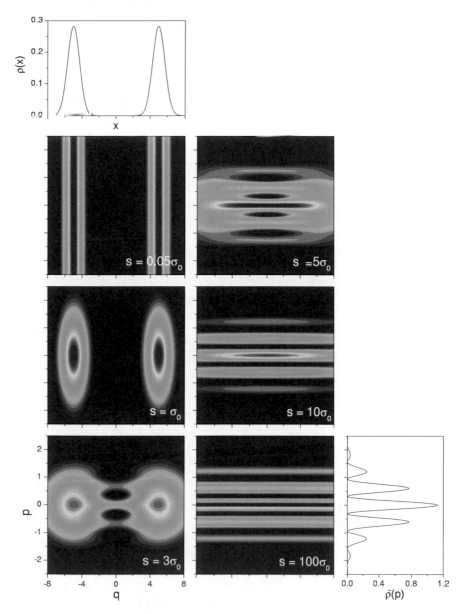

Fig. 3.2 Set of Husimi distributions from the family associated with the wave packet superposition (3.90) and generally specified by (3.11a) and (3.11b). To compare with, at the *left top* and the *right bottom* $\rho(x)$ (note that the same scale is used for both x and q, since both denote a configuration space coordinate) and $\tilde{\rho}(p)$ are respectively displayed

$\tilde{\rho}(p) = |\tilde{\Psi}(p)|^2$ (note that in those limits the prefactors inside the corresponding integrals become δ-functions).

In the particular case of the examples given above in Sect. 3.3.1, one finds that the Husimi distribution for the Gaussian state (3.88) is

$$\rho_H(q, p) = \frac{1}{\pi \hbar} \left(\frac{\sigma_0 s}{\sigma_0^2 + s^2} \right) e^{-(q-x_0)^2/2(\sigma_0^2 + s^2)} e^{-2(p-p_0)^2/(\sigma_0^{-2} + s^{-2})\hbar^2}, \quad (3.110)$$

while the Husimi distribution associated with the superposition (3.90) is

$$\rho_H(q, p) = \frac{N^2 \sigma_0 s}{\pi \hbar (\sigma_0^2 + s^2)} e^{-2p^2/(\sigma_0^{-2} + s^{-2})\hbar^2} \left\{ e^{-(q+x_0)^2/2(\sigma_0^2 + s^2)} \right.$$

$$\left. + e^{-(q-x_0)^2/2(\sigma_0^2 + s^2)} + 2 e^{-(q^2 + x_0^2)/2(\sigma_0^2 + s^2)} \cos \left[\frac{2 p x_0}{\hbar} \left(\frac{s^2}{\sigma_0^2 + s^2} \right) \right] \right\}.$$

$$(3.111)$$

In Fig. 3.2 it is observed a set of Husimi distributions of the family associated with the wave packet superposition (3.90) and given generally by (3.111). More specifically, as s increases, a gradual transition from a Husimi distribution, whose projection along the q axis is $\rho(q)$, to the opposite case, when $s \to \infty$, whose projection along the p axis reproduces the momentum probability density, $\tilde{\rho}(p)$, is patent. Contrary to the Wigner distribution, this type of distributions is in all cases positive definite.

3.4 Feynman's Path Integrals

Together with the wave and matrix formulations of quantum mechanics, started by Schrödinger and Heisenberg, respectively, around the first quarter of the twentieth century, the path integral formulation developed by Feynman in 1948 [53, 54] constitutes the third great approach (based on the wave function) to quantum mechanics. The idea behind of the path integral formulation apparently comes originally from Dirac and it was incorporated in the second edition of his famous book [55]. This approach has been a very important interpretational and computational tool since its inception, being particularly used to deal with statistical problems involving many degree-of-freedom systems in gas and condensed phases [56–58]. Its starting point consists of expressing the formal solution to the time-dependent Schrödinger equation in the integral form

$$\Psi(q, t) = \int K[q, t; q_0, t_0] \Psi(q_0, t_0) dq_0, \quad (3.112)$$

where the *kernel* or *propagator*, K, connects the initial and final states—as will be seen in Chap. 4, in the way how (3.112) is expressed, it keeps a very close relation with

Huyghen's principle and the so-called *Huygens' construction* [7, 9] (see Sect. 4.3.1). The propagator K also obeys Schrödinger's equation as well as the boundary condition

$$\lim_{t \to t_0^+} K[q, t; q_0, t_0] = \delta(q - q_0), \tag{3.113}$$

which allows us to interpret it as a time-evolved δ-function initially centered at q_0. From a classical perspective, this initial condition represents a uniform p_0-momentum distribution on the surface $q = q_0$, the same initial ensemble represented by the (classical) Jacobian determinant in configuration space [59]. Actually, in the classical limit, at t_0, the propagator approaches the Jacobian determinant, known as the *Van Vleck determinant* [60–62] within the context of the semiclassical approximation (see below). Therefore, this determinant and the classical action (which rules the interference effects) constitute the basic building block of almost any semiclassical approximation.

In order to construct the propagator K, Feynman proceeded [53, 54] considering the path-integral approach a way to reach quantum mechanics—though the inverse way [57] turns out to be an elegant, alternative to derive the path-integral formalism from quantum mechanics. Thus, consider the state of the system at t_0 and t_f is described by the coordinates q_0 and q_f, respectively, which can be connected by an infinity of paths, all of them satisfying the boundary conditions $q_{0,\alpha} = q_0$ and $q_\alpha = q_f$. A probability amplitude is then associated with every individual path, as

$$\Phi_\alpha(q_f, t_f) \propto e^{i S_\alpha(q_f, t_f)/\hbar}, \tag{3.114}$$

where S_α is given by (1.1) evaluated along the path α from $t_a = t_0$ to $t_b = t_f$. The propagator is now obtained by summing up over all possible paths, i.e.,

$$K[q_f, t_f; q_0, t_0] = \sum_\alpha \Phi_\alpha(q_f, t_f), \tag{3.115}$$

with the proportionality constant in (3.114) being such that (3.115) is properly normalized. From an interpretational viewpoint, (3.115) is already helpful (for example, it is clear from it that quantum effects arise from the *interference* among the probability amplitudes associated with each path), however it is not in the form of a path integral yet. To do so, first both space coordinates and time are discretized, as in Fig. 3.3, and then use of the multiplicative property satisfied by the propagator is made,

$$K[q_f, t_f; q_0, t_0] = \int K[q_f, t_f; q_k, t_k] K[q_k, t_k; q_0, t_0] dq_k, \tag{3.116}$$

where the subscript k denotes an intermediate state, such that $t_0 \leq t_k \leq t_f$. Then, considering the infinitesimal discreteness of every path and summing up over all paths joining q_0 with q_f [53, 54], one obtains the general and standard expression

Fig. 3.3 Discretization
process of both space and
time (polygonal trajectory)
to evaluate the propagator
(3.115). This process,
analogous to carrying out a
Riemann integral, gives rise
to the path-integral
expression (3.117) for the
propagator K

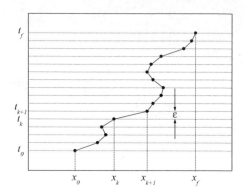

$$K[x_f, t_f; q_0, t_0] = \int e^{i S[q_f, t_f; q_0, t_0]/\hbar} \, \mathcal{D}q, \qquad (3.117)$$

known as the *path integral* [54]. The notation $\mathcal{D}q$ denotes "sum over histories" and explicitly illustrates the physics behind this sum over all paths connecting the initial and final states. That is, it emphasizes the fact that (within Feynman's framework) the evolution of a quantum system can always be understood as a combination of classical-like paths and the superposition principle.

A critical aspect of the path integral approach to quantum mechanics comes from the way how path integrals are solved. Most of times, the Riemann integral is used since it displays a nice convergence no matter what intermediate point in the intervals is selected to evaluate the corresponding approximating sums. This ideal situation is broken when, for example, a magnetic field, derivable from a vector potential, is included in the Hamiltonian [53, 57]. For this case, the midpoint rule of the interval should be adopted to obtain consistent results with the Schrödinger equation. This problem was already found in the theory of stochastic processes by Itô and Stratonovich when discussing the Brownian motion (see Appendix B). This midpoint rule is the choice due to Stratonovich.

Among the former applications of Feynman's path integrals is found the so-called *semiclassical theory* [63–74] (see Sect. 3.5.2), started by the earlier 1970s and further developed in the following decades. The basic idea here is, starting from the path integral approach, considering the classical limit, where only the paths close to the stationary one which makes S an extremum contribute to the transition probability obtained from the expression [63, 64]

$$\langle \Psi(q_f, t_f) | e^{-i \hat{H}(t_f - t_0)/\hbar} | \Psi(q_0, t_0) \rangle \sim e^{i S_{\rm cl}[q_f, t_f; q_0, t_0]/\hbar}, \qquad (3.118)$$

where $\hat{H}$ is the system Hamiltonian and $S_{\rm cl}$ is the *classical action*, i.e., the value of S when (1.1) is evaluated along the classical trajectory with boundary conditions q_0 at t_0 and q_f at t_f. In this way, although the evolution of both the Van Vleck determinant and the classical action are determined classically, due to the construction of (3.118), which allows for interference, observables can be computed approximately (i.e., semiclassically). This theoretical approach was initially developed to study both inelastic and reactive molecular collisions, and shortly after it was also adapted [75, 76] and applied [77–81] to atom–surface scattering. Within this particular field, the semiclassical theory resulted from an interpretation point of view advantageous because of the classical picture of scattering phenomena it provides. However, this theory can only be exactly applied to elastic scattering, being necessary the use of additional approximations in the inelastic case [80, 81]. Moreover, comparisons with other approximate quantum methods have shown [78, 79] that the semiclassical approach only provides better results for low surface corrugation. For highly corrugated surfaces, a case of interest in the study of surface resonances [82], there is a serious drawback that complicates the root search process involved in the calculation of the (semiclassical) S-matrix: the appearance of classically chaotic dynamics [83, 84]. This makes that a large number of (chaotic) trajectories connect the initial and final states [85].

Recently, work related to the tunnel effect within the path integral and semiclassical formulations has been reviewed by Ankerhold [34].

3.5 The Semiclassical Route to Quantum Mechanics

3.5.1 The Ehrenfest Theorem

Though incoming fluxes are usually described by means of plane waves, this is only an approximation which allows us to develop analytical treatments (e.g., the partial wave analysis used in scattering theory). However, when describing particles (whatever the meaning of "particle"), it is more common to consider wave packets, i.e., wave functions which have a certain extension in the corresponding configuration space. Accordingly, the classical limit is regarded as the dynamical regime at which the space variations of the forces (or, equivalently, potential functions) acting on the wave packet are relatively large in comparison with the effective extension of the latter. In other words, if the wave packet remains relatively localized, very little influenced, for sufficiently long times (namely, the *Ehrenfest time*), classical mechanics provides a good description of quantum processes. In such cases, the wave packet displays a classical-like behavior. For example, in bound systems, Ehrenfest's time should be larger than the associated classical period while, in scattering problems, this time should be larger than the typical (classical) times involved in collisions or diffraction.

Such remarks on the classicality of quantum systems are summarized within the so-called *Ehrenfest theorem* [86], which constitutes a way to obtain the time-evolution of expectation or average values, according to

$$\frac{d\langle \hat{A} \rangle}{dt} = \frac{i}{\hbar}\langle [\hat{H}, \hat{A}] \rangle + \left\langle \frac{\partial \hat{A}}{\partial t} \right\rangle, \tag{3.119}$$

where $\hat{A}$ is a quantum-mechanical operator and $\langle \hat{A} \rangle$ denotes its expectation or average value. Note that this expression is analogous to the classical one

$$\frac{dA}{dt} = \{H, A\} + \frac{\partial A}{\partial t}, \qquad (3.120)$$

written in terms of the Poisson bracket and where A is a certain classical variable (see Chap. 1). However, if $\hat{A}$ is the density matrix, $\hat{\rho}$, (3.120) will read as

$$\frac{d}{dt} \langle \hat{\rho} \rangle = -\frac{i}{\hbar} \langle [\hat{H}, \hat{\rho}] \rangle, \qquad (3.121)$$

if $\hat{\rho}$ does not depend explicitly on time, which happens for systems where probability is conserved, i.e., for quantum systems that are *closed* and described by pure states. The corresponding time evolution for *open* quantum systems will be discussed in Chap. 5.

If $\hat{H}$ and $\hat{\rho}$ commute, then $d\langle \hat{\rho} \rangle / dt = 0$ which is equivalent to the classical condition (1.55), thus (3.121) resembling (1.56) or, equivalently, (1.57). However, these classical counterparts are for the distribution function itself and not for its averaged values. The passage to the classical limit just consists in finding an equation of motion for the quantum expectation values themselves rather than having an equation of motion relating different expectation values ($d\langle \hat{A} \rangle / dt$, $\langle \partial \hat{A} / \partial t \rangle$ and $\langle [\hat{H}, \hat{A}] \rangle$). That is, if the wave packet is expected to display classical-like features, the expectation values for its position and momentum,

$$\frac{d\langle \hat{q} \rangle}{dt} = \left\langle \frac{\partial \hat{H}}{\partial \hat{p}} \right\rangle = \frac{\langle \hat{p} \rangle}{m}, \qquad (3.122a)$$

$$\frac{d\langle \hat{p} \rangle}{dt} = -\left\langle \frac{\partial \hat{H}}{\partial \hat{q}} \right\rangle = -\left\langle \frac{\partial \hat{V}}{\partial \hat{q}} \right\rangle, \qquad (3.122b)$$

should essentially behave as a classical trajectory and be described by

$$\frac{dq}{dt} = \frac{p}{m}, \qquad (3.123a)$$

$$\frac{dp}{dt} = -\frac{\partial V}{\partial q}. \qquad (3.123b)$$

The condition now for (3.122) to behave like their classical counterparts (3.123) is that $\Delta \hat{q}$ and $\Delta \hat{p}$ remain small during times of the order of the typical timescales involved in the system dynamics, for this would imply $\langle \hat{H}(\hat{q}, \hat{p}, t) \rangle \approx \hat{H}(\langle \hat{q} \rangle, \langle \hat{p} \rangle, t)$ and therefore (3.122) would read as

$$\frac{d\langle\hat{q}\rangle}{dt} = \frac{\partial H(\langle\hat{q}\rangle, \langle\hat{p}\rangle, t)}{\partial\langle\hat{p}\rangle}, \tag{3.124a}$$

$$\frac{d\langle\hat{p}\rangle}{dt} = -\frac{\partial H(\langle\hat{q}\rangle, \langle\hat{p}\rangle, t)}{\partial\langle\hat{q}\rangle}, \tag{3.124b}$$

which are much closer to (3.123). In order to find the condition leading to this result, let us define the force vector operator,

$$\hat{F}(\hat{q}) = -\frac{\partial\hat{H}}{\partial\hat{q}}, \tag{3.125}$$

representing the forcing acting on the quantum particle. The Taylor series expansion of this operator around $\langle q\rangle(t)$ is given by

$$\hat{F}(\hat{q}) = \hat{F}(\langle\hat{q}\rangle) + (\hat{q} - \langle\hat{q}\rangle) \cdot \nabla\hat{F}\Big|_{\hat{q}=\langle\hat{q}\rangle}$$

$$+ \frac{1}{2}\sum_{i,j}(\hat{q}_i - \langle\hat{q}_i\rangle)(\hat{q}_j - \langle\hat{q}_j\rangle)\frac{\partial^2\hat{F}}{\partial\hat{q}_i\partial\hat{q}_j}\Big|_{\hat{q}=\langle\hat{q}\rangle} + \cdots \tag{3.126}$$

Substituting now (3.126) into (3.122b) yields

$$\frac{d\langle\hat{p}\rangle}{dt} \simeq \hat{F}(\langle\hat{q}\rangle) + \frac{1}{2}\sum_{i,j}\Delta_{ij}\frac{\partial^2\hat{F}}{\partial q_i\partial q_j}\Big|_{\hat{q}=\langle\hat{q}\rangle}, \tag{3.127}$$

where $\Delta_{ij} \equiv \langle\hat{q}_i\hat{q}_j\rangle - \langle\hat{q}_i\rangle\langle\hat{q}_j\rangle$. If $\Delta_{ij} \approx \sigma^2$, where σ is the effective size or width of the wave packet, the nonclassical term in (3.127) can be neglected and the center of the wave packet will only be affected by the first term of the Taylor series (3.126), thus displaying a classical-like motion. A more quantitative criterion is

$$\left|\sigma^2\frac{\partial^3 V/\partial q_i\partial q_j\partial q_k}{\partial V/\partial q_l}\right| \ll 1, \tag{3.128}$$

i.e., whenever the spreading of the wave packet remains sufficiently small in comparison with the spatial variations of the potential acting on it, the wave packet will behave like a classical particle and its evolution will be describable by means of the classical laws. Note that although (3.128) is independent of $\hbar$, it is required that the wave packet be sufficiently narrow and potentials vary slowly with q. On the other hand, also notice that the criterion (3.128) is not valid for potentials which are polynomials of second or less degree, for (3.126) does not provide any correcting term. In such cases, $\langle\hat{q}\rangle$ and $\langle\hat{p}\rangle$ coincide with their classical counterparts at any time and, therefore, it is not possible to speak about an Ehrenfest-like classical limit, but other criteria (if any) have to be chosen.

3.5.2 The JWKB Approximation

The roots of semiclassical mechanics lie in the old quantum theory [33, 34, 87]. It represents a short wavelength link between classical and quantum mechanics, similar to that found between wave optics and geometric optics (see Sect. 7.3). It is known as the *JWKB* or *WKB approximation* (as it is more generally known), developed by Jeffreys [88], Wentzel [89], Kramers [90] and Brillouin [91, 92]. Jeffreys was a mathematician who developed a general method of approximating solutions to linear second-order differential equations. On the other hand, Wentzel, Kramers and Brillouin, apparently unaware of Jeffreys' work, used a similar method for the resolution of Schrödinger equation, discovered two years later.

As it can be inferred from Sect. 3.4, a good understanding of the emerge of the classical limit within Feynman's approach also constitutes a key point to derive a semiclassical expression for the propagator. As seen before, every path contributes to the transition probability with the same probability ($|\Phi_\alpha|^2 = 1$) independently of whether it corresponds to a classical trajectory or not. On the other hand, in the classical limit one expects that the corresponding action S_α becomes very large in comparison to $\hbar$, this translating into very fast oscillations in the waves associated with every path (see (3.114)). Interference among waves associated with paths far from the classical trajectory will be destructive and they will barely contribute to the propagator, while the phase difference for paths in the vicinity of the classical trajectory will be more important (due to the stationarity of S) and interferences will be constructive. Thus, if q_0 and q are joined by a single classical trajectory, because of stationarity to first order the phase associated with the paths closer to the classical trajectory will be S_{cl} for all of them, thus becoming the global phase factor for the propagator. Meanwhile, second-order phase differences give a Gaussian integral, whose modulus can be interpreted as a measure of how many neighboring paths contribute to the transition probability in the classical limit.

The derivation of the semiclassical propagator can be carried out by means of the JWKB approximation, using the one-dimensional, time-independent JWKB ansatz [33, 93]

$$\psi(q) = \rho^{1/2}(q)e^{i\phi(q)/\hbar}. \tag{3.129}$$

This approximation is also called the *primitive semiclassical approximation* [94], where ρ is the probability density and ϕ a real-valued phase. Since (3.129) is assumed to be a formal solution of the time-independent Schrödinger equation at an energy E,

$$\frac{\partial^2 \psi(q)}{\partial q^2} + \frac{4\pi^2}{\lambda^2(q)}\psi(q) = 0, \tag{3.130}$$

where $\lambda(q) \equiv 2\pi\hbar/\sqrt{2m[E - V(q)]}$ is a *local* or *effective* de Broglie wavelength, after substituting (3.129) into (3.130) is found

$$\nabla \cdot \mathbf{J} = 0, \tag{3.131a}$$

$$\frac{p^2}{2m} = \frac{(\nabla \phi)^2}{2m} - \frac{\hbar^2}{2m} \frac{\nabla^2 \rho^{1/2}}{\rho^{1/2}}, \tag{3.131b}$$

where $p = \hbar/\lambda$, $\mathbf{J} = \rho(\nabla\phi)/m$ and the second term in the right-hand side of (3.131b) is the so-called *quantum potential* (see Chap. 6),

$$Q_P \equiv -\frac{\hbar^2}{2m} \frac{\nabla^2 \rho^{1/2}}{\rho^{1/2}}. \tag{3.132}$$

As can be noticed, (3.131a) tells us that the quantum flow is stationary (time independent)—since ρ does not depend on time, there cannot be any source or sink of quantum flow and therefore the quantum probability current density has to be stationary. On the other hand, according to (3.131b) the local quantum momentum field depends on the variations of the phase and also on another component with a purely quantum origin. In classical mechanics, the momentum depends only on the action through its gradient, so if the classical limit is defined when $\phi \approx S_{cl}$, then

$$\frac{p^2}{2m} \approx \frac{(\nabla \phi)^2}{2m} \gg \left| -\frac{\hbar^2}{2m} \frac{\nabla^2 \rho^{1/2}}{\rho^{1/2}} \right|. \tag{3.133}$$

This criterion for classicality is similar to the standard one that the JWKB approximation consists of dropping terms containing $\hbar^2$ in (3.131b) [93].

From (3.133), if it holds, ϕ then corresponds to the classical action evaluated along a trajectory $q(t)$ at an energy E. Therefore, a solution for the continuity equation (3.131a) along this trajectory can be readily found,

$$\rho(q) = \left| \frac{p(q_0)}{p(q)} \right| \rho(q_0), \tag{3.134}$$

where q_0 is the initial condition for $q(t)$ and $\rho(q_0)$ is the density distribution evaluated at that point. Putting these results together one reaches the well-known semiclassical solution

$$\psi_{sc} \propto \sqrt{\lambda(q)} e^{i S_{sc}(q)/\hbar}, \tag{3.135}$$

valid whenever the potential varies slowly over many wavelengths. This can be seen from (3.133), which can be recast as

$$\left| \frac{V'(q)}{V'''(q)} \right| \gg \frac{\lambda^2(q)}{4} \tag{3.136}$$

taking into account (3.134) and that $V'' \simeq 0$ to assure the smoothness of the potential [95] (primes indicate the order of the derivatives with respect to q). This condition

of classicality is the same as the one found through Ehrenfest's theorem, given by
(3.128), if it is assumed the width of the wave packet (σ) is of the order of $\lambda(q)/2$.

For potentials that are polynomials of degree less than or equal to two the condition
(3.136) is always satisfied ($\lambda(q) \ll \infty$) provided one does not approach the classical turning points, where $p = 0$ and the semiclassical approximation breaks down
($\lambda \to \infty$). Actually, this is not only for this kind of potentials, but in general. Therefore, the semiclassical wave function ψ_{sc} is only of *local* validity. Nevertheless, this
problem can be overcome by finding a *uniform approximation* to the wave function
[94, 96, 97], i.e., expressing it as a combination of primitive semiclassical forms with
validity in each region of phase space. This can be carried out by using a technique
developed by Maslov [96] that takes advantage of the possibility to represent the wave
function in both the configuration and the momentum representations. According to
this technique, one computes the semiclassical wave function from the initial point
to another one close to the turning point where the wave function is still valid. From
this new point to its inverted (i.e., the same point but after passing through the turning
point) the evaluation of the wave function is made in the momentum representation,
where there is no breaking down. Once in the inverted point, the configuration representation is used again. This process, repeated for any turning point that appears until
reaching the final point q_f, leads to a good estimation of the wave function.

The step to switch from the configuration to the momentum representation is
carried out by making use of the *stationary phase method*, well-known in geometric
optics [98] (this method is also known as the steepest descent or saddle point method).
This method consists of approximating the phase in multidimensional integrals like

$$\mathcal{I} = \int A(q)\, e^{i\Upsilon(q)/\hbar} dq \tag{3.137}$$

by their expansion around a point $\tilde{q}$ to second order in $\delta q_n \equiv |q_n - \tilde{q}_n|$, i.e.,

$$\Upsilon(q) \approx \Upsilon(\tilde{q}) + \frac{1}{2}\frac{\partial^2 \Upsilon}{\partial q_k \partial q_l}\bigg|_{q=\tilde{q}} \delta q_k \delta q_l, \tag{3.138}$$

and the amplitude by its value at $\tilde{q}$ if it varies slower than the phase over many
oscillations of the latter. Taking this into account and integrating by using the optical
Fresnel-integral formula, (3.137) becomes

$$\mathcal{I} = \frac{(2\pi\hbar)^{N/2}}{|\Delta(\mathcal{M})|^{1/2}}A(\tilde{q})e^{i\Upsilon(\tilde{q})/\hbar + i\pi\,\mathrm{sgn}\,(\mathcal{M})/4}, \tag{3.139}$$

where $\Delta(\mathcal{M})$ and $\mathrm{sgn}(\mathcal{M}) \equiv \Delta(\mathcal{M})/|\Delta(\mathcal{M})|$ represent, respectively, the deter-
minant and the signature (positive minus negative eigenvalues) of the matrix $\mathcal{M}$,
defined as

$$\mathcal{M} \equiv \frac{\partial^2 \Upsilon(q)}{\partial q_k \partial q_l}\bigg|_{q=\tilde{q}}. \tag{3.140}$$

If $\Delta(\mathcal{M}) = 0$, the stationary phase approximation fails; the next order in the expansion of $\Upsilon(q)$ is then considered. This problem, equivalent to solving the Schrödinger equation for a linear potential (the linearized approximation of $V(q)$), involves the use of Airy functions [97] and allows the semiclassical wave function to penetrate into the classical forbidden region.

So far a time independent analysis of semiclassical wave functions has been developed. However, it can also be extended to time-dependent problems by considering the ansatz

$$\Psi(q, t) = \rho^{1/2}(q, t)e^{i\Phi(q,t)/\hbar}. \tag{3.141}$$

Proceeding as before, (3.131) become

$$\frac{\partial \rho}{\partial t} + \nabla \cdot \mathbf{J} = 0, \tag{3.142a}$$

$$\frac{\partial \Phi}{\partial t} + \frac{(\nabla \Phi)^2}{2m} + V + Q_P = 0, \tag{3.142b}$$

respectively, where now $\mathbf{J} = \rho(\nabla \Phi)/m$. That is, (3.142a) is the continuity equation for the conservation of the probability density, while (3.142b) is a *quantum Hamilton–Jacobi equation*, where Q_P is now time-dependent due to the dependence on time of ρ. As before, the semiclassical approximation also arises when neglecting Q_P in (3.142b), according to (3.133). This makes (3.142b) to become the classical Hamilton–Jacobi equation and $\Phi \approx S_{\text{sc}}$ Hamilton's *principal function* [13] discussed in Chap. 1. Then, a solution of (3.142a) along the trajectory $q(t)$ can be found, reading as

$$\rho(q(t), t) = |\Delta(\mathcal{J})| \rho(q_0, t_0), \tag{3.143}$$

where

$$\mathcal{J} \equiv \frac{\partial q_0}{\partial q} \tag{3.144}$$

is the *Jacobian matrix*.

At sufficiently small times, Φ remains single-valued and the semiclassical wave function (3.141) can be then written as

$$\Psi_{\text{sc}}(q, t) = |\Delta(\mathcal{J})|^{1/2} e^{i S_{\text{sc}}[q,t;q_0,t_0]/\hbar} \Psi(q_0, t_0). \tag{3.145}$$

However, as time goes on, the wave function (3.145) might no longer be valid since Φ is not necessarily unique because more than one trajectory with different values of S_{sc} may connect q_0 with $q(t)$. These points belong to *caustics* [85], which are the

projection onto configuration space of all those points for which the Jacobian determinant, $\Delta(\mathcal{J})$, vanishes (thus the amplitude of (3.145) becoming infinity). Under
this condition, one can overcome the problem by finding a uniform approximation
near the caustic, which is analogous to finding it close to the turning point in time-
independent problems. In scattering problems, for example, this procedure consists
in considering separately: the incident part of the trajectories as well as the outgoing
part of those trajectories that do not reach the caustic; the scattered part of the trajectories after "touching" the caustic; and the part corresponding to the singular points
on the caustic. This leads to a division of the configuration space in nonoverlapping
domains where the globally valid approximate solution is well defined on the boundaries of such domains and corresponds to the different piecewise local approximate
solutions in every domain.

   Unless one needs to evaluate the semiclassical expression of the wave function on a point of a caustic, it is not necessary to consider the momentum representation. However, it is important to take into account that since there might be
different trajectories connecting q_0 with q at t, one has to consider the evaluation
of $\Delta(\mathcal{J}_\alpha)$ and $S_{sc}^{(\alpha)}$ for each particular trajectory α. On the other hand, since the
orientation of the Lagrangian manifold $(q, \nabla\Phi)$ at t will vary with respect to the
initial manifold, the eigenvalues of $\mathcal{J}$ will also change their sign each time that the
corresponding trajectory touches a caustic. This is expressed by writing the Jacobian
determinant as

$$\Delta(\mathcal{J}_\alpha) = e^{-i\pi\mu_\alpha(q(t),q_0)} |\Delta(\mathcal{J}_\alpha)|, \tag{3.146}$$

where $\mu_\alpha(q(t), q_0)$ is the *Maslov index* associated with the trajectory α, which counts
the number of times that the sign of the Jacobian determinant changes along such a
trajectory when going from q_0 to $q(t)$. Taking all these remarks into account, one
obtains the correct form for the evolved semiclassical wave function,

$$\Psi_{sc}(q,t) = \int \left\{ \sum_\alpha |\Delta(\mathcal{J}_\alpha)|^{1/2} e^{i S_{sc}^{(\alpha)}[q,t;q_0,t_0]/\hbar - i\pi\mu_\alpha(q(t),q_i)/2} \right\} \Psi(q_0, t_0) dq_0, \tag{3.147}$$

which, as can be seen, includes the contribution from all possible trajectories that
connect any point q_0 with $q(t)$.

   The expression between the curly brackets in (3.147) is precisely the expression
for the propagator that maps the initial ansatz $\Psi(q_0, t_0)$ into $\Psi_{sc}(q,t)$. Observe,
however, that there is a small subtlety in this expression: the propagator in (3.147)
does not satisfy the boundary condition (3.113). This is because this propagator has
been obtained under the assumption that particles have a well-defined momentum at q,
what clearly contradicts the fact that for $q = q_0$ its amplitude is infinity and its phase
is not well-defined—and therefore particles can have any momentum. Notice that this
is also connected to the fact that the calculation of the semiclassical wave function
has been carried out by predetermining a certain trajectory $q(t)$. In order to obtain
the correct expression for the propagator, one has to consider the expression between
the curly brackets in (3.147) and the one corresponding to the short-time evolution

of the propagator [97]. Putting both expressions together results in the semiclassical or *Van Vleck* propagator:

$$K_{sc}(q, t) = \sum_{\alpha} \frac{1}{(2\pi i \hbar)^{N/2}} |\Delta(\mathcal{V}_{\alpha})|^{1/2} \, e^{i S_{sc}^{(\alpha)}[q, t; q_0, t_0]/\hbar - i\pi \mu_{\alpha}(q(t), q_0)/2}, \quad (3.148)$$

where

$$\mathcal{V}_{\alpha} \equiv \frac{\partial p_0}{\partial q} = -\frac{\partial^2 S_{sc}^{(\alpha)}}{\partial q \partial q_0}. \quad (3.149)$$

Two important points about the use of the semiclassical propagator are worth stressing. First, the use of the semiclassical approximation requires the search for all classical trajectories connecting q_0 with $q(t)$; this constitutes a difficult task. Indeed, under the presence of chaotic dynamics, although several alternatives can be followed, the problem is apparently intractable. One of the techniques that is receiving much attention, specially within the chemical physics community, is the *initial value representation* (IVR) of the semiclassical propagator [34, 63, 64, 99–103]. The best known IVR propagator is the so-called *Hermann–Kluk propagator* although a variety of alternative IVR propagation schemes have been proposed [34]. Second, the semiclassical propagator (3.149) satisfies the property (3.116) only approximately. Thus, the multidimensional integral

$$K_{sc}[q_f, t_f; q_0, t_0] \approx \int K_{sc}[q_f, t_f; q_k, t_k] K_{sc}[q_k, t_k; q_0, t_0] dq_k \quad (3.150)$$

can only be evaluated out approximately by using the stationary phase method integrating over regions near points q_k where the phase is stationary. Classical trajectories will contribute whenever the final momentum from q_0 to q_k and the initial momentum from q_k to q_f coincide.

Quantum tunneling and escape processes, in general, have been some of the main problems studied in the semiclassical approximation. Two powerful thermodynamics methods (the bounce and the instanton methods) are also widely used to perform thermal averages over tunneling rates in a dissipationless regime within the energy and time domains. Recently, this theoretical scheme has been extended to density matrices by a unified semiclassical approach [34].

When dealing with genuine quantum processes such as, for example, barrier or dynamical tunneling, interference and diffraction and the zero point motion, a classical analysis has to be carried out with great care. The time-evolution of many molecular processes involving elastic and inelastic collisions, reactive scattering, dissociation problems or diabatic transitions, to just cite a few of them, is usually analyzed by running classical trajectories where some sort of quantization at the beginning (initial conditions) or at the end (statistical treatments) or in between (hope probabilities), or even all of them, need to be implemented. Furthermore, complex classical trajectories are also very often performed. This gave rise to a large variety of methods which were applied with great success. One of the main reasons why such "quasi-classical" methods were used is the high dimensionality of the process under consideration where quantum mechanical treatments become prohibitive. Nowadays, a very important number of "hybrid" (combination of classical and quantum elements) methods more sophisticated can be found in the literature. To cite all of them with their advantages and disadvantages is a huge effort out of the scope of this monograph.

Finally, a very active field of research is also the role played by periodic orbits and, in general, classically chaotic systems in building wave functions. An introduction to this field can be found in [104].

3.5.3 The Eikonal Approach

Semiclassical approximations, like the JWKB discussed above, are often inspired from optics. In this regard, another interesting semiclassical approach is the so-called *eikonal approximation* [14, 98, 105, 106]. This approach readily applies to systems with many degrees of freedom, e.g., elastic atom–surface scattering assuming simple corrugated hard-wall potentials [107], or atomic and molecular collisions including any kind of potential function [108, 109].

As before, the starting point here is also the ansatz

$$\Psi(\mathbf{q}) = \chi(\mathbf{q})e^{iS(\mathbf{q})/\hbar}, \tag{3.151}$$

though the functions χ and S are now complex and real, respectively (actually, in general, χ could be a complex column matrix) and $\mathbf{q}$ represents the position of the particle in three dimensions. After substitution of (3.151) into the time-independent Schrödinger equation,

$$\left[\frac{1}{2m}(-i\hbar\nabla + \nabla S)^2 + V - E\right]\chi = 0. \tag{3.152}$$

From this expression, one can define the *local* total energy of the system, as

$$\begin{aligned}
E(\mathbf{q}) &\equiv \text{Re}\left[\frac{\Psi^*(\mathbf{q})\hat{H}\Psi(\mathbf{q})}{|\Psi(\mathbf{q})|^2}\right] \\
&= \frac{(\nabla S)^2}{2m} + V - \frac{\hbar^2}{4m|\chi|^2}\left(\chi^*\nabla^2\chi + \chi\nabla^2\chi^*\right) - \frac{i\hbar(\nabla S)}{4m|\chi|^2}\left(\chi^*\nabla\chi - \chi\nabla\chi^*\right).
\end{aligned} \tag{3.153}$$

Although S is not determined at this stage, proceeding as in classical mechanics, the momentum is defined again as

$$\mathbf{p} \equiv \nabla S, \tag{3.154}$$

which allows to rewrite the energy (3.153) as

$$E = \frac{\mathbf{p}^2}{2m} + \tilde{V}(\mathbf{q}, \mathbf{p}), \tag{3.155}$$

and where $\tilde{V}$ is the so-called effective quantum potential,

$$\tilde{V} = V - \frac{\hbar^2}{4m|\chi|^2} \left(\chi^* \nabla^2 \chi + \chi \nabla^2 \chi^* \right) - \frac{i\hbar p}{4m|\chi|^2} \left(\chi^* \nabla \chi - \chi \nabla \chi^* \right). \tag{3.156}$$

since terms containing $\hbar$ are included, apart from the V potential. This allows to express again the system Hamiltonian as

$$\hat{H}(\mathbf{q}, \mathbf{p}) = \frac{\mathbf{p}^2}{2m} + \tilde{V}(\mathbf{q}, \mathbf{p}), \tag{3.157}$$

which is similar to a classical Hamiltonian, with $\mathbf{q}$ and $\mathbf{p}$ being the system variables.

Taking advantage of the Hamiltonian-like form acquired by the initial time-independent Schrödinger equation after considering the ansatz (3.151) and assuming now that both q and p depend parametrically on t, the evolution of the latter are assumed to follow Hamilton's equation of motion,

$$\frac{d\mathbf{q}}{dt} = \frac{\mathbf{p}}{m} + \frac{\partial \tilde{V}}{\partial \mathbf{p}}, \tag{3.158a}$$

$$\frac{d\mathbf{p}}{dt} = -\frac{\partial \tilde{V}}{\partial \mathbf{q}}. \tag{3.158b}$$

The corresponding trajectories are not classical since the effective quantum potential rules the underlying dynamics. From this type of *quantum trajectories*, one can then calculate S and the transition integrals by projection on the final states [109] by

$$T_f = \int e^{i\mathbf{p}_f \cdot \mathbf{q}/\hbar} V(\mathbf{q}) \chi(\mathbf{q}) e^{iS(\mathbf{q})/\hbar} d\mathbf{q}. \tag{3.159}$$

Nonetheless, note that one has to evaluate χ, which can be done by substituting (3.154) into (3.152). Due to the explicit dependence of $\tilde{V}$ on χ and S, this type of calculation is quite involved. In the limit of high energy collisions, however, (3.158a) and (3.158b) can be solved in an easier manner. Observe that in this limit, the (effective) de Broglie wavelength, $\lambda = 2\pi\hbar/|\mathbf{p}|$, becomes small and one can then take into account the conditions leading to the optical eikonal approximation [14]. That is, the spatial variations of χ are slower than those of S, and therefore

$$|\nabla^2 \chi| \ll \left| \frac{\mathbf{p}}{\hbar} \nabla \chi \right| \ll \left| \left(\frac{\mathbf{p}}{\hbar} \right)^2 \chi \right|. \tag{3.160}$$

Under these conditions, the effective quantum potential, $\tilde{V}$, reduces to the classical one, V, and (3.158a) and (3.158b) become the usual classical equations of motion.

References

1. Jammer, M.: The Conceptual Development of Quantum Mechanics, 2nd edn. American Institute of Physics, New York (1989)
2. Jammer, M.: The Philosophy of Quantum Mechanics. The Interpretations of Quantum Mechanics in Historical Perspective. Wiley-Interscience, New York (1974)
3. Cushing, J.T.: Quantum Mechanics Historical Contingency and the Copenhagen Hegemony. The University of Chicago Press, Chicago (1994)
4. Sánchez-Ron, J.M.: Historia de la Física Cuántica. El período fundacional. Drakontos, Barcelona (2001)
5. Selleri, F.: El debate de la teoría cuántica. Alianza Editorial, Madrid (1986)
6. Bub, J.: Interpreting the Quantum World. Cambridge University Press, Cambridge (1999)
7. Margenau, H., Murphy, G.M.: The Mathematics of Physics and Chemistry, 2nd edn. Van Nostrand, New York (1956)
8. Schrödinger, E.: An undulatory theory of the mechanics of atoms and molecules. Phys. Rev. **28**, 1049–1070 (1926)
9. Courant, R., Hilbert, D.: Methods of Mathematical Physics. vol. 1. Wiley, New York (1953)
10. Schrödinger, E.: Quantisierung als Eigenwertproblem (series of papers). Ann. Phys. (Leipzig) **79**, 361–376, 489–527 (1926)
11. Schrödinger, E.: Quantisierung als Eigenwertproblem (series of papers). Ann. Phys. (Leipzig) **80**, 437–490 (1926)
12. Schrödinger, E.: Quantisierung als Eigenwertproblem (series of papers). Ann. Phys. (Leipzig) **81**, 109–139 (1926)
13. Goldstein, H.: Classical Mechanics. Addison-Wesley, Reading (1980)
14. Born, M., Wolf, E.: Principles of Optics. Pergamon Press, Oxford (1980)
15. de Broglie, L.: Recherche sur la théorie des quanta. Ann. Physique **3**, 22–128 (1925)
16. Sommerfeld, A., Runge, I.: Anwendung der Vektorrechnung auf die Grundlagen der geometrischen Optik. Ann. Phys. (Leipzig) **35**, 290–298 (1911)
17. Sommerfeld, A.: Mechanics. Lectures on Theoretical Physics, vol. 1, pp. 229–230. Academic Press, London (1964)
18. Morse, P.M., Feshbach, H.: Methods of Theoretical Physics. McGraw-Hill, New York (1953)
19. Heisenberg, W.: Über den anschaulichen Inhalt der quantentheoretischen Kinematik und Mechanik. Z. Phys. **43**, 172–198 (1927)
20. Folland, G.B., Sitaram, A.: The uncertainty principle: a mathematical survey. J. Fourier Anal. Appl. **3**, 207–238 (1997)
21. Pinsky, M.: Introduction to Fourier Analysis and Wavelets. Brooks/Cole, Pacific Grove (2002)
22. Plancherel, M.: Contribution à l'étude de la représentation d'une fonction arbitraire par des intégrales définies. Rendiconti del Circolo Matematico di Palermo **30**, 289–335 (1910)
23. Schiff, L.I.: Quantum Mechanics, 3rd edn. McGraw-Hill, Singapore (1968)
24. Born, M.: Zur Quantenmechanik der Stoßvorgänge (series of papers). Z. Phys. **37**, 863–867 (1926)
25. Born, M.: Zur Quantenmechanik der Stoßvorgänge (series of papers). Z. Phys. **38**, 803–840 (1927)
26. Born, M.: Physical aspects of quantum mechanics. Nature **119**, 354–357 (1927)
27. Born, M.: Quantenmechanik und Statistik. Naturwissenschaften **15**, 238–242 (1927)
28. Ballentine, L.E.: The statistical interpretation of quantum mechanics. Rev. Mod. Phys. **42**, 358–381 (1970)
29. Ballentine, L.E.: Quantum Mechanics. A Modern Development. World Scientific, Singapore (1998)
30. Tarozzi, G., van der Merwe, A. (eds.): Open Questions in Quantum Physics. Reidel, Dordrecht (1985)
31. Wheeler, J.A., Zurek, W.H. (eds.): Quantum Theory and Measurement. Princeton University Press, Princeton (1983)

32. von Neumann, J.: Mathematische Grundlagen der Quantenmechanik. Springer, Berlin (1932). Translated into English by Beyer, R.T.: Mathematical Foundations of Quantum Mechanics. Princeton University Press, Princeton (1955)
33. Razavy, M.: Quantum Theory of Tunneling. World Scientific, Singapore (2003)
34. Ankerhold, J.: Quantum Tunneling in Complex Systems. Springer, Berlin (2007)
35. Main, I.G.: Vibrations and Waves in Physics, 3rd edn. Cambridge University Press, Cambridge (1993)
36. Fowler, R.H., Nordheim, L.W.: Electron emission in intense electric fields. Proc. R. Soc. Lond. A **119**, 173–181 (1928)
37. Gamow, G.: Zur Quantentheorie des Atomkernes. Z. Phys. **51**, 204–212 (1928)
38. Gurney, R.W., Condon, E.U.: Wave mechanics and radioactive disintegration. Nature **122**, 439–439 (1928)
39. Wood, R.W.: A new form of cathode discharge and the production of X-rays, together with some notes on diffraction. Preliminary communication. Phys. Rev. (Ser. I) **5**, 1–10 (1897)
40. Elster, J., Geitel, H.F.: Über einige zweckmässige Abänderungen am Quadrantelectrometer. Ann. Phys. (Leipzig) **300**, 680–684 (1898)
41. Fornel, F.: Evanescent Waves. From Newtonian Optics to Atomic Optics. Springer, Berlin (2001)
42. Zachos, C.K., Fairlie, D.B., Curtright, T.L. (eds.): Quantum Mechanics in Phase Space. An Overview with Selected Papers. World Scientific, Singapore (2005)
43. Wigner, E.P.: Quantum mechanical distribution functions revisited. In: Yourgrau, W., van der Merwe, A. (eds.) Perspectives in Quantum Theory, pp. 25–36. Dover, New York (1971)
44. Wigner, E.P.: On the quantum correction for thermodynamic equilibrium. Phys. Rev. **40**, 749–759 (1932)
45. Husimi, K.: Some formal properties of the density matrix. Proc. Phys. Math. Soc. Jpn **22**, 264–283 (1940)
46. Lee, H.-W.: Theory and application of the quantum phase–space distribution functions. Phys. Rep. **259**, 147–211 (1995)
47. Scully, M.O., Zubairy, M.S.: Quantum Optics. Cambridge University Press, Cambridge (1997)
48. Moyal, J.E.: Quantum mechanics as a statistical theory. Math. Proc. Camb. Philos. Soc. **45**, 99–124 (1949)
49. Groenewold, H.J.: On the principles of elementary quantum mechanics. Physica **12**, 405–460 (1946)
50. Fletcher, P.: The uniqueness of the Moyal algebra. Phys. Lett. B **248**, 323–328 (1990)
51. Strachan, I.A.B.: The Moyal bracket and the dispersionless limit of the KP hierarchy. J. Phys. A **28**, 1967–1975 (1995)
52. Lee, H.-W., Scully, M.O.: Wigner phase–space description of a Morse oscillator. J. Chem. Phys. **77**, 4604–4610 (1982)
53. Feynman, R.P.: Space–time approach to non-relativistic quantum mechanics. Rev. Mod. Phys. **20**, 367–387 (1948)
54. Feynman, R.P., Hibbs, A.R.: Quantum Mechanics and Path Integrals. McGraw-Hill, New York (1965)
55. Dirac, P.A.M.: The Principles of Quantum Mechanics, 2nd edn, p. 125. Clarendon Press, Oxford (1935)
56. Feynman, R.P.: Statistical Mechanics. W. A. Benjamin, Reading (1972)
57. Schulman, L.S.: Techniques and Applications of Path Integrals. Wiley, New York (1981)
58. Ceperly, D.M.: Path integrals in the theory of condensed helium. Rev. Mod. Phys. **67**, 279–355 (1995)
59. Gaspard, P.: Chaos, Scattering and Statistical Mechanics. Cambridge University Press, Cambridge (1998)
60. van Vleck, J.H.: The correspondence principle in the statistical interpretation of quantum mechanics. Proc. Natl. Acad. Sci. USA **14**, 178–188 (1928)

61. Morette, C.: On the definition and approximation of Feynman's path integrals. Phys. Rev. **81**, 848–852 (1951)
62. DeWitt, B.S.: Dynamical theory in curved spaces. I. A review of the classical and quantum action principle. Rev. Mod. Phys. **29**, 377–397 (1957)
63. Miller, W.H.: Semiclassical theory of atom–diatom collisions: path integrals and the classical S matrix. J. Chem. Phys. **53**, 1949–1959 (1970)
64. Miller, W.H.: Classical S matrix: numerical applications to inelastic collisions. J. Chem. Phys. **53**, 3578–3587 (1970)
65. Miller, W.H.: Classical-limit quantum mechanics and the theory of molecular collisions. Adv. Chem. Phys. **25**, 69–177 (1974)
66. Miller, W.H.: The classical S-matrix in molecular collisions. Adv. Chem. Phys. **30**, 77–136 (1975)
67. Miller, W.H.: Classical S matrix for rotational excitation: quenching of quantum effects in molecular collisions. J. Chem. Phys. **54**, 5386–5397 (1971)
68. Doll, J.D., Miller, W.H.: Classical S-matrix for vibrational excitation of H_2 by collision with He in three dimensions. J. Chem. Phys. **57**, 5019–5026 (1972)
69. Marcus, R.A.: Theory of semiclassical transition probabilities (S-matrix) for inelastic and reactive collisions. J. Chem. Phys. **54**, 3965–3979 (1971)
70. Marcus, R.A.: Theory of semiclassical transition probabilities for inelastic and reactive collisions. V. Uniform approximation in multidimensional systems. J. Chem. Phys. **57**, 4903–4909 (1972)
71. Marcus, R.A.: Semiclassical S matrix theory. VI. Integral expression and transformation of conventional coordinates. J. Chem. Phys. **59**, 5135–5144 (1973)
72. Connor, J.N.L., Marcus, R.A.: Theory of semiclassical transition probabilities for inelastic and reactive collisions. II. Asymptotic evaluation of the S-matrix. J. Chem. Phys. **55**, 5636–5643 (1971)
73. Marcus, R.A.: Theory of semiclassical transition probabilities for inelastic and reactive collisions. III. Uniformization using exact trajectories. J. Chem. Phys. **56**, 311–320 (1972)
74. Marcus, R.A.: Theory of semiclassical transition probabilities (S-matrix) for inelastic and reactive collisions. Uniformation with elastic collision trajectories. J. Chem. Phys. **56**, 3548–3550 (1972)
75. Doll, J.D.: Semiclassical theory of atom–solid surface collisions: elastic scattering. Chem. Phys. **3**, 57–264 (1974)
76. Doll, J.D.: Semiclassical theory of atom–solid-surface collisions: application to the He–LiF diffraction. J. Chem. Phys. **61**, 954–957 (1974)
77. McCann, K.J., Celli, V.: A semiclassical treatment of atom–surface scattering: He+LiF(001). Surf. Sci. **61**, 10–24 (1976)
78. Masel, R.I., Merrill, R.P., Miller, W.H.: A semiclassical model for atomic scattering from solid surfaces: He and Ne scattering from W(112). J. Chem. Phys. **64**, 45–56 (1976)
79. Masel, R.I., Merrill, R.P., Miller, W.H.: Atomic scattering from a sinusoidal hard wall: comparison of approximate methods with exact quantum results. J. Chem. Phys. **65**, 2690–2699 (1976)
80. Hubbard, L.M., Miller, W.H.: Application of the semiclassical perturbation (SCP) approximation to diffraction and rotationally inelastic scattering of atoms and molecules from surfaces. J. Chem. Phys. **78**, 1801–1807 (1983)
81. Hubbard, L.M., Miller, W.H.: Application of the semiclassical perturbation approximation to scattering from surfaces. Generalization to include phonon inelasticity. J. Chem. Phys. **80**, 5827–5831 (1984)
82. Guantes, R., Borondo, F., Jaffé, C., Miret-Artés, S.: Diffraction of atoms from stepped surfaces: a semiclassical chaotic S-matrix study. Phys. Rev. B **53**, 14117–14126 (1996)
83. Smilansky, U.: The classical and quantum theory of chaotic scattering. In: Giannoni, M.-J., Voros, A., Zinn-Justin, J. (eds.) Proceedings of the Les Houches Summer School in Chaos and Quantum Physics, Elsevier, Amsterdam (1992)

84. Eckhardt, B.: Irregular scattering. Physica D **33**, 89–98 (1988)
85. Guantes, R., Sanz, A.S., Margalef-Roig, J., Miret-Artés, S.: Atom–surface diffraction: a trajectory description. Surf. Sci. Rep. **53**, 199–330 (2004)
86. Ehrenfest, P.: Bemerkung über die angenäherte Gültigkeit der klassischen Mechanik innerhalb der Quantenmechanik. Z. Phys. **45**, 455–457 (1927)
87. Child, M.S.: Semiclassical Mechanics with Molecular Applications. Clarendon Press, Oxford (1991)
88. Jeffreys, H.: On certain approximate solutions of linear differential equations of the second order. Proc. Lond. Math. Soc. **23**(2), 428–436 (1925)
89. Wentzel, G.: Eine Verallgemeinerung der Quantenbedingungen für die Zwecke der Wellenmechanik. Z. Phys. **38**, 518–529 (1926)
90. Kramers, H.A.: Wellenmechanik und halbzahlige Quantisierung. Z. Phys. **39**, 828–840 (1926)
91. Brillouin, L.: La mécanique ondulatoire de Schrödinger; une méthode générale de résolution par approximations successives. Comptes Rendus **183**, 24–26 (1926)
92. Brillouin, L.: Sur un type général de problèmes, permettant la séparation des variables dans la mécanique ondulatoire de Schrödinger. Comptes Rendus **183**, 270–271 (1926)
93. Messiah, A.: Quantum Mechanics. Dover, New York (1999)
94. Delos, J.B.: Semiclassical calculation of quantum mechanical wavefunctions. Adv. Chem. Phys. **65**, 161–214 (1985)
95. Gottfried, K.: Quantum Mechanics, vol. 1. W.A. Benjamin, New York (1966)
96. Maslov, V.P., Fedoriuk, M.V.: Semiclassical Approximations in Quantum Mechanics. D. Reidel, Boston (1981)
97. Cvitanović, P., Artuso, R., Mainieri, R., Tanner, G., Vattay, G. (eds.): Classical and quantum chaos. http://www.nbi.dk/ChaosBook/. Niels Bohr Institute, Copenhagen (2002)
98. Eckart, C.: The approximate solution of the one-dimensional wave equations. Rev. Mod. Phys. **20**, 399–417 (1948)
99. Miller, W.H., Jansen op de Haar, B.M.D.D.: A new basis set method for quantum scattering calculations. J. Chem. Phys. **86**, 6213–6220 (1987)
100. Zhang, J.Z.H., Chu, S.-I., Miller, W.H.: Quantum scattering via the S-matrix version of the Kohn variational principle. J. Chem. Phys. **88**, 6233–6239 (1988)
101. Sepúlveda, M.A., Grossmann, F.: Time-dependent semiclassical mechanics. Adv. Chem. Phys. **96**, 191–304 (1996)
102. Miller, W.H.: Spiers memorial lecture quantum and semiclassical theory of chemical reaction rates. Faraday Discuss. **110**, 1–21 (1998)
103. Miller, W.H.: The semiclassical initial value representation: a potentially practical way for adding quantum effects to classical molecular dynamics simulations. J. Phys. Chem. A **105**, 2942–2955 (2001)
104. Gutzwiller, M.C.: Chaos in Classical and Quantum Mechanics. Springer, Berlin (1990)
105. Glauber, R.J.: High energy collision theory. In: Brittin, W.E., Dunham, L.G. (eds.) Lectures in Theoretical Physics, vol. 1, p. 315. Interscience, New York (1959)
106. Weinberg, S.: Eikonal method in magnetohydrodynamics. Phys. Rev. **126**, 1899–1909 (1962)
107. Garibaldi, U., Levi, A.C., Spadacini, R., Tommei, G.E.: Quantum theory of atom–surface scattering: diffraction and rainbow. Surf. Sci. **48**, 649–675 (1975)
108. Micha, D.A.: A self-consistent eikonal treatment of electronic transitions in molecular collisions. J. Chem. Phys. **78**, 7138–7145 (1983)
109. Cohen, J.M., Micha, D.A.: Electronically diabatic atom–atom collisions: a self-consistent eikonal approximation. J. Chem. Phys. **97**, 1038–1052 (1992)

Chapter 4
Optics and Quantum Mechanics

4.1 Introduction

Wave theory can be considered one of the most important mathematical models in physics. It has been successfully applied to describe many different physical phenomena and processes [1, 2]. Hence it is not strange to find different branches of physics where the same or similar concepts are considered. This is precisely the case of wave optics[1] and quantum mechanics, where historically theoretical treatments and concepts developed within the former were later on considered to explain some phenomena described by the latter [3, 4]. Actually, the fact that both formulations—indeed, the conception of light and matter—have followed a somewhat similar and, to some extent, intertwined evolution cannot be neglected. As it would happen with quantum mechanics (see Chap. 3), the wave theory of light found many troubles in being accepted since the former wave-based models proposed by Huygens [5] in 1678. By that time, there was already a strongly settled theory: Newton's corpuscle theory of light [6], based on assuming that light consists of different kinds of tiny particles which carry the colors and propagate in straight lines. Newton's scientific authority at the moment made that only a few scientists, like Euler [7–9] and Franklin [8, 9], dared by the end of the eighteenth century to reject this viewpoint and adhered to Huygens' conception.

   Newton's approach was based on the everyday (direct) experience of light. This leads in a natural way to the concept of rays propagating along straight lines, reflecting or refracting according to very simple and well-defined geometrical laws, i.e., to *geometric optics* (see Chap. 7). Without appealing to wave aspects of light, he was also able to successfully describe phenomena such as the formation of the so-called Newton's rings—an interference phenomenon formerly discovered by Hooke in 1664

---

[1] The terms wave optics, physical optics and undulatory optics are usually considered in the literature to denote the same: the part of optics described by the wave equation derived from Maxwell's equations of electromagnetism. Throughout this monograph, the first term will be used preferably, since it allows a more direct conceptual connection with Schrödinger's wave mechanics or quantum mechanics.

A. S. Sanz and S. Miret-Artés, *A Trajectory Description of Quantum Processes.*
I. Fundamentals, Lecture Notes in Physics 850, DOI: 10.1007/978-3-642-18092-7_4,
© Springer-Verlag Berlin Heidelberg 2012

[10]—, or birefringence—first described by Erasmus Bartholin in 1669 [11, 12], who observed it in calcite crystals. However, there are also optical processes where the shape of the obstacles met by light during its propagation plays an important role, giving rise not only to interference, but to *diffraction phenomena*. These phenomena cannot be tackled by means of geometric optics: according to this theory rays should "bend", which can only occur in nonhomogeneous media (those with a varying refractive index), as it can be inferred from Fermat's principle (see Chap. 7). In this sense, it is worth mentioning the crucial experiment carried out by Arago to prove in 1818 Poisson's hypothesis: if Fresnel's wave theory of light was correct, a light spot should be observed at the centre of the geometric shadow caused by a circular obstacle [13, 14]—this effect being first observed by Maraldi in 1723 [14, 15]. Contrary to Poisson, a Newtonian himself, the spot was observed and the wave theory of light succeeded. This experiment and the failure of Newton's theory in explaining diffraction thus led to the acceptance and rise of Huygens' theory during the nineteenth century by the hand of Fresnel, Arago, Foucault, Young, Fraunhofer, or Kirkchoff, among others, reaching its climax by the end of the century with Maxwell's unified formulation of electromagnetism, which constitutes the core of *wave optics*. Of course, also by the end of the nineteenth century, just when everything seemed to be very well established in physics, new physical effects appeared, such as the photoelectric effect or the discrete levels observed in absorption spectra, which would bring some troubles to the Maxwellian electromagnetism and would lead to considering again Newton's former idea of light constituted by particles.

4.2 Maxwell's Equations and the Wave Equation

In wave optics, light is conceived as a perturbation or *wave* caused in vacuum or a material medium by an electromagnetic field—as well as an effect of the propagation of electromagnetic energy through those media. This particular type of electromagnetic radiation is characterized by wavelengths (λ) ranging from 380 to 750 nm or, equivalently, frequencies (ν) between 400 and 790 THz (remember that $\lambda\nu = c$, where c denotes the speed of light in vacuum), within the so-called *visible* region of the electromagnetic spectrum. The way how light propagates or interacts with matter is described by the well-known Maxwell equations of (classical) electromagnetism [16, 17],

$$\nabla \cdot \mathbf{E} = \frac{\rho_e}{\varepsilon_0}, \tag{4.1a}$$

$$\nabla \cdot \mathbf{H} = 0, \tag{4.1b}$$

$$\nabla \times \mathbf{E} = -\mu_0 \frac{\partial \mathbf{H}}{\partial t}, \tag{4.1c}$$

$$\nabla \times \mathbf{H} = \mathbf{J}_e + \varepsilon_0 \frac{\partial \mathbf{E}}{\partial t}, \tag{4.1d}$$

which are applicable to any type of electromagnetic radiation, regardless of its wavelength or frequency. In other words, this set of differential equations allows us to establish the link between the behavior of the electromagnetic field—specified by its electric (**E**) and magnetic (**H**) components—in space and time with the causes or sources that produce and modify it (electric charge densities, ρ_e, current densities, $\mathbf{J}_e$ and some boundary conditions). Locally, in space regions with absence of electric and/or magnetic sources, (4.1) can also be recast in a more compact form as two wave equations,

$$\frac{\partial^2 \mathbf{E}}{\partial t^2} - c^2 \nabla^2 \mathbf{E} = 0, \qquad (4.2a)$$

$$\frac{\partial^2 \mathbf{H}}{\partial t^2} - c^2 \nabla^2 \mathbf{H} = 0. \qquad (4.2b)$$

These equations make more apparent the connection between the electromagnetic field and its wave description. In this case, the particular form of the electromagnetic field will be determined by the boundaries imposed on it as well as the properties of the medium through which it propagates or it is confined.

The solutions of (4.2) are therefore *electromagnetic waves*, where **E** and **H** are not independent one another due to Maxwell's equations (4.1c) and (4.1d). The simplest solutions to (4.2) are the harmonic solutions,[2] which are characterized by a periodic time-dependence with frequency ω ($\omega = 2\pi \nu$). These are also called monochromatic fields because of their dependence on a single frequency. In complex form [16, 17], these solutions read as

$$\mathbf{E}(\mathbf{r}, t) = \mathbf{E}_0(\mathbf{r}) e^{-i\omega t}, \quad \mathbf{H}(\mathbf{r}, t) = \mathbf{H}_0(\mathbf{r}) e^{-i\omega t}, \qquad (4.3)$$

where $\mathbf{E}_0(\mathbf{r})$ and $\mathbf{H}_0(\mathbf{r})$ are in general also complex fields with magnitude and phase that change with position—the actual electric and magnetic fields are then obtained from their respective components (4.3) by taking their real parts. Together with the corresponding boundary conditions (at a given time), these fields obey the time-independent partial differential equations

$$\nabla^2 \mathbf{E}_0 + k^2 \mathbf{E}_0 = 0, \qquad (4.4a)$$

$$\nabla^2 \mathbf{H}_0 + k^2 \mathbf{H}_0 = 0, \qquad (4.4b)$$

where $k = \omega/c$ is the associated wave vector. *Plane waves* traveling along the **k**-direction ($\mathbf{k} = k\mathbf{n}$, with **n** being a unitary vector pointing along the propagation direction) are simple solutions to these equations, for which the modulus of both $\mathbf{E}_0$

---

[2] Of course, other solutions can also be devised. It is sufficient to assume a particular form (initial condition) for **E** and **H** satisfying the corresponding boundaries at a given time and then solving (4.2), which will give us the time-evolution of these fields.

and $\mathbf{H}_0$ is constant and their associated phase is given by $e^{i\mathbf{k}\cdot\mathbf{r}}$. With this, the fields (4.3) read as

$$\mathbf{E}(\mathbf{r}, t) = E_0 e^{i(\mathbf{k}\cdot\mathbf{r}-\omega t)}\mathbf{u}_e, \quad \mathbf{H}(\mathbf{r}, t) = H_0 e^{i(\mathbf{k}\cdot\mathbf{r}-\omega t)}\mathbf{u}_h, \tag{4.5}$$

where $\mathbf{u}_e$ and $\mathbf{u}_h$ denote the oscillation direction of these fields, respectively. By considering Maxwell's equations (4.2), one readily finds that $\mathbf{u}_e \cdot \mathbf{u}_h = \mathbf{u}_e \cdot \mathbf{n} = \mathbf{u}_h \cdot \mathbf{n} = 0$ as well as the relationship $E_0/H_0 = \mu_0 \omega/k = \sqrt{\mu_0/\varepsilon_0}$.

In many situations of physical interest, both the electric and magnetic fields can be described in terms of a scalar field, $\Psi(\mathbf{r}, t)$, from which they are derived [16, 17]. In particular, the electromagnetic field will arise from the solution Ψ and its first derivatives, but also their components fit this scalar description. This scalar field satisfies the scalar wave equation

$$\nabla^2 \Psi - \frac{1}{c^2}\frac{\partial^2 \Psi}{\partial t^2} = 0. \tag{4.6}$$

Assuming that the space and time parts of Ψ are separable, as before, and then using again the method of separation of variables, (4.6) gives rise to the well-known *Helmholtz equation* [1, 2],

$$\nabla^2 \psi + k^2 \psi = 0, \tag{4.7}$$

where ψ is time-independent. This general equation of the wave theory allows us to establish a very important link between the solutions found in wave optics and those from quantum mechanics [3, 4] (see Sect. 4.6). Nevertheless, Helmholtz's equation is much more general and appears in the description of any time-independent wave phenomenon, regardless of its nature.

4.3 Interference and Diffraction

4.3.1 The Huygens–Fresnel Principle

Any partial differential equation which does not contain any nonlinear term of its solution, such as the wave or the Helmholtz equations, gives rise to solutions satisfying an interesting (and useful) mathematical property: they can be linearly combined in order to provide us with other alternative solutions of the same equation. This simple mathematical idea is known as the *superposition principle* (see Chap. 1). From a practical viewpoint, this principle is very important to obtain general solutions to the wave equation, but it is also the mechanism that allows us to explain the observation of *interference* and *diffraction* phenomena. Although there is not a fundamental, physical distinction between these two phenomena, they will be considered as different throughout this monograph for pedagogical purposes. The criterion followed is the same found in different textbooks on optics [18]. Thus, here interference will refer to any situation where several wavefronts coming from different, separate sources coalesce on a space region; when the number of such wavefronts becomes so large that no spatial distinction between two different sources is possible, the phenomenon will be considered to be diffraction. Accordingly, note that interference patterns will only depend on the overlapping of waves, while diffraction patterns will manifest a strong dependence on the relationship between the wavelength of the incoming wave and the (relative) size of the diffracting object. In this way, the pattern originated by a periodic grating can be considered as an *interference pattern* when the emphasis is made on the overlapping of the different (diffracted) outgoing wavefronts, while it is a *diffraction pattern* if the emphasis is put on the particular features of the pattern in relation to the grating properties.

Bearing in mind such a distinction, consider now the issue of the "physical" grounds of the superposition principle, namely the *Huygens–Fresnel principle*. By the end of the seventeenth century, Huygens proposed [5] a mechanism to explain the propagation of light which consisted of assuming it behaves like the waves on a water surface. Accordingly, he established the following principle [14]:

> Every point on a propagating wavefront serves as the source of spherical secondary wavelets, such that the wavefront at some later time is the envelope of these wavelets.

Moreover, "if the propagating wave has a frequency v, and is transmitted through the medium at a speed v, then the secondary wavelets have the same frequency and speed". By plotting sections of circles or spheres (secondary wavelets) of radius vt centered at different points along a wavefront, and then tracing the envelope common to all of them, one obtains the new wavefront. This very simple method, called the *Huygens' construction* (see also Sects. 1.2.1 and 3.4), allows us to describe the evolution of the wave across a medium, explaining the straight-line propagation of light as well as its reflection, refraction and passage through birefringent media. This method, however, has an important drawback: it neglects backward-traveling contributions, only considering the forward-traveling ones [19]. At the level of description considered by Huygens, i.e., the onwards propagation of wavefronts, this did not constitute a major issue. Actually, his model does not present any trace of the principle of wave interference and, therefore, cannot account for diffraction phenomena. Now, it is clear that two opposite traveling wavefronts have to be considered within a more refined model, which is precisely the result arising from the wave equation. As will be seen below, the refinement necessary to account for these effects was introduced later on by Fresnel and Kirchoff.

According to the previous discussion, Huygens' principle only focusses on the part of the secondary wavelets common to the enveloping wavefront, without providing any description of what happens near the borders of diffracting objects or how the shape of the latter influences the shape of the emerging wavefronts. Consider, for example, an aperture of width w illuminated by a wave with wavelength λ. According to Huygens' principle, the emerging wavefront will look the same regardless of the relative size between w and λ, except for its spatial extension, which is limited by the size of the aperture, w. The region beyond the (geometric) projections of the obstacle with respect to the propagation direction of the wave is the so-called *geometric shadow* region, which is always "dark" according to Huygens' principle. However, waves "bend over" around objects, this effect being more important as the size of the object becomes closer to the wavelength of the incoming wave (see, for example, the experiments in ripple tanks displayed in [20]). In other words, the formation of shadows depends on the wavelength of the incoming wave—a tree gives rise to shadows when illuminated by light, but it does not produce such shadows for radiofrequencies. Hence it is possible to observe light in regions of geometric shadow, with the well-known Arago-Poisson spot constituting the most remarkable experimental fact illustrating this effect [14]. The concept of geometric shadow is only strictly valid in the limit $\lambda \to 0$.

In order to overcome this problem, in the nineteenth century Fresnel added a postulate to Huygens' principle introducing the concept of interference between secondary wavelets as follows [14]:

> Every unobstructed point of a wavefront, at a given instant, serves as a source of spherical secondary wavelets (with the same frequency as that of the primary wave). The amplitude of the optical field at any point beyond is the superposition of all these wavelets (considering their amplitudes and relative phases).

This generalization of Huygens' principle, which allows us to describe diffraction phenomena, is known as the *Huygens–Fresnel principle*. This principle constitutes the core of the scalar theory of light, as shown by Kirchhoff, who proved that it is a consequence of the scalar differential wave equation.

In a more formal fashion, the Huygens–Fresnel principle can be formulated as follows. For the sake of simplicity and without loss of generality, consider a point-like source radiating at point P_s in vacuum [1]. The outgoing solutions to the corresponding wave equation are given by spherical oscillatory waves,

$$\Psi(r_s, t) = \frac{A}{r_s} e^{i(kr_s - \omega t)}, \tag{4.8}$$

where r_s denotes the radial distance from P_s to any point in space. Now, consider there is a closed surface Σ between P_s and another point P_o where the effect of the source after the wave has met the surface is going to be measured. The surface Σ coincides with a wavefront of radius r_s and centered at P_s. The value of the perturbation at P_o is denoted by Ψ_o. Following the Huygens–Fresnel principle, Ψ_o arises from the interference of the secondary wavelets originated at each element of area $d\Sigma$ (on Σ) when Ψ reaches the surface Σ. In order to determine the contribution from each of these elements at P_o, it can be reasonably argued that they will be proportional to:

1. The so-called *obliquity factor* (also called *transmission function* when dealing with apertures), $f(\theta_s, \theta_o)$, which accounts for the angular dependence of the secondary wavelets (on θ_s and θ_o).
2. The value of (4.8) at P_Σ, where the wavefront emerging from P_s, at a distance r_s, meets the surface element $d\Sigma$ at a time t'.

3. The spherical factor associated with each (point-like) secondary source, also given by (4.8) but evaluated at r_o (the distance between P_Σ and P_o) and $t - t'$ (the time elapsed to cover the distance r_o).
4. The area element $d\Sigma$.

That is,

$$d\Psi_o \propto \underbrace{f(\theta_s, \theta_o)}_{(1)} \underbrace{\left(\frac{A}{r_s}\, e^{i(kr_s - \omega t')}\right)}_{(2)} \underbrace{\left(\frac{1}{r_o}\, e^{i[kr_o - \omega(t - t')]}\right)}_{(3)} \underbrace{d\Sigma}_{(4)}$$

$$= Kf(\theta_s, \theta_o)\frac{A}{r_s r_o} e^{i[(k(r_s + r_o) - \omega t]}d\Sigma, \qquad (4.9)$$

where K is a proportionality constant with the dimensions of inverse length and then proportional to $1/\lambda$. Therefore, the (total) wave amplitude reaching P_o is obtained by integrating (4.9) over the region of the surface Σ (hereby denoted as $\Delta\Sigma$) which is unobstructed,

$$\Psi_o = K A e^{-i\omega t} \int_{\Delta\Sigma} f(\theta_s, \theta_o) \frac{e^{ik(r_s + r_o)}}{r_s r_o} d\Sigma. \qquad (4.10)$$

In other words, this is equivalent to integrating over the whole space, but taking into account the boundary conditions imposed on the transmission by the presence of a possible obstacle.

In general, the obliquity factor is not known and therefore (4.10) cannot be directly solved. However, some analytical results can be obtained when considering harmonic solutions to the Helmholtz equation together with Green's theorem [1, 16]. In particular, this procedure leads to the *Helmholtz–Kirchhoff integral theorem*, which gives us $\Psi_0(t)$ as an exact time-harmonic scalar wave. By applying the so-called *Kirchhoff boundary conditions* [1, 16], (4.10) acquires the form

$$\Psi_o(t) = -\frac{iA}{\lambda} \int_{\Delta\Sigma} \left[\left(\frac{\cos\theta_s + \cos\theta_0}{2}\right) + \frac{i\lambda}{4\pi}\left(\frac{\cos\theta_s}{r_s} + \frac{\cos\theta_0}{r_0}\right)\right]$$

$$\times \frac{e^{i[k(r_s + r_o) - \omega t]}}{r_s r_o} d\Sigma, \qquad (4.11)$$

where $K = -i/\lambda$, which is in agreement with the hypothesis above about the wavelength dependence of K. One can further proceed and assume the limit $r_s, r_o \gg \lambda$, i.e., both the source and observation points at relatively far compared with the light wavelength. In this case, the second term in the integrand of (4.11) vanishes and

$$\Psi_o = K A e^{-i\omega t} \int_{\Delta\Sigma} \left(\frac{\cos\theta_s + \cos\theta_0}{2}\right) \frac{e^{ik(r_s + r_o)}}{r_s r_o} d\Sigma, \qquad (4.12)$$

with $f(\theta_s, \theta_o) = (\cos\theta_s + \cos\theta_o)/2$. This expression is known as *Fresnel–Kirchhoff diffraction formula*. Although there are no full analytic solutions for (4.12) due to the complexity involved in the calculation along Σ, some approximations can be obtained in two limiting cases:

- *Near-field* or *Fresnel diffraction*.
- *Far-field* or *Fraunhofer diffraction*.

For example, consider an aperture of width w. If z_o denotes the distance between the center of the aperture and the position of P_o perpendicular with respect to the aperture, then the Fresnel approximation is applied for $z_0 \gtrsim w$ and the Fraunhofer one for $z_0 \gg w^2/\lambda$ (see below in Sect. 4.3.4); for $z_0 \gtrsim \lambda/2$, (4.12) cannot be used, but (4.11). In any case, as can be seen, the integral is strongly dependent on the phase factor inside the integrand, which will give rise to constructive and destructive interference and therefore to the typical diffraction fringe patterns.

4.3.2 Interference Phenomena

When two or more wavefronts coalesce on a space region, a series of alternate dark and light fringes appear when the coalescing wavefronts are *coherent*, i.e., all of them have the same or nearly the same wavelength. From now on, this process will be referred to as *interference* and to the associated pattern as the *interference pattern*. The dark fringes arise when the wavefronts cancel among themselves, while light fringes appear where their combination is *constructive*. In the latter case, the intensity observed can be greater than the intensities associated with each separated wavefront.

In order to provide a more quantitative view of interference phenomena, consider two point-like sources, S_1 and S_2, coherently emitting monochromatic, linearly polarized light in vacuum with wavelength λ. They are separated a distance $a \gg \lambda$. Moreover, let us assume that P_o in this case is so far from S_1 and S_2 that the incident wavefronts from each source can be assumed as plane waves. With these assumptions, the corresponding electric fields can be expressed as

$$\mathbf{E}_1(\mathbf{r}, t) = \mathbf{E}_{1,0} e^{i(\mathbf{k}_1 \cdot \mathbf{r} - \omega t + \phi_1)}, \tag{4.13a}$$

$$\mathbf{E}_2(\mathbf{r}, t) = \mathbf{E}_{2,0} e^{i(\mathbf{k}_2 \cdot \mathbf{r} - \omega t + \phi_2)}. \tag{4.13b}$$

These fields are both solutions of (4.2a), with $|\mathbf{k}_1| = |\mathbf{k}_2| = k = \omega/c$, though pointing along different space directions ($\hat{\mathbf{k}}_1 = \mathbf{k}_1/|\mathbf{k}_1|$ and $\hat{\mathbf{k}}_2 = \mathbf{k}_2/|\mathbf{k}_2|$). Due to their monochromaticity, the space-dependent part of these fields also satisfies the Helmholtz equation (4.7). To observe interference, (4.13) are now substituted into the time-averaged expression for the intensity,

$$I = \frac{1}{2} \, \mathrm{Re}\, \{\langle \mathbf{E}^2 \rangle\}, \tag{4.14}$$

which yields

$$I = \frac{1}{2} \, \mathrm{Re} \left\{ \langle \mathbf{E}^2 \rangle \right\} = \frac{1}{2} \, \mathrm{Re} \left\{ \langle \mathbf{E}_1^2 \rangle \right\} + \frac{1}{2} \, \mathrm{Re} \left\{ \langle \mathbf{E}_2^2 \rangle \right\} + \mathrm{Re} \left\{ \langle \mathbf{E}_1 \cdot \mathbf{E}_2 \rangle \right\}$$

$$= \frac{1}{2} E_{1,0}^2 + \frac{1}{2} E_{2,0}^2 + \mathbf{E}_{1,0} \cdot \mathbf{E}_{2,0} \cos \delta, \tag{4.15}$$

where $E_{1,0} = |\mathbf{E}_{1,0}|$ and

$$\delta = \Delta \mathbf{k} \cdot \mathbf{r} + \Delta \phi = (\mathbf{k}_1 - \mathbf{k}_2) \cdot \mathbf{r} + \phi_1 - \phi_2. \tag{4.16}$$

The first two contributions in (4.15) represent the intensity of each separate electric field at P_o (I_1 and I_2), while the third one (I_{12}) is the *interference term*.

As it can be inferred from (4.16), interference depends on three factors: the direction of both fields through the scalar product $\hat{\mathbf{k}}_1 \cdot \hat{\mathbf{k}}_2$, the dephasing associated with the path difference $\Delta \mathbf{k} \cdot \mathbf{r}$, and the initial phase difference $\Delta \phi$. If both fields are perpendicular, i.e., $\hat{\mathbf{k}}_1 \cdot \hat{\mathbf{k}}_2 = 0$, the interference term vanishes and the total intensity at P_o is given by the sum of the partial intensities, $I = I_1 + I_2$, thus satisfying the law of addition of intensities. On the other hand, when the sources are so far from P_o that the fields can be assumed as parallel, $\hat{\mathbf{k}}_1 \cdot \hat{\mathbf{k}}_2 = 1$ and therefore

$$I = I_1 + I_2 + 2\sqrt{I_1 I_2} \cos \delta. \tag{4.17}$$

Interference thus depends on the relative position of P_o with respect to the sources and the initial phase of the fields. Its maximum is reached when $\delta = \pm 2n\pi$ (with $n = 0, 1, 2, \ldots$), since

$$I_{\max} = I_1 + I_2 + 2\sqrt{I_1 I_2} = \frac{1}{2}(E_{1,0} + E_{2,0})^2. \tag{4.18}$$

In this case, both fields (waves) are *in phase* and there is *total constructive interference*; for any δ such that $1 > \cos \delta > 0$, interference is still constructive but not total, since the waves are partly *out of phase* and $I_{\max} > I > I_1 + I_2$. The opposite case is *total destructive interference*, which is reached when the fields are totally out of phase or in *opposite phase*, $\delta = \pm(2n + 1)\pi$, leading to

$$I_{\min} = I_1 + I_2 - 2\sqrt{I_1 I_2} = \frac{1}{2}(E_{1,0} - E_{2,0})^2. \tag{4.19}$$

If the phase difference is not totally opposite, there is partial destructive interference and $I_1 + I_2 > I > I_{\min}$. The intermediate case between total constructive and destructive interference, $I = I_1 + I_2$, is reached for $\delta = \pm(2n + 1)\pi/2$, which is a similar situation to having perpendicular fields. For example, consider that $\mathbf{E}_{1,0} = \mathbf{E}_{2,0}$. Then, $I_1 = I_2 = I_0$ and (4.17) reduces to

$$I = 4I_0 \cos^2 \delta/2. \tag{4.20}$$

In this particular case, the maximum value of the intensity is four times larger ($I_{\max} = 4I_0$) than the intensity of each separated field, while the minimum value just consists of the total suppression of the intensity ($I_{\min} = 0$) by destructive interference.

4.3.3 Young's Two-Slit Experiment

When speaking about interference, Young's two-slit experiment has to be mentioned. Not only it was crucial in the development of wave optics, but also for quantum mechanics, about 100 years later than the former. The reason why this experiment has become so famous relies on the fact that, though very simple, it contains the essence of the wave nature of both light and matter. In order to display such features, consider a simple system of three screens, one (S_1) with a slit that produces a narrow wave or beam; another (S_2) containing the two slits (a_1 and a_2), which gives rise to the splitting of the incoming beam into two (outgoing) ones; and, finally, the third screen (S_3), at a large enough distance L from S_2, is used to scan the pattern formed with the interference of the two beams. Unlike the case of interference between two different wavefronts analyzed before, here the interference is due to the recombination of a wavefront previously separated (coherently): the incoming wavefront is split up by means of S_2 and then recombined again, thus originating an interference pattern by *self-interference* (light coming from the same source interferes with itself).

Consider the thickness of the screens is negligible and the slits are very narrow along the x-direction and very large along the y-direction (perpendicular to this page), so that the outgoing waves can be assumed to be cylindrical wavefronts. Due to cylindrical symmetry, the problem can then be reduced to the XZ-plane. The slit at S_1 acts like the source—obviously, there is a light source behind. If this slit is centered at $x = 0$ and the source radiates homogeneously, there is radial symmetry and the phase δ depends only on the path difference, i.e., $\delta = \mathbf{k}_1 \cdot \mathbf{r}_1 - \mathbf{k}_2 \cdot \mathbf{r}_2 = -k(r_2 - r_1)$, where r_1 and r_2 are the distances from a_1 and a_2 to P_o, respectively, and $|\mathbf{k}_1| = |\mathbf{k}_2| = k = 2\pi/\lambda$ because the outgoing wavefronts are coherent. The observation angles from a_1 and a_2 are then

$$\sin\theta_1 = \frac{r - r_1}{d/2}, \quad \sin\theta_2 = \frac{r_2 - r}{d/2}, \tag{4.21}$$

respectively, where r is the distance from the center between the two slits to P_o, and d is the inter-slit distance. Assuming $x \ll L$, where x is the position of P_o along the x-axis,

$$\sin\theta_1 \approx \tan\theta_1 = \frac{x}{L}, \quad \sin\theta_2 \approx \tan\theta_2 = \frac{x + d/2}{L}. \tag{4.22}$$

Combining now the relations (4.21) and (4.22), we find

$$r_2 - r_1 \approx \frac{xd}{L} + \frac{d^2}{4L} \approx \frac{xd}{L}, \tag{4.23}$$

where the assumption $d \ll x \ll L$ has been considered in the last step. From (4.23), the phase difference reads as

$$\delta(x) = -k(r_2 - r_1) \approx -\frac{2\pi dx}{\lambda L}, \tag{4.24}$$

from which one infers that the separation between two consecutive points with the same intensity is

$$\Delta x = x_1 - x_2 = \frac{\lambda L}{d}. \tag{4.25}$$

This result indicates that the spacing between the dark and light fringes of the interference pattern is constant, with the fringes being parallel to the y-axis. Moreover, by equating (4.24) to $2n\pi$, total constructive interference will happen for those points such that their path difference, $r_2 - r_1$, is an integer number of wavelengths, while it will be destructive if it corresponds to a half integer.

Here, it is assumed that the distance L between S_3 and S_2 has to be very large compared with the distance x between P_o and the z-axis. Within this approximation, the phase of any elementary wave reaching S_3 can be assumed to be a linear function of the coordinates defining the observation plane (i.e., a plane wave), as it can be inferred from (4.24). As will be seen below, this is known as the *far field* or *Fraunhofer* approximation within the context of diffraction.

4.3.4 Fresnel and Fraunhofer Diffraction

When looking at the edge of a knife illuminated by sunlight, it is possible to observe a spot of light just as if there was a notch on the edge. What is happening, indeed, is that light is getting diffracting, i.e., light rays are "bending" around the obstacle rather than passing straight ahead. In order to describe diffraction phenomena, of which this is just a common example, the relationship between the wavelength of the incoming wavefront and the size of the obstacles that it meets has to be considered. In general, there are no analytical solutions for diffraction problems, as mentioned at the end of Sect. 4.3.1. However, there are two limits which allow us to obtain some approximated solutions, as mentioned above: near-field or Fresnel diffraction and far-field or Fraunhofer diffraction. In particular, for far-field diffraction is very useful Babinet's principle which states that the diffraction patterns from two complementary objects are identical (by complementary objects is meant that the opaque parts of one correspond to the transparent parts of the other).

The limit of *Fresnel diffraction* is defined when the distances from the obstacle to the source and to the observation point are both relatively small compared to the size of such an obstacle (or the space variations of its particular shape). In this way, changes of the wavefront phase near the obstacle can be assumed to be a quadratic perturbation of the coordinates on the obstacle. This leads to relatively complicated analytical treatments: any displacement of the observation point means considering a new integration region on the obstacle, since the curvature of the wavefronts cannot be neglected.

In order to briefly analyze this type of diffraction with some more detail, consider an aperture, as in Fig. 4.1. Within this approximation, not all points on the aperture contribute equally to the intensity at P_o, but only those located in a neighborhood of the intersection between the plane containing the aperture and the straight line joining the source with P_o. The intersection point is called the *stationary point* and the generating straight line is the *ray* described by geometric optics (see Chap. 7). In this way, the contributions to the intensity at P_o come from points on the aperture such that their associated rays vary slightly with respect to the geometric optics ray— this does not mean the intensity at P_o will coincide exactly with the value predicted by geometric optics.[3] Thus, consider the Huygens–Fresnel integral (4.12) expressed [18] as

$$\Psi_o = \frac{i\alpha}{\lambda} \int_{\Delta\Sigma} f(x, y) \frac{e^{-i\mathbf{k}\cdot(\mathbf{r}_s+\mathbf{r}_o)}}{r_s r_o} dx dy, \qquad (4.26)$$

where r_s and r_o are, respectively, the distances from the source and the observer to a point P on the aperture near the stationary point $\bar{P}$, as seen in Fig. 4.1. Taking into account the stationarity condition (i.e., the distances from the source and the observer to the aperture, z_s and z_o, are much larger than the distance between P and P_s), these distances can be expressed as

$$r_s = \sqrt{(x_s - x)^2 + (y_s - y)^2 + z_s^2} \approx z_s + \frac{(x_s - x)^2 + (y_s - y)^2}{2z_s}, \qquad (4.27a)$$

$$r_o = \sqrt{(x - x_o)^2 + (y - y_o)^2 + z_o^2} \approx z_o + \frac{(x - x_o)^2 + (y - y_o)^2}{2z_o}, \qquad (4.27b)$$

which explicitly show the quadratic dependence on the aperture coordinates characteristic of Fresnel diffraction. Replacing r_s and r_o in the denominator of (4.26) by z_s and z_o, respectively, does not affect much the amplitudes of the spherical waves due to the small variations in the corresponding paths with respect to the stationary one. However, the same does not hold for the phase factor: small variations in the phase may lead to dramatic changes when going from one point on the aperture to another. This phase factor is proportional to

$$r_s + r_o \approx z_s + z_o + \left[\frac{(x_s - x)^2}{2z_s} + \frac{(x - x_o)^2}{2z_o} \right] + \left[\frac{(y_s - y)^2}{2z_s} + \frac{(y - y_o)^2}{2z_o} \right]. \qquad (4.28)$$

This change makes the resulting integral very complicated analytically. So, consider the parameters

---

[3] In quantum mechanics, this is somehow similar to what happens within Feynman's path-integral formulation [21] (see Sect. 3.4). In this case, the geometric optics ray is substituted by a classical (stationary) trajectory and the secondary wavelets are replaced by imaginary exponential functions, with their argument being the classical action evaluated along the corresponding varied (with respect to the classical trajectory) paths. As one approaches the *classical limit*, mainly contributions in a neighborhood of the classical trajectories are relevant.

Fig. 4.1 Scheme employed to analyze the Fresnel diffraction produced by an aperture. The *dashed line* denotes the stationary path, namely the *optical ray*, between a source point P_s and the point P_o where the effect is observed

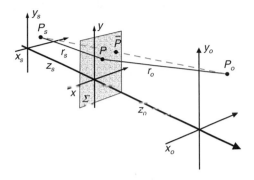

$$\rho = \frac{z_s z_o}{z_o + z_s}, \quad \bar{x} = \frac{z_s x_o + z_o x_s}{z_s + z_o}, \quad \bar{y} = \frac{z_s y_o + z_o y_s}{z_s + z_o}, \tag{4.29}$$

where $\bar{x}$ and $\bar{y}$ are the coordinates of the stationary point $\bar{P}$. After substitution of these parameters into (4.28),

$$r_s + r_o \approx z_s + z_o + \left[\frac{(x_o - x_s)^2 + (y_o - y_s)^2}{2(z_s + z_o)} \right] + \left[\frac{(x - \bar{x})^2 + (y - \bar{y})^2}{2\rho} \right]. \tag{4.30}$$

Also, expressing the distance between P_s and P_o as

$$L = z_o + z_s + \frac{(x_o - x_s)^2 + (y_o - y_s)^2}{2(z_s + z_o)}, \tag{4.31}$$

one obtains $z_s z_o \approx \rho L$. Taking into account these relation, (4.26) can be finally expressed as

$$\Psi_o = \frac{i\alpha}{\lambda \rho L} e^{-ikL} \int_{\Delta\Sigma} f(x, y) e^{-ik[(x-\bar{x})^2 + (y-\bar{y})^2]/2\rho} dx dy. \tag{4.32}$$

According to this expression, the wave at P_o is the spherical wave e^{-ikL}/L that would be observed without the aperture. However, due to the presence of the aperture, both the amplitude and phase of this wave are modulated by the integral in (4.32), which contains a phase depending quadratically on the aperture coordinates.

If P_o is progressively moved further away from the aperture, a gradual, smooth variation of the diffraction pattern is observed until reaching a certain distance. Beyond this distance, the features of the corresponding pattern remain stable, i.e., independent of the distance to the diffracting object. This is the limit of the *Fraunhofer diffraction*, where the dominant term in the Huygens–Fresnel integral is the linear one. Nevertheless, once the Fraunhofer region is reached (i.e., the space region where Fraunhofer diffraction takes place), in principle Fresnel-like patterns can be recovered by decreasing the wavelength of the incident wave and, in the limit case $\lambda \to 0$, the patterns corresponding to the geometric optics are obtained. Quantitative studies of Fraunhofer diffraction are thus based on two assumptions:

1. The distances considered as well as the size of the obstacle are large enough with respect to the light wavelength.
2. The obstacle angular size, when observed from either P_s or P_o, is relatively small. This is the so-called *paraxial approximation*.

These assumptions allow us to consider (locally) the wave detected at P_o as a plane wave.

Making use of the paraxial approximation for the incident wave, consider an aperture which, as in Fig. 4.2, is illuminated by a plane wave parallel to the aperture. Instead of (4.26), only

$$\Psi_o = \frac{i\alpha}{\lambda} \int_{\Delta\Sigma} f(x, y) \frac{e^{-ik\cdot r_o}}{r_o} dx dy \tag{4.33}$$

needs to be evaluated. Here, as before,

$$r_o = \sqrt{(x - x_o)^2 + (y - y_o)^2 + z_o^2} \tag{4.34}$$

describes the distance between the point P on the aperture and P_o. Equation (4.33) can be expressed as

$$\Psi_o = \frac{i\alpha}{\lambda} e^{-ik\cdot r_o'} \int_{\Delta\Sigma} f(x, y) \frac{e^{-ik\cdot(r_o - r_o')}}{r_o} dx dy, \tag{4.35}$$

where

$$r_o' = \sqrt{x_o^2 + y_o^2 + z_o^2} \tag{4.36}$$

is the distance from the center of the aperture to P_0. As seen in Fig. 4.2, $\mathbf{r} = \mathbf{r}_o' - \mathbf{r}_o$. Since P_o is very far from the aperture, the angle φ between $\mathbf{r}_o'$ and $\mathbf{r}_o$ is negligible and, therefore,

$$r = \sqrt{r_o'^2 + r_o^2 - 2r_o' r_o \cos\varphi}$$
$$\approx r_o' - r_o = \frac{r_o'^2 - r_o^2}{r_o' + r_o} = \frac{2(x x_o + y y_o) - (x^2 + y^2)}{r_o' + r_o}, \tag{4.37}$$

and

$$k|r_o' - r_o| \ll k r_o'. \tag{4.38}$$

This condition implies the Huygens–Fresnel integral does not vanish, ensuring that all Huygens secondary wavelets produced between the center of the aperture and the position denoted by r have similar phases. Thus, contributing constructively to the interference at P_o (i.e., with nonzero amplitude).

Fig. 4.2 Scheme employed to analyze the Fraunhofer diffraction produced by an aperture. Within the Fraunhofer limit, the optical ray essentially coincides with the line joining P and P_o

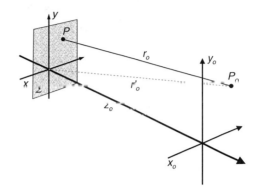

Using the relation

$$\frac{1}{r_0' + r_o} = \frac{1}{2r_o'}\left(1 + \frac{r_o - r_o'}{2r_o'}\right)^{-1} \tag{4.39}$$

as well as (4.38), (4.37) can be expressed as

$$r \approx \left(\frac{xx_o + yy_o}{r_o'} - \frac{x^2 + y^2}{2r_o'}\right)\left(1 - \frac{r_o' - r_o}{2r_o'}\right)^{-1}. \tag{4.40}$$

According to (4.38), the second factor of (4.40) has to be approximately 1, which indicates the aperture is relatively small when seen from P_o. Substituting (4.40) into (4.35) yields

$$\Psi_o = \frac{iAe^{-i\mathbf{k}\cdot\mathbf{r}_o'}}{\lambda r_o'}\int_{\Delta\Sigma} f(x, y)e^{ik(xx_o+yy_o)/r_o'-ik(x^2+y^2)/2r_o'}dxdy, \tag{4.41}$$

where A is the amplitude of the plane wave irradiated by the aperture. In (4.41), the phase depends linearly on the P_o coordinates and quadratically on the aperture coordinates. However, since P_o is assumed to be far away, this second, quadratic contribution can be neglected provided the condition

$$r_o' \gg \frac{x^2 + y^2}{4\pi\lambda} \sim \frac{d^2}{4\pi\lambda} \tag{4.42}$$

satisfies, where d is some measure of the area covered by the aperture. This condition, namely the far-field criterion, defines a lower bound or distance, L_R, after which the far-field or Fraunhofer approximation starts to be valid.

In the phase factor of the integrand in (4.41), $x_o/r_o' = \cos\theta_x$ and $y_o/r_o' = \cos\theta_y$ are the directional cosines defining the point P_o with respect to the geometrical center of the aperture. Thus, by defining the space frequencies

$$\kappa_x = -\frac{2\pi}{\lambda}\cos\theta_x = -\frac{2\pi}{\lambda}\sin\theta_o^x, \quad \kappa_y = -\frac{2\pi}{\lambda}\cos\theta_y = -\frac{2\pi}{\lambda}\sin\theta_o^y, \tag{4.43}$$

along these directions or with respect to the normal, (4.41) can be alternatively expressed as

$$\Psi_o(\kappa_x, \kappa_y) = \frac{i A e^{-i\mathbf{k}\cdot\mathbf{r}_o'}}{\lambda r_o'} \int_{\Delta\Sigma} f(x, y) e^{-i(\kappa_x x + \kappa_y y)} dx dy. \qquad (4.44)$$

According to our conventional interpretation of the Huygens–Fresnel integral, the diffracted wave (4.44) can be understood as a coherent superposition of plane waves leaving the aperture along the directions marked by the directional cosines, whose contributions are modulated by $f(x, y)$. However, (4.44) can also be interpreted as a two-dimensional Fourier transform of the aperture transmission function. In other words, a Fraunhofer pattern is simply the space-frequency spectrum associated with the obstacle (or, equivalently, its transmission function).

4.3.5 Diffraction by Gratings

A *diffracting grating* is a periodic array consisting of N identical and equally space diffracting elements, where diffraction can occur either by transmission through or reflection from each one of such elements. The crystal structure of solids is an example of diffracting grating. By analyzing the Fraunhofer diffraction pattern formed when a beam of X-rays passes through the material (in terms of the X-rays wavelength), one can determine the structure of the solid, namely the *lattice struc-*ture*, and the distribution of its constituents (atoms or molecules) within the lattice sites or *unit cells*. Of course, something similar can be done with slits gratings: by illuminating the grating with light, one can obtain information about the number of slits per length unit, their width and any imperfection. In this case, the unit cell is called the *period* of the grating.

As seen in the previous Section, Fraunhofer diffraction patterns can be understood as the Fourier transform of the obliquity factor associated with the grating. In the case of periodic structures, this gives rise to a very interesting and important result. Consider a one-dimensional grating consisting of N identical slits. If the transmission function of one of these slits is $f(x')$ in the local coordinate x' of the slit, the transmission function for a slit centered around a position x_k (with respect to some fixed position along the grating) can be expressed as

$$f(x - x_k) = \int \delta(x' - x_k) f(x - x') dx'. \qquad (4.45)$$

The total transmission function for the full grating can be then given as

$$F(x) = \sum_{k=1}^{N} w_k f(x - x_k) = \int \left(\sum_{k=1}^{N} w_k \delta(x' - x_k) \right) f(x - x') dx', \qquad (4.46)$$

where w_k is the weight (in general, a complex number) with which the kth slit contributes to the diffraction pattern. Expression (4.46) can be written as a convolution function $F(x) = [g*f](x)$, where g refers to the sum of δ-functions in (4.46). According to (4.44), the associated Fraunhofer diffraction pattern is given by the Fourier transform of (4.46),

$$\Psi(\kappa_x) = \int F(x)e^{-i\kappa_x x} dx. \tag{4.47}$$

In virtue of the *convolution theorem* [22], (4.47) can be written as the product of the Fourier transforms of the convoluted functions in (4.46),

$$\Psi(\kappa_x) = \mathcal{F}\{f\}\mathcal{F}\{g\} = \Psi_{\text{slit}}(\kappa_x) \sum_{k=1}^{N} w_k e^{-i\kappa_x x_k}, \tag{4.48}$$

where $\mathcal{F}$ denotes the Fourier transform of the function between the square brackets and

$$\Psi_{\text{slit}} \sim \int f(x)e^{-i\kappa_x x} dx \tag{4.49}$$

is the Fraunhofer diffraction amplitude associated with one of the slits. Equation (4.48) formally summarizes the so-called *array theorem* [23], which states:

> The field distribution of Fraunhofer diffraction from an array of similarly oriented, identical apertures is the product of the field distribution of Fraunhofer diffraction from any one of the apertures with the Fourier transform of the set of delta functions distributed in the same manner (random or otherwise) as the apertures of the array.

Accordingly, the first term in (4.48) can be associated with a pure diffraction process (diffraction by an aperture), for it provides information about the diffractive properties of the slits and therefore modulates the total diffraction pattern. The second term, however, is related to a pure interference process involving N identical point-like sources. Because the interference pattern (i.e., the interference fringes) depends on how the position and weight of the slits within the array, this term is called the *array factor, grating function* or *form factor*. By playing around with positions and weights one can vary the transmission performance of the grating, which can be optimized in order to achieve some desirable properties—this is well known in antenna theory [24], where the array can be steered (change the direction of maximum radiation or reception) by changing the weights. On the other hand, in the fields of condensed matter physics, solid state physics or crystallography the array factor is known as the *structure factor* and describes how a material scatters an incident beam of, for example, X-rays, electrons, neutrons or rare-gas atoms. Meanwhile, the weights w_k are the equivalent to the *atomic form factor* or *atomic scattering factor*, which is the particular way how each atom of the unit cell responds to the incident beam.

An application of the array theorem to the particular case of a *Ronchi grating*—a slit array where the distance between two consecutive slits is equal to their width—will be seen in more detail in Volume 2 in the case of matter waves.

4.4 Quantization in Bound Optical Systems: Waveguides

When light crosses the boundary that separates two media with different refractive indices, it is partly refracted at the boundary surface and partly reflected. This is well-known from geometric optics, where it is explained in terms of rays of light (see Chap. 7), although the more rigorous explanation is obtained from Maxwell's equations. Refraction is described by the Snell or Snell–Descartes law, which states that the ratio of the sines of the incidence (θ_i) and refraction (θ_r) angles is equal to the ratio of the speed of light in the two media. Since the speed of light in a medium can be expressed as $v = c/n$, where n is the medium refractive index, this law can also be expressed in its more well-known form

$$n_1 \sin \theta_1 = n_2 \sin \theta_2. \tag{4.50}$$

As it can be inferred from this expression, if $n_1 > n_2$ (i.e., the second medium is less optically dense than the former), as the incidence angle θ_1 increases, there is a "critical" value, namely the *critical angle*, after which there is no refraction, but all light is reflected back into the first medium. The critical angle, for which $\theta_2 = \pi/2$, is given by the relation

$$\theta_c = (\sin)^{-1} \left(\frac{n_2}{n_1} \right). \tag{4.51}$$

Whenever $\theta_1 > \theta_c$, one can observe the so-called *total internal reflection* phenomenon. Optical fibers and waveguides constitute well-known applications of this phenomenon. As is well-known, nowadays optical fibers are highly involved in our everyday life due to their properties to transmit information over longer distances and at higher bandwidths than other forms of communications, allowing signal traveling with less losses, being immune to electromagnetic interference, or carrying images through very narrow channels. However, the guidance of light by refraction was first demonstrated long ago by Colladon [25] and Babinet [26] in 1842. Later on Tyndall [27, 28] explained the phenomenon in terms of total internal reflection for light transmitted within a stream of water (this is the so-called Tyndall effect).

In order to understand how an optical fiber works at a basic level, consider a light ray incident on an end of a fiber with refractive index n_f, which is higher than the refractive index of the surrounding medium, n_s. The critical angle leading to total internal refraction inside the fiber is $\theta_c = (\sin)^{-1}(n_s/n_f)$. According to (4.50), this means that the maximum incidence angle ensuring that a ray penetrating into the fiber will remain inside it is given by

$$\sin \theta_{max} = \frac{\sqrt{n_f^2 - n_s^2}}{n_s}. \tag{4.52}$$

This angle defines the so-called *acceptance cone*: any ray with incidence angle θ_i such that $|\theta_i| \leq \theta_{max}$ will keep propagating along the fiber. The distance or path length, ℓ, traveled by the ray bouncing back and forth along the fiber is then given by

$$\ell = \frac{L}{\cos \theta_f} = \frac{n_f L}{\sqrt{n_f^2 - n_g^2 \sin^2 \theta_g}}, \tag{4.53}$$

where L is the fiber length.

To avoid any leakage of light through *frustrated total internal reflection* (see Sect. 4.5) or when a large number of fibers are packed very close, it is common to cover the guiding material with a thin layer of another material with lower refractive index (but still higher than the refractive index of the surrounding medium). The layer is called the *cladding* and the enshrouded material is the *core*, both being parts of the same fiber. In this case, instead of (4.52), the maximum angle of incidence is given by

$$\sin \theta_{\max} = \frac{\sqrt{n_f^2 - n_c^2}}{n_s}, \tag{4.54}$$

where n_c is the refractive index of the cladding. The quantity $n_s \sin \theta_{\max}$, given by (4.54), is called the *numerical aperture* and its square gives us a measure of the light-gathering power of the system, i.e., the amount of incident light that can be transmitted through the guide by total internal reflection. This is related to the fact that, depending on the launch angle into the fiber and the core diameter, there can be a myriad of different paths by which energy can propagate along the fiber core. This is the case of *multimode fibers*. When the core is very narrow, only one ray can travel parallel to the central axis, this being the *single-mode fibers*.

Up to now, a ray-based description of optical fibers has been considered. However, the notion of modes propagating inside them leads us back to the wave conception of light and the generalized concept of *waveguide*, i.e., a physical structure that guides electromagnetic waves (optical fibers guide light in the visible region of the electromagnetic spectrum). As happens with optical fibers, although the technology based on waveguides is very common nowadays, the first waveguide was proposed in 1893 by Thomson [29] and experimentally verified in 1894 by Lodge [30]. The mathematical analysis of the propagating modes within a hollow cylinder was first performed in 1897 by Lord Rayleigh [31]. This analysis starts with the vector wave equations (4.2) for the electric and magnetic fields, seeking for non-plane-wave solutions fitting the boundary conditions prescribed by the shape of the waveguide. Moreover, these solutions are required to satisfy Gauss' law (4.1a) with $\rho_e = 0$ (i.e., $\nabla \cdot \mathbf{E} = 0$) inside the waveguide. Solving analytically the vector wave equations (4.2) is relatively difficult, so it is more effective to consider the scalar wave equation (4.7). If $\hat{\mathbf{u}}$ is a fixed-direction unit vector and $\Psi (\mathbf{r}, t)$ is a solution of (4.7) with the corresponding boundary conditions, the vector field $\hat{\mathbf{u}} \Psi (\mathbf{r}, t)$ will be a solution of the vector wave equation, for $\hat{\mathbf{u}}$ commutes with the derivative operators ∇^2 and $\partial^2 / \partial t^2$. In terms of the electric field, the two only possible (non redundant) solutions which satisfy the vanishing divergence condition [4] are

$$\mathbf{E} \sim \nabla \times (\hat{\mathbf{u}} \Psi), \tag{4.55a}$$

$$\mathbf{E}' \sim \nabla \times \nabla \times (\hat{\mathbf{u}} \Psi). \tag{4.55b}$$

Any linear combination of these two solutions is sufficient to provide a general solution for the vector wave equation (4.2a). Moreover, due to the relationship between the electric and magnetic fields through Maxwell's equations, $\mathbf{H} \sim \nabla \times \mathbf{E}$, note that if one of these represents the electric field, the other one will be proportional to the magnetic field.

Consider a time-harmonic scalar field of frequency ω traveling parallel to the waveguide axis, which will be assumed to be oriented along the z-direction. The wave scalar field can be described as

$$\Psi(\mathbf{r}, t) = \phi(x, y)e^{i(k_z z - \omega t)}, \tag{4.56}$$

where the cylindrical symmetry of the problem allows us to assume $\hat{\mathbf{u}} = \mathbf{k}$ is along the z-direction. If this solution is substituted into (4.55a), $\mathbf{E}$ has vanishing z-component ($E_z = 0$), while the z-component of its associated $\mathbf{H}$ field (proportional to (4.55b), according to Maxwell's equations) is a function of the Laplacian of ϕ, $\nabla^2 \phi$. If this Laplacian vanishes, then both electric and magnetic fields are orthogonal or *transverse* to the propagation direction. The corresponding wave is called *transverse electromagnetic* (TEM) wave (or mode). However, in the general case, if $\nabla^2 \phi \neq 0$ and therefore $B_z \neq 0$, the solutions are called as *transverse electric* (TE) waves or H-waves, because only the electric field is transverse to the propagation direction (while the magnetic field propagates in all directions). In other words, the electric field is confined in the waveguide cross-section plane. On the other hand, similarly (4.55a) can be associated with the magnetic field, obtaining analogous results. Thus, the case of TEM waves arises again for $\nabla^2 \phi = 0$, while in the general case $\nabla^2 \phi \neq 0$ ($E_z \neq 0$) only the magnetic field is transverse to the propagation direction (now it is this field the one confined in the waveguide cross-section plane). These are called *transverse magnetic* (TM) waves or E-waves. Of course, general solutions can also be obtained by considering superpositions of TE and TM waves.

The problem now consists of finding the function $\phi(x, y)$. By introducing (4.56) into the scalar wave equation (4.7), it can be seen that ϕ satisfies the two-dimensional Helmholtz equation,

$$\nabla_\perp^2 \phi + k_c^2 \phi = 0, \tag{4.57}$$

with the appropriate boundary conditions, where

$$\nabla_\perp^2 = \frac{\partial^2}{\partial x^2} + \frac{\partial^2}{\partial y^2} = \frac{1}{r} \frac{\partial}{\partial r} \left(r \frac{\partial}{\partial r} \right) + \frac{1}{r^2} \frac{\partial^2}{\partial \theta^2} = \cdots \tag{4.58}$$

is defined as the two-dimensional transverse Laplacian operator ($\nabla^2 \equiv \nabla_\perp^2 + \partial^2/\partial z^2$) and

$$k_c^2 = \frac{\omega^2}{c^2} - k_z^2, \tag{4.59}$$

which results after applying the second time and z-dependent derivatives on Ψ. Equation (4.57) will render discrete or *quantified* values for k_c, which will depend on the boundaries (shape and size) of the waveguide cross section, and will lead to discrete values or *modes* for the electric and magnetic fields inside the waveguide. Equation (4.59) is a *dispersion relation*, since it gives the relationship between the wave number k_z (or wavelength $\lambda = 2\pi/k_z$) of the wave propagating inside the waveguide to the frequency ω of this wave. Taking this into account, if (4.59) is rewritten as

$$k_z^2 = \frac{\omega^2}{c^2} - k_c^2, \tag{4.60}$$

note that for $\omega < ck_c$ the wave attenuates and dies out exponentially; this is called an *evanescent wave* (see Sect. 4.5). Thus, in order to have waves propagating along the waveguide, their frequency has to be greater than the *cutoff frequency*,

$$\omega_c = ck_c. \tag{4.61}$$

Relation (4.60) can be alternatively expressed in terms of a *cutoff wavelength*, as

$$\lambda_g = \frac{\lambda_0}{\sqrt{1 - (\lambda_0/\lambda_c)^2}}, \tag{4.62}$$

where $\lambda_g = 2\pi/k_z$ is the wavelength of the bounded wave traveling down the waveguide, $\lambda_0 = 2\pi c/\omega$ is the wavelength of a plane wave of frequency ω traveling in vacuum, and $\lambda_c = 2\pi/k_c$ is a parameter determined by the shape and size of the waveguide cross-section.

As an example, consider a waveguide with rectangular cross-section, with dimensions a and b along the x and y directions, respectively. After setting up the appropriate boundary conditions,

$$\phi(x, y) = A \cos(k_x x) \cos(k_y y), \tag{4.63}$$

where $k_x = l\pi/a$ and $k_y = m\pi/b$, with $l, m = 0, 1, 2, \ldots$, satisfy the constraint $k_x^2 + k_y^2 = k_c^2$. Taking this into account, the (l, m)-mode cutoff frequency is

$$\omega_c = \pi c \sqrt{\left(\frac{l}{a}\right)^2 + \left(\frac{m}{b}\right)^2} \tag{4.64}$$

and the cutoff wavelength,

$$\lambda_c = \frac{1}{\sqrt{(l/2a)^2 + (m/2b)^2}}. \tag{4.65}$$

According to (4.65), for a TE mode either l or m can be zero, but not both at the same time; for a TM mode it can be shown that both l and m must be greater than zero in order to ensure non-vanishing solutions.

According to this description, given the size of a waveguide, there is always necessarily a particular mode of lowest cutoff frequency, such that waveguide propagation below this cutoff will not be possible (except for the case of evanescent waves). Above this cutoff and below the cutoff frequency corresponding to the second lowest mode, only one mode can propagate down the waveguide. This lowest mode is known as the *dominant mode*. Waveguides with a size such that only one mode can propagate at a given frequency are called *dominant-mode waveguides* (for that frequency), as happens with single-mode optical fibers, which are prepared in such a way that the core can only support one propagating mode. If the size is larger and several modes can propagate, the waveguide is called *oversized*. This is the case of multi-mode optical fibers, for example. Dominant-mode waveguides have the advantage that small discontinuities do not convert energy to other modes in an uncontrolled manner, thus enabling the bending or twisting of the waveguide without inducing reflections provided the deformation takes place gradually relative to a wavelength. Otherwise coupling terms may appear in the corresponding scalar equation between the different directions, which leads to the internal exchange of energy between modes and then to the passage from one mode to another.

4.5 Evanescent Waves and Optical Tunneling

As mentioned in the previous Section, total internal reflection causes the reflection of light when it reaches a medium with smaller refractive index at an incidence angle larger than the critical angle. However, although no energy transmits through the second medium, an *evanescent wave* [32] with an exponentially damped amplitude (or intensity) with the distance from the boundary surface at which it was formed appears (i.e., on average, no energy is transmitted along the direction perpendicular to the boundary surface). Evanescent waves arise from any kind of wave equation, thus being independent of the kind of wave considered in the description (waves propagating in strings or membranes, sound waves, electromagnetic waves or quantum-probability waves). They form at the boundary between two media with different properties regarding the wave motion, being more intense within one-third of the wavelength from the boundary, where they appear to travel along the boundary between the two media.

To understand the appearance of evanescent waves, consider that the wave vector of the wave transmitted to a medium 2 is given by

$$\mathbf{k}_2 = k_2 \sin(\theta_2)\hat{\mathbf{x}} + k_2 \cos(\theta_2)\hat{\mathbf{z}}. \qquad (4.66)$$

As seen in Sect. 4.4, if $n_1 > n_2$, for angles larger than the critical one, $\sin(\theta_2) = (n_1/n_2)\sin(\theta_1) > 1$. Therefore, $\cos(\theta_2)$ becomes a complex number which can be expressed as

$$\cos(\theta_2) = i\sqrt{\left(\frac{n_1}{n_2}\right)^2 \sin^2(\theta_1) - 1}. \tag{4.67}$$

If the transmitted electric field, for example, is given by a harmonic plane wave, $\mathbf{E}_2 = \mathbf{E}_0 \exp[i(\mathbf{k}_2 \cdot \mathbf{r} - \omega t)]$, then

$$\mathbf{E}_2 = \mathbf{E}_0 e^{-\kappa z} e^{i(kx - \omega t)}, \tag{4.68}$$

where

$$\kappa = k_2 |\cos(\theta_2)| = k_0 \sqrt{n_1^2 \sin^2(\theta_1) - n_2^2} \tag{4.69}$$

is the *attenuation constant* and

$$k = k_2 \sin(\theta_2) = k_0 n_1 \sin(\theta_1) \tag{4.70}$$

is the *propagation constant* (here, $v_{\text{prop}}^2 = c/n_2 = \omega/k_2$ and $k_0 = \omega/c$). Physically, the damping undergone by the transmitted wave arises as a consequence from the fact that electric and magnetic fields cannot be discontinuous at a boundary surface, which would be the case if there was no evanescent wave. This is the direct analog of the exponentially decaying wave functions in quantum mechanics: because the wave function and its first derivative cannot be discontinuous at the turning point of a potential, the solution inside the potential has to be an exponentially decaying function. In electromagnetism, the attenuation constant depends on the refractive index of the second medium; in quantum mechanics, it depends on the effective energy of the particle. Thus, a certain relationship can already be established between optical refractive indices and quantum particle energies.

It is clear that a damping of the wave occurs due to the presence of the second medium, with smaller refractive index. The *penetration distance*, i.e., the effective distance that a light ray can pursue before getting totally damped, can be defined in terms of the inverse of the attenuation constant, $\ell \sim 1/\kappa$. Thus, if the second medium is "sandwiched" between two media with the same refractive index n_1 (or, at least, similar and larger than n_2) and separated a distance of the order of ℓ or smaller, there should be an emerging wave exiting through the third medium. This phenomenon is called *frustrated total internal reflection*. Accordingly, the evanescent wave *couples* two media in which traveling waves are allowed. This thus enables the energy transfer from one of these media to the other one, although no traveling-wave solutions are allowed within the sandwiched region. This is the direct quantum-mechanical analog of *tunneling*, due to coupling of evanescent waves. This effect can be nicely observed by sandwiching a think layer of air (a transparent, low refractive index material) between two prisms. By means of this simple device, it can be observed how the incident beam will "tunnel" through from one prism to the next one. The excitation that the evanescent wave causes on the second prism, thus causing the appearance of a power-carrying transmitted wave, leads to a decrease in the power carried by the reflected wave in the first prism.

There are different phenomena related to evanescent waves which have interesting counterparts in quantum mechanics. For example, on the one hand, there is the *Goos–Hänchen* [33–37] and *Imbert–Fedorov effects* [38, 39, 40]. The Goos–Hänchen effect consists of a small shift undergone by linearly polarized light when it is totally internally reflected. This shift takes place along the propagation direction and is parallel to the surface—i.e., in the plane containing the incident and reflected beams. This *longitudinal* shift is unexpected from a purely geometric treatment unless a certain width (non-monochromaticity) is associated with the incident beam. In the case of elliptically or circularly polarized light, apart from this shift a *transversal* shift is also observed. This is the Imbert–Fedorov effect, which consequently is perpendicular to the plane containing the incident and reflected beams. On the other hand, another interesting effect related to optical tunneling (but also with its quantum-mechanical counterpart) is the *Hartman effect* [41, 42]. This effect consists of the independence of the delay-time undergone by a tunneling beam under frustrated total internal reflection conditions with respect to the thickness of the barrier tunneled. In other words, according to this effect, firstly described by Thomas Hartman in 1962 [41], the tunneling time tends to a constant for relatively large barriers [42]. Consider two prisms sandwiching a medium with higher refractive index, as before. If the prisms are in contact, light will pass straight through. However, when a gap is introduced in between, the light will get refracted, which can be frustrated by increasing the incidence angle above the critical angle by total internal reflection. For larger gaps between the prisms the tunneling time required to cross from one prism to the other one approaches a constant, which can be mistakenly interpreted as light transmitting with a superluminal speed [43]. Nevertheless, a careful analysis of this effect shows [44] that it cannot be used to violate relativity by transmitting signals faster than c, since tunneling times should not be linked to a velocity, for evanescent waves do not propagate.

Finally, it is also worth mentioning that in modern material science one finds the so-called *metamaterials*, with very interesting electromagnetic properties, such as negative refractive indices [45]—nevertheless, areas of active research in this kind of materials are also connected with other types of waves, such as acoustic or seismic. In this regard, it is remarkable how it is possible to create materials such that light gradually becomes extinct inside them, giving rise to a phenomenon which is known as *optical black holes* [46–50], in analogy to the effect caused by gravitational black holes on light. That is, this phenomenon essentially consists of taking advantage of the possibility of reducing the light speed inside dielectric materials [51] to later on recreate dielectric analogs of astronomical effects. Thus, though still under research (see references before), in optical black holes slow light is passed through a fast-spinning Bose–Einstein condensate (faster than the local speed of light). In analogy to gravitational black holes, this rotation would give rise to a vortex capable of trapping the light behind an event horizon [46]. Nevertheless, these optical black holes analog would still present a series of problems (e.g., it cannot display Hawking radiation or the related quantum effects cannot be reproduced), thus breaking a full analogy with their gravitational counterparts.

4.6 Wave Optics and Schrödinger Equation

4.6.1 Schrödinger-like Formulation of Electromagnetism

The description of wave optics presented up to here is based on the standard approach to electromagnetism, grounded on Maxwell's equations. Since this monograph is concerned with Bohmian mechanics [52–56], a hydrodynamic picture of quantum mechanics, it is interesting to show that electromagnetism—and therefore wave optics—also admits a hydrodynamic picture or reformulation. In this case, as shown by Bialynicki-Birula [57–60] and Sipe [61], a Schrödinger-like equation ruling the behavior of the electromagnetic field emerges from Maxwell's equations. This formulation arises after assuming that, as happens with matter particles, the electromagnetic fields can also be described by a well-defined wave function.[4] More specifically, following the same argumentation that led Dirac to the relativistic equation for the electron, Bialynicki-Birula [57–60] reaches Maxwell's equations. As summarized in Table 1.5 from the book of Scully and Zubairy *Quantum Optics* [67], "semi-classically" electromagnetic fields and massive particles are treated by means of the Maxwell and Schrödinger equations, respectively. In the case of matter, it is common to talk about *first quantization*, with $\hbar$ appearing as the key element which establishes the difference with the description of matter particles in classical mechanics. However, in the case of electromagnetic fields, although there is also a wave-like behavior associated with the fields (just as the wave function Ψ describes matter particles in the Schrödingerian counterpart), $\hbar$ does not appear in Maxwell equations and therefore one can still talk about "classical" electromagnetism. It is when going to second quantization in both cases (i.e., to quantum field theory), by means of the Schwinger and Dirac equations for matter and electromagnetic field, respectively, that both are treated on the same footing. Thus, at the "semiclassical" level, there is a sort of asymmetric consideration for matter and electromagnetic fields [67, 68].

Such an asymmetry can be avoided [67, 68], however, if one starts from the theoretical framework provided by the least action principle (see Chap. 7 for details). Though the quantization scheme remains the same for matter particles within this framework, in the case of electromagnetic fields there is an important difference: the starting point is not the Maxwell equations, but Fermat's principle and geometric optics. As it will be further analyzed in Sect. 7.2, Fermat's principle is the direct analog of Hamilton's least action principle (indeed, it predates the latter). Accordingly, within the Fermatian formulation of optics, the topology displayed by optical rays is influenced by the properties of the refractive index traversed by light during its propagation (just as the topology of Newtonian trajectories for matter particles

---

[4] In particular, the problem of finding a wave function for the photon is an issue which has received much attention in the literature [57–73]. It is remarkable that recently it has been shown experimentally [74] that such a wave function can be directly measured, which might open a very fruitful and interesting debate of unexpected implications.

depends on external forces or potentials). In this way, the passage from Fermat's ray formulation (i.e., geometric optics) to fields (waves) in terms of Maxwell's equations would constitute in electromagnetism a somewhat first quantization. In this regard, if the electromagnetic field is confined (for example, inside a waveguide), one can readily observe the appearance of discrete modes (frequencies), just as discrete energy levels appear in the problem of a particle confined in a box in quantum mechanics—of course, this analogy can be extended to any type of wave, although here this instance is stressed in order to make clearer the correspondence between electromagnetic fields and matter waves. Proceeding along this direction, an analog of the spin in the case of matter particles can also be found in the polarization vector for electromagnetic fields—actually, both spin and polarization states admit a similar algebraic description in terms of Pauli matrices.

The similarity between the descriptions in terms of complex wave fields applicable to electromagnetic fields and matter waves can be noted through the so-called *Riemann–Silberstein complex electromagnetic vector* [75–77],

$$\boldsymbol{\Xi}(\mathbf{r}, t) = \frac{1}{\sqrt{2}} [\sqrt{\epsilon_0} \, \mathbf{E}(\mathbf{r}, t) + i \sqrt{\mu_0} \, \mathbf{H}(\mathbf{r}, t)], \qquad (4.71)$$

where $\mathbf{E}$ and $\mathbf{H}$ are the time-dependent, real electric and magnetic fields, respectively, satisfying Maxwell's equations (4.1). After substitution of

$$\mathbf{E} = \frac{1}{\sqrt{2\epsilon_0}} \left(\boldsymbol{\Xi} + \boldsymbol{\Xi}^* \right), \quad \mathbf{H} = \frac{1}{i \sqrt{2\mu_0}} \left(\boldsymbol{\Xi} - \boldsymbol{\Xi}^* \right) \qquad (4.72)$$

into Maxwell's equations in the absence of electric charges and charge densities, one finds

$$i \frac{\partial \boldsymbol{\Xi}}{\partial t} = c \nabla \times \boldsymbol{\Xi}, \qquad (4.73)$$

with

$$\nabla \cdot \boldsymbol{\Xi} = 0. \qquad (4.74)$$

Equation (4.73) is the analog for the complex electromagnetic field $\boldsymbol{\Xi}$ of Schrödinger's equation for matter waves. Moreover, if $\rho = \Psi^* \Psi$ describes the quantum-mechanical probability density, the electromagnetic analogous, the energy density, can be expressed as

$$\mathcal{U} = \frac{1}{2} \left(\epsilon_0 \, \mathbf{E} \cdot \mathbf{E} + \mu_0 \, \mathbf{H} \cdot \mathbf{H} \right) = \boldsymbol{\Xi} \cdot \boldsymbol{\Xi}^*. \qquad (4.75)$$

In Sect. 7.6, it will also be seen how this analogy extends even to the energy flux.

Notwithstanding, there is a difference between (4.73) and its quantum-mechanical homologous: it is a vector equation, while Schrödinger's equation is scalar. This is related to the fact that electromagnetic fields are characterized by a polarization state, which is absent in matter particles unless the spin state is considered (at the level of first quantization), which leads to a vectorial description of the wave function Nevertheless, in those cases independent of the polarization state (e.g., if linearly polarized light is being used), it is possible to reach a scalar description by applying the operator $-i\partial/\partial t$ on both sides of (4.73). This yields the wave equation

$$\frac{\partial^2 \Xi}{\partial t^2} = c^2 \nabla^2 \Xi, \tag{4.76}$$

which is a more compact form of expressing (4.2). Now, consider for example, a diffraction problem in vacuum, which is essentially a boundary-condition problem, as seen in Sect. 4.3.4). In this case, one can further proceed with (4.76) and assume a separable solution. Accordingly, if Ξ is decomposed as

$$\Xi(\mathbf{r}, t) = \Xi_0(\mathbf{r})\varphi(t) = \Xi_0(\mathbf{r})\, e^{-i\omega t}, \tag{4.77}$$

the time-dependent part of the Riemann–Silberstein vector field is described by the equation

$$\frac{\partial^2 \varphi(t)}{\partial t^2} = -\omega^2 \varphi(t), \tag{4.78}$$

with $c = \omega/k$, while its space-dependent part, Ξ_0, will satisfy Helmholtz's equation,

$$\nabla^2 \Xi_0(\mathbf{r}) + k^2 \Xi_0(\mathbf{r}) = 0. \tag{4.79}$$

This equation is the direct analog of the time-independent Schrödinger equation, which immediately emerges if k is expressed in terms of the refractive index characterizing the medium where the electromagnetic field is confined (analogy with bound problems) or introducing the corresponding boundary conditions (analogy with matter wave diffraction problems).

This alternative formulation of electromagnetism in terms of the complex Riemann–Silberstein vector field has only been briefly discussed here (it will appear again in Chap. 7) in order to show that the electromagnetic field also admits a similar description to matter waves—actually, the Riemann–Silberstein vector field could be regarded as a sort of semiclassical wave function for the photon [57–61]. However, it should be noticed that this approach has also been considered from a practical viewpoint in recent years. In particular, it has been used to analyze different problems from condensed matter physics and solid state physics [78–80], developing the corresponding numerical tools, which keep certain resemblance to those in the standard propagation of quantum-mechanical wave packets.

4.6.2 Paraxial Approximation and Schrödinger Equation

The formulation presented in previous Section is general and allows us to understand the similarity between optics and quantum mechanics from a different perspective (but based on similar equations of motion). At a different level, but also very
interesting, another way to reach a description similar to that provided by quantum
mechanics for matter particles arises when appealing to the paraxial approximation
[16]. As seen in Sect. 4.3.4, this is a *small-angle* approximation, which allows us
to replace sine and tangent functions by the value of their arguments (in radians),
indicating (within the viewpoint of geometric optics) that rays deviate little from
the optical axis of a system—more specifically, for angles $\simeq 10°$. This approximation is very common in geometric optics in relation to the analysis of lenses, but
also in Gaussian optics (paraboloidal waves and Gaussian beams), laser light and
waveguides (optical guides), all connected.

In the case of wave optics, in order to set up the paraxial approximation, if the
optical axis is oriented along the z-axis, one can assume that the time-independent
electric and magnetic fields can be expressed as

$$\psi(\mathbf{r}) = \phi(\mathbf{r})e^{ik_z z}. \tag{4.80}$$

Substituting this expression into Helmholtz's equation (4.7), the latter can be
recast as

$$2ik_z \frac{\partial \phi}{\partial z} + \frac{\partial^2 \phi}{\partial z^2} = -\nabla_\perp^2 \phi + (k_z^2 - k^2)\phi, \tag{4.81}$$

where $\nabla_\perp^2$ is the transverse part of the Laplacian (for example, in Cartesian coordinates, it reads as $\nabla_\perp^2 = \partial^2/\partial x^2 + \partial^2/\partial y^2$). Now, making use of the paraxial
approximation, the longitudinal space variations of ϕ are neglected when compared
with the value of this function—$\partial^2 \phi/\partial z^2 \approx 0$—, (4.81) can be expressed as

$$2ik_z \frac{\partial \phi}{\partial z} = -\nabla_\perp^2 \phi + (k_z^2 - k^2)\phi. \tag{4.82}$$

As can be readily noticed, this equation is very similar to the Schrödinger one,
except for the evolution, which is not in time, but along the z-coordinate, acting here
as an "evolution" parameter. Equation (4.82) has been used, for example, to study the
design of waveguides with optimal conditions of light transmission [81–83]: based on
the direct relationship between (4.82) and Schrödinger's equation, analogous optimal
control techniques can be found in the case of waveguides.

References

1. Elmore, W.C., Heald, M.A.: Physics of Waves. Dover, New York (1969)
2. Main, I.G.: Vibrations and Waves in Physics, 3rd edn. Cambridge University Press, Cambridge (1993)
3. Morse, P.M., Feshbach, H.: Methods of Theoretical Physics, vol. 1. McGraw-Hill, New York (1953)
4. Morse, P.M., Feshbach, H.: Methods of Theoretical Physics, vol. 2. McGraw-Hill, New York (1953)
5. Huygens, C.: Treatise on Light. Dover, New York (1962)
6. Newton, I.: Opticks. Dover, New York (1952)
7. Home, R.W.: Leonhard Euler's "anti-Newtonian" theory of light. Ann. Sci. **45**, 521–533 (1988)
8. Eliasmith, C., Thagard, P.: Waves, particles and explanatory coherence. Br. J. Phil. Sci. **48**, 1–19 (1997)
9. Iliffe, R.: Philosophy of science, in The Cambridge History of Science. In: Porter, R. (ed) Eighteenth-Century Science, vol. 4, pp. 267–284. Cambridge University Press, Cambridge (2003)
10. Hooke, R.: Micrographia or Some Physiological Descriptions of Minute Bodies Made by Magnifying Glasses with Observations and Inquiries Thereupon, 1st edn. J. Martyn and J. Allestry, London, (1665). Preprinted by Dover Publications (Dover Phoenix Editions, New York, 2003)
11. Bartholin, E.: Experimenta crystalli islandici disdiaclastici quibus mira et infolita refractio detegitur. Daniel Pauli, Copenhagen (1669). Translated into English by Archibald, T.: Experiments on birefringent Icelandic crystal through which is detected a remarkable and unique refraction. Danish National Library of Science and Medicine, Copenhagen (1991); Acta Historica Scientiarium Naturalium et Medicinialium 40 (1991)
12. Bartholin, E.: An account of sundry experiments made and communicated by That Learn'd mathematician, Dr Erasmus Bartholin, upon a Chrystal-like body, sent to him out of Island. Phis. Trans. Roy. Soc. Lond. **5**, 2039–2048 (1670)
13. Fresnel, A.: Œuvres Complètes. Imprimerie Impèriale, Paris, t.1, pp. 254–255 (1866). Translated into English by Crew, H.: The Wave Theory of Light: Memoirs o Huygens, Young and Fresnel. American Book Company, New York (1900)
14. Hecht, E.: Optics. Addison-Wesley, Reading (1987)
15. Harvey, J.E., Forgham, J.L.: The spot of Arago: new relevance for an old phenomenon. Am. J. Phys. **52**, 243 (1984)
16. Born, M., Wolf, E.: Principles of Optics. Electromagnetic Theory of Propagation, Interference and Diffraction of Light, 7th edn. Cambridge University Press, Cambridge (1999)
17. Jackson, J.D.: Classical Electrodynamics, 3rd edn. Wiley, New York (1998)
18. Guenther, R.: Modern Optics. Wiley, New York (1990)
19. Daniels, J.M.: An explanation of a difficulty with Huygens' secondary wavelets. Can. J. Phys. **74**, 236–239 (1996)
20. Haber-Schaim, U., Cross, J.B., Dodge, J.H., Walter, J.A.: PSSC Physics, 3rd edn. D. C. Heath, Lexington, p. 122 (1971)
21. Feynman, R.P., Hibbs, A.R.: Quantum Mechanics and Path Integrals. McGraw-Hill, New York (1965)
22. Arfken, G.: Mathematical Methods for Physicists, 3rd edn. Academic Press, San Diego (1985)
23. Sharma, K.K.: Optics. Principles and Applications. Academic Press, Amsterdam (2006)
24. Balanis, C.A.: Antenna Theory. Analysis and Design, 2nd edn. Wiley, New York (1997)
25. Colladon, J.-D.: Sur les réflexions d'un rayon de lumière à l'interieur d'une veine liquide parabolique. Comptes Rendus **15**, 800–802 (1842)
26. Babinet, J.: Note sur la transmission de la lumière par des canaux sinueux. Comptes Rendus **15**, 802–802 (1842)

27. Tyndall, J.: On some phenomena connected with the motions of liquids. Proc. R. Inst. G. B. **1**, 446–448 (1854)

28. Tyndall, J.: Notes of a Course of Nine Lectures on Light. The Royal Institution of Great Britain, London, pp. 20–22 (1870)

29. Thomson, J.J.: Notes on Recent Researches in Electricity and Magnetism. Clarendon Press, Oxford (1893)

30. Lodge, O.J.: The Work of Hertz and Some of His Succesors, London (1894)

31. Lord Rayleigh, J.W.S.: On the passage of electric waves through tubes or the vibrations of dielectric cylinders. Phil. Mag. **43**, 125–132 (1897)

32. de Fornel, F.: Evanescent waves: from Newtonian mechanics to atomic optics. Springer, Berlin (2001)

33. Goos, F., Hänchen, H.: Ein Neuer und Fundamentaler Versuch zur Totalreflexion. Ann. Phys. (Leipzig) **436**, 333–346 (1947)

34. Goos, F., Lindberg-Hänchen, H.: Neumessung des Strahlversetzungeffktes bei Totalreflexion. Ann. Phys. (Leipzig) **436**, 251–252 (1949)

35. Lotsch, H.K.V.: Beam displacement at total reflection: the Goos–Hänchen effect (series of papers). Optik **32**, 116–137, 189–204 (1970)

36. Lotsch, H.K.V.: Beam displacement at total reflection: the Goos–Hänchen effect (series of papers). Optik **32**, 299–319, 553–569 (1971)

37. Delgado, M., Delgado, E.: Evaluation of a total reflection set-up by an interface geometric model. Optik **113**, 520–526 (2003)

38. Fedorov, F.I.: Theory of total reflection. Dokl. Akad. Nauk. SSR **105**, 465–468 (1955)

39. Imbert, C.: L'effet inertial de spin du photon: Théorie et preuve expérimentale. Nouv. Rev. Opt. Appl. **3**, 199–208 (1972)

40. Pillon, F., Gilles, H., Girard, S.: Experimental observation of the Imbert–Fedorov transverse displacement after a single total reflection. App. Opt. **43**, 1863–1869 (2004)

41. Hartman, T.E.: Tunneling of a wave packet. J. Appl. Phys. **33**, 3427–3433 (1962)

42. Martínez, J.C., Polatdemir, E.: Origin of the Hartman effect. Phys. Lett. A **351**, 31–36 (2006)

43. Nimtz, G., Stahlhofen, A.A.: Macroscopic violation of special relativity (2007). arXiv: 0708.0681

44. Winful, H.: Tunneling time, the Hartman effect and superluminality: a proposed resolution of an old paradox. Phys. Rep. **436**, 1–69 (2006)

45. Shelby, R.A., Smith, D.R., Schultz, S.: Experimental verification of a negative index of refraction. Science **292**, 77–79 (2001)

46. Leonhardt, U., Piwnicki, P.: Relativistic effects of light in moving media with extremely low group velocity. Phys. Rev. Lett. **84**, 822–825 (2000)

47. Unruh, W.G., Schützhold, R.: On slow light as a black hole analogue. Phys. Rev. D **68**, 024008(1–14) (2003)

48. Philbin, T.G., Kuklewicz, Ch., Robertson, S., Hill, S., König, F., Leonhardt, U.: Fiber-optical analog of the event horizon. Science **319**, 1367–1370 (2008)

49. Narimanov, E.E., Kildishev, A.V.: Optical black hole: broadband omnidirectional light absorber. Appl. Phys. Lett. 95:41106(1–3) (2009)

50. Cheng, Q., Cui, T.J., Jiang, W.X., Cai, B.G.: An omnidirectional electromagnetic absorber made of metamaterials. New J. Phys. **12**, 063006(1–10) (2010)

51. Hau, L.V., Harris, S.E., Dutton, Z., Behroozi, C.H.: Light speed reduction to 17 metres per second in an ultracold atomic gas. Nature **397**, 594–598 (1999)

52. Madelung, E.: Quantentheorie in hydrodynamischer Form. Z. Phys. **40**, 322–326 (1926)

53. Bohm, D.: A suggested interpretation of the quantum theory in terms of "hidden" variables I. Phys. Rev. **85**, 166–179 (1952)

54. Bohm, D.: A suggested interpretation of the quantum theory in terms of "hidden" variables II. Phys. Rev. **85**, 180–193 (1952)

55. Takabayashi, T.: On the formulation of quantum mechanics associated with classical pictures. Prog. Theor. Phys. **8**, 143–182 (1952)

56. Takabayashi, T.: Remarks on the formulation of quantum mechanics with classical pictures and on relations between linear scalar fields and hydrodynamical fluids. Prog. Theor. Phys. **9**, 187–222 (1953)
57. Bialynicki-Birula, I.: On the wave function of the phonon. Acta Phys. Pol. A **86**, 97–107 (1994)
58. Bialynicki-Birula, I.: The photon wave function. In: Eberly, J.H., Mandel, L., Wolf, E. (eds.) Coherence and Quantum Optics, vol. 7. Plenum Press, New York, pp. 313–322 (1996)
59. Bialynicki-Birula, I.: Photon wave function. Prog. Opt. **36**, 245–294 (1996)
60. Bialynicki-Birula, I.: Hydrodynamics of relativistic probability flows. In: Infeld, E., Zelazny, R., Galkowski, A. (eds.) Nonlinear Dynamics, Chaotic and Complex Systems. Cambridge University Press, Cambridge, pp. 64–71 (1997)
61. Sipe, J.E.: Photon wave functions. Phys. Rev. A **52**, 1875–1883 (1995)
62. Landau, L., Peierls, R.: Quantenelektrodynamik im Konfigurationsraum. Z. Phys. **62**, 188–200 (1930)
63. Dirac, P.A.M.: The Principles of Quantum Mechanics. Clarendon Press, Oxford (1958)
64. Cook, R.J.: Photon dynamics. Phys. Rev. A **25**, 2164–2167 (1982)
65. Cook, R.J.: Lorents covariance of photon dynamics. Phys. Rev. A **26**, 2754–2760 (1982)
66. Inagaki, T.: Quantum-mechanical approach to a free photon. Phys. Rev. A **49**, 2839–2843 (1994)
67. Scully, M.O., Zubairy, M.S.: Quantum Optics. Cambridge University Press, Cambridge (1997)
68. Kobe, D.H.: A relativistic Schrödinger-like equation for a photon and its second quantization. Found Phys. **29**, 1203–1231 (1999)
69. Berry, M.V.: Riemann–Silberstein vortices for paraxial waves. J. Opt. A **6**, S175–S177 (2004)
70. Holland, P.R.: Hydrodynamic construction of the electromagnetic field. Proc. R. Soc. A **461**, 3659–3679 (2005)
71. Raymer, M.G., Smith, B.J.: The Maxwell wave function of the photon. Proc. SPIE **5866**, 293–297 (2005)
72. Smith, B.J., Raymer, M.G.: Photon wave functions: wave-packet quantization of light and coherence theory. New J. Phys. **9**, 414(1–37) (2007)
73. Zhi-Yong, W., Cai-Dong, X., Ole, K.: The first-quantized theory of photons. Chin. Phys. Lett. **24**, 418–420 (2007)
74. Lundeen, J.S., Sutherland, B., Patel, A., Stewart, C., Bamber, C.: Direct measurement of the quantum wavefunction. Nature **474**, 188–191 (2011)
75. Silberstein, L.: Elektromagnetische Grundgleichungen in bivectorieller Behandlung. Ann. Phys. (Leipzig) **22**, 579–586 (1907)
76. Silberstein, L.: Nachtrag zur Abhandlung über "Elektromagnetische Grundgleichungen in bivectorieller Behandlung". Ann. Phys. (Leipzig) **24**, 783–784 (1907)
77. Bateman, H.: The Mathematical Analysis of Electrical and Optical Wave Motion on the Basis of Maxwell's Equations. Dover, New York (1955)
78. Borisov, A.G., Shabanov, S.V.: Lanczos pseudospectral method for initial-value problems in electrodynamics and its applications to ionic crystal gratings. J. Comp. Phys. **209**, 643–664 (2005)
79. Borisov, A.G., García de Abajo, F.J., Shabanov, S.V.: Role of electromagnetic trapped modes in extraordinary transmission in nanostructured materials. Phys. Rev. B 71:075408(1–7) (2005)
80. Borisov, A.G., Shabanov, S.V.: Wave packet propagation by the Faber polynomial approximation in electrodynamics of passive media. J. Comp. Phys. **216**, 391–402 (2006)
81. Pant, D.K., Coalson, R.D., Hernández, M.I., Campos-Martínez, J.: Optimal control theory for the design of optical waveguides. J. Lightwave Technol. **16**, 292–300 (1998)
82. Pant, D.K., Coalson, R.D., Hernández, M.I., Campos-Martínez, J.: Optimal control theory for optical waveguides design: application to Y-branch structures. Appl. Opt. **38**, 3917–3923 (1999)
83. Campos-Martínez, J., Coalson, R.D.: The wide-angle equation and its solution through the short-time iterative Lanczos method. Appl. Opt. **42**, 1732–1742 (2003)

Chapter 5
Dynamics of Open Quantum Systems

5.1 Introduction

Quantum dissipation constitutes a broad and active field of research within quantum mechanics. One of the crucial problems in the early days of quantum mechanics was to explain the stability of matter [1]. As it was very well known, classically electrons should emit radiation when they move in closed orbits around the nucleus. Therefore, they should lose energy, this leading to a *radiation damping* and eventually to the collapse of electrons on nuclei. Bohr's atomic model and then quantum mechanics provided a solution to this problem: electrons remain in stationary *orbitals* and only emit or absorb radiation when they "jump" from one orbital to another (quantum transitions).

Strictly speaking, real physical systems do not exist in complete isolation in Nature. All physical systems are open systems, since the interaction with their environment can never be totally neglected. This interaction gives rise to a strong correlation or *entanglement* between system and environment which eventually leads the former to become a statistical mixture. The so-called theory of open quantum systems [2] encompasses a series of formalisms and approaches developed to deal with this kind of problems. Nowadays, the corresponding quantum dynamics is much more developed and it can be considered as an interdisciplinary field, where very broad branches of physics, chemistry and biology meet together to describe processes that are ubiquitous in Nature [2–6]. Nevertheless, finding good quantum analogues of classical dissipative systems constitutes a very difficult task and still remains an open problem, because of the involvement of elements such as commutation rules, time ordering and symmetrization, which cannot be neglected when dealing with the quantum world.

A quite natural way to tackle this problem is starting from the information acquired and processed in classical mechanics. It is well known that the form of Lagrangian and Hamiltonian functions is not unique for conservative and nonconservative systems. Hence different wave equations corresponding to the same classical equations of motion can be obtained. Nevertheless, the fact that physical or mathematical incon-

A. S. Sanz and S. Miret-Artés, *A Trajectory Description of Quantum Processes.*
I. Fundamentals, Lecture Notes in Physics 850, DOI: 10.1007/978-3-642-18092-7_5,
© Springer-Verlag Berlin Heidelberg 2012

sistencies can be found in the different quantization procedures makes that most of such wave equations can be disregarded. Furthermore, for dissipative systems the Hamiltonian no longer represents the total energy of the system.

In general, it could be said there are two main routes to quantization. First, appealing to arguments based on the correspondence principle, one can focus on the classical equations of motion and try to find a quantized version of them, passing from classical variables to homologous quantum operators subject to certain commutation rules. This is the case, for example, of the so-called *Yang–Feldman method* [7]. Also, the Heisenberg equations of motion have their analog in classical mechanics. Nevertheless, the formal connection between the equations of motion in classical and quantum mechanics arises from Ehrenfest's theorem, though in general it does not discriminate Hamiltonians coming from different orderings in the dynamical variables [6]. The rate of change in time of the expectation value associated with an arbitrary Hermitian operator follows the classical equation of motion. Such an expectation value has two interpretations: either it describes a property of a wave packet or the (same) averaged property for a particle ensemble. Several attempts have also been addressed to quantize systems with dissipative forces that depend quadratically (or even higher powers) on velocity as well as those displaying quadratic damping [6]. The equation of motion for the reduced density matrix (once a tracing over the environment degrees of freedom is carried out) or any system observable can be obtained following several quantum approaches worth mentioning [8–12]: the Langevin equation, c-number equations, the Fokker–Planck equation, the linear response theory, the Redfield theory (with the secular approximation similar to the rotating wave approximation used in spectroscopy), the Agarwal master equation, the optical Bloch equations, the Floquet–Markov master equation, the Linblad equation, the quantum phase space distributions based on the Wigner representation of quantum mechanics [13–15] (see Sect. 3.3.1), the Yan–Mukamel phase space distribution, the path integral approach (see Sect. 3.4), the quantum-classical hybrid formalisms and the quantum state diffusion. All these approaches have been used in different branches of physics, such as quantum optics, condensed matter physics, astrophysics, chemical physics or mathematical physics.

Sometimes quantum processes have no analog in the classical world and therefore there is not a unique way to reach the corresponding Hamiltonians, quantization rules, etc., and some intuition is needed for such purposes. In any case, the only check is comparison with experiment.

In quantum mechanics, if the total Hamiltonian is also split according to the system-plus-bath Hamiltonian model used in the classical context (see Chap. 2), dissipation is easier to tackle and understand. Both system and reservoir are in continuous interaction and the effects—quantum coherence loss or decoherence, population transfer, and/or (system-environment) energy exchange—arising from that interaction will depend to a greater or a lesser extent on the coupling strength and its intrinsic nature. Under these conditions, one speaks about the environment induced decoherence (see Appendix B). An interesting and very important process is vibrational dephasing of small molecules immersed in a rare gas in liquid phase [11, 16–20]. Fluctuations occur because of random collision events. Due to its interaction with the environment, the system usually behaves quite different with respect to its behavior in isolation. Its time-evolution is not unitary and therefore cannot be described in terms of the Schrödinger equation. In these cases, it is necessary to resort to statistical quantum methods invoking, for example, the density matrix and Langevin formalisms and/or introducing, in general, quantum stochasticity into the time-evolution equations: quantum master equations, quantum Langevin type equations, and so on. The energy transfer from the system to the environment is termed quantum *relaxation* or *damping*. If there is no chance for the energy to move backwards into the system, the unidirectional energy flow into the reservoir is then called quantum *dissipation*. On short timescales, the distinction between quantum relaxation and dissipation is obviously unclear. Only when the environment has a small number of degrees of freedom, the energy moves backwards into the system; this phenomenon is called a *recurrence*. Nonetheless, even for large systems, quantum *noise* arises since the reservoir distributes some of its energy back into the system. Under certain conditions the duration of the reservoir correlations is very short compared to the dynamical evolution of the system. This leads to a total memory loss of the bath dynamics that gives rise to a subsequent irreversible loss of coherence and energy (or population) relaxation in the system. This is called a *Markovian regime*. Within this regime, the time-evolution of the system does only depend on the present state of the system; this is called a *Markovian process*. As will be seen, when this happens, the system dynamics can be characterized by (relatively) simple Markovian master equations, where one does not need to take into account the reservoir dynamics and its effects on the system are described by means of certain operators. Quantum stochastic methods in the Markovian and non-Markovian regimes have been widely developed in quantum optics [5]. In particular, the non-Markovian dynamics of open quantum systems has being developed very fast in the last years. It has been finally established that there is no quantum Onsager regression theorem and the correct generalization of the Onsager hypothesis is the fluctuation-dissipation theorem [21].

Irreversible quantum processes take place when dealing with extended systems or, more precisely, with systems displaying an infinite number of degrees of freedom (photon or phonon fields, for example). Quantum noise has its origin in our inability to specify each of all the infinite modes of a given field when considering an environment around a system; noise appears when the reaction of the field back on the system is not neglected. Furthermore, spontaneous emission can also be considered quantum noise since this emission is random with respect to the incoming signal adding noise to the detector.

Another interesting aspect comes from the measurement process itself. A measure apparatus can be understood as a sort of environment; the measurement can then be identified with an outcome arising from the coupling between this apparatus and a physical system. As the coupling strength reduces, the precision of the measurement also decreases. This drawback, though, can be surmounted by averaging over many measurements. Consequently, subsequent strong ("standard") measurements of a complementary observable will not be disturbed in the limit of vanishing coupling. In this regard, the initial state of the quantum system is preselected, while the final state is post-selected, this allowing to define the weak measurement of a certain property A as $A_w = \langle \Psi_f | A | \Psi_i \rangle / \langle \Psi_f | \Psi_i \rangle$. The complex weak value A_w of a given observable thus characterizes the observed outcomes of weak measurements. This post-selection is a common tool in quantum information processing. The average of the so-called *weak measurement* [22–25] is a simple expectation value of the property to be measured when the final and initial states are not the same. This type of measurements are in contrast with quantum state tomography [26, 27], which does not allow to measure both complementary properties at once. These weak measurements have other interesting properties. For example, a stochastic interpretation of quantum mechanics in complex space has been proposed [28] based on an interpretation of the weak value associated with the position operator as a conditional expectation value. Another example is that, if repeated measurements take place on a given state of the system, Heisenberg's uncertainty principle applies, giving rise under certain conditions to the so-called *quantum Zeno* and *anti-Zeno effects* [2, 5, 29–32].

In analogy to open classical systems (see Chap. 2), three main different approaches to deal with quantum dissipative dynamics can also be considered here:

1. Effective time-dependent Hamiltonians.
2. Nonlinear Schrödinger equation.
3. The system-plus-bath model within a conservative scenario.

Although there are links among them, they will be discussed separately in order to make more apparent such connections afterwards. This theoretical scheme is valid for both dissipative and stochastic dynamics; their corresponding formalisms are common to both cases and only those related somehow to trajectories will be reported here. As an illustrative example, adsorbate diffusion on flat surfaces will be considered, which can be handled analytically. This example will also provide the theoretical background to later on study in detail wave-packet stochastic dynamics in Volume 2.

5.2 The Quantization Problem: Standard Theoretical Approaches

A natural way to find the quantum version of a classical dissipative system consists of starting from the classical equation of motion for a given dynamical variable[1] $A(q, p, t)$ [6],

$$\frac{dA}{dt} = \{A, H\} + \frac{\partial A}{\partial p} F(q, p), \qquad (5.1)$$

---

[1] Throughout this chapter, to simplify the notation, quantum operators will be represented as dynamical variables, i.e., without the *hat* symbol on top (e.g., O instead of $\hat{O}$). Depending on the context, the reader will be able to identify easily whether a given symbol is acting either as a variable or as an operator.

where $F(q, p)$ is a friction force. Then, the Poisson bracket $\{A, H\}$ is replaced by the commutator of the two operators A and H according to the correspondence rule

$$\{A, H\} \quad \longrightarrow \quad \frac{1}{i\hbar}[A, H]. \tag{5.2}$$

This leads to the well-known Heisenberg equation,

$$i\hbar\frac{dA}{dt} = [A, H] + i\hbar\frac{\partial A}{\partial p}F(q, p), \tag{5.3}$$

where an appropriate symmetrization of the second term in the right-hand side should be taken into out. Furthermore, if A is considered to be q and p, from the time derivative of $[q, p] = i\hbar$ and (5.3), it follows that

$$[q, F] = 0 \tag{5.4}$$

and therefore F can only be a function of q.

As stated by some authors, quantum mechanics deals with operators following an "awkward" noncommutative algebra. This algebra is also followed for the same operator at different times. This fact is crucial when trying to quantize a dynamical system from a classical description. This problem is known as the time ordering problem. The ordered operator algebra allows us to associate c-number functions with them and transform quantum problems to equivalent "classical" or c-number problems where theoretical manipulation is easier. This is very much related to the correspondence principle where the prescription is to carry out the limiting procedure, $\hbar \rightarrow 0$, and/or by considering Ehrenfest's theorem.

An alternative route to quantization is the so-called Yang–Feldman method [7] in quantum field theory, where the idea is to quantize the equation of motion (or its solution) directly in the Heisenberg picture instead of starting from the canonical formalism. This starting point was postulated and used by Vineyard [33], for example, within the Langevin equation formalism in the scattering of slow neutrons by liquids.

Since noise is a time-dependent (stochastic) process, its autocorrelation function at two different times gives its spectrum of frequencies (see Appendix B). The quantum treatment was originally developed by Callen and Welton starting from the classical Nyquist–Shannon sampling theorem [34, 35]—though credit should also be given to Whittaker [36] and Kotel'nikov [37]. However, a better treatment was carried out by different authors by taking into account the time ordering of the corresponding operators. In general, as in the classical case, and in presence of environments, when the quantum equations of motion are considered without averaging, one finds quantum fluctuations. Dissipation and fluctuations have the same physical origin and it is due to the coupling with the environment. Both phenomena are related by the fluctuation-dissipation theorem [38]. A new aspect to deal with is decoherence or the transition from a pure state into a mixed one again due to the coupling to the environment.

5.2.1 Time-Dependent (Effective) Hamiltonians

An effective description of dissipation can be achieved through explicitly time-dependent Hamiltonians. As mentioned in Chap. 2, one of the former Hamiltonian models is the so-called Caldirola–Kanai (CK) Hamiltonian [39, 40],

$$H_{CK} = \frac{\bar{p}^2}{2m} e^{-\gamma t} + V(\bar{q}) e^{\gamma t}, \tag{5.5}$$

which comes from the Lagrangian

$$L_{CK} = \left[\frac{\dot{\bar{q}}^2}{2m} - V(\bar{q}) \right] e^{\gamma t}, \tag{5.6}$$

with the canonical variables

$$\bar{p} = p e^{\gamma t} = m \dot{q} e^{\gamma t}, \qquad \bar{q} = q. \tag{5.7}$$

The transformation from (q, p) (physical variables) to $(\bar{q}, \bar{p})$ is non-canonical. Originally, it was seen as a system with variable mass, $m(t) = m e^{-\gamma t}$, when working with the physical variables. As said in Chap. 2, this Hamiltonian leads to the correct equation of motion in configuration space, but not in phase space. The damped harmonic oscillator, where $V(q) = m\omega_0^2 q^2 / 2$, represents the paradigmatic example. It is straightforward to show that

$$[\bar{q}, \bar{p}] = i\hbar e^{\gamma t}, \tag{5.8}$$

which fulfills the commutation relation only at $t = 0$. As a consequence, the uncertainty principle is violated. Even more, the Schrödinger equation with H_{CK} applied to the harmonic oscillator leads to the so-called *loss-energy states*. However, such a Hamiltonian is not the energy of the system and it does not give the correct equation for the time-evolution of the momentum operator [6, 41–43].

Several attempts to circumvent such problems can be found in the literature. For example, for the damped harmonic oscillator, Schuch [44] has shown that there is a canonical transformation between H_{CK} and the harmonic oscillator Hamiltonian,

$$H_Q = \frac{P^2}{2m} + \frac{1}{2} m\Omega^2 Q^2, \tag{5.9}$$

in terms of what is called expanding variables,

$$Q = \bar{q} e^{\gamma t}, \tag{5.10a}$$

$$P = \bar{p} e^{-\gamma t} + m\gamma \bar{q} e^{\gamma t}, \tag{5.10b}$$

with $\Omega = \sqrt{\omega_0^2 - \gamma^2/4}$. In the underdamped motion, $\omega_0 \gg \gamma$ and thus Ω is real. Obviously, with these new coordinates, the uncertainty principle is not violated since

$[Q, P] = i\hbar$. Equation (5.9) represents the initial total energy and the correct classical equation for the corresponding damped harmonic oscillator is obtained in terms of q. Schuch also showed that H_Q can be obtained from a non-canonical transformation between (q, p) and (Q, P) from

$$H_{qp} = \left(\frac{p^2}{2m} + \frac{1}{2}m\omega_0^2 q^2 + \frac{1}{2}\gamma\, qp \right) e^{\gamma t}. \tag{5.11}$$

This Hamiltonian, which yields the correct equation of motion, expresses the fact that the initial energy is dissipated into an implicit environment.

The Schrödinger equation for $\Psi(Q, t)$ is formally written as

$$i\hbar\frac{\partial\Psi(Q, t)}{\partial t} = H_Q\Psi(Q, t), \tag{5.12}$$

carrying out the quantum calculation in the expanding system. Even more, the corresponding Ehrenfest equations for $\langle P \rangle$ and $\langle Q \rangle$ are also fulfilled. A non-unitary transformation to the physical variables or, in particular, to $\Psi(q, t)$ can finally be performed to reach observables. As will be shown later on, this formalism can be related to a logarithmic nonlinear Schrödinger equation.

Coherent and squeezed states of a damped harmonic oscillator have also been described fulfilling the uncertainty principle and the usual commutation relations [6] as well as using Dekker's Hamiltonian [45]. In the case of a general interaction potential, $V(q)$, and many particles, it should be stressed that it is better to proceed by following the dynamics of a conservative system, i.e., using system-plus-environment Hamiltonians. This approach is much more natural and successful for such purposes.

5.2.2 Nonlinear Hamiltonians

In the sixties, Senitzky [46] and Ford et al. [47] showed that a system of coupled harmonic oscillators could model a heat bath. The corresponding Brownian motion was then studied both classically and quantum-mechanically. In this model, a Langevin equation was obtained as a result of reducing the dimensionality of the full problem by tracing out over the bath variables. The heat bath was thus only described by two parameters: the friction coefficient and the temperature through a random force. This random force was shown to be a Gaussian stochastic process. The variables in the Langevin equation were assumed to be operators with order preserved according to the Heisenberg picture. Different orderings are available, such as the symmetric rule, Weyl's rule, etc., in order to construct Hermitian operators. The normal product of the random force was shown to be a Gaussian process, but not Markovian. This point is critical when compared to the classical counterpart. For an Ohmic or constant friction coefficient, the classical Gaussian process is also Markovian. The denomination of Ohmic friction comes from the fact that for a constant field of force F, the average value of the momentum reads like Ohm's law from the Langevin equation, i.e., $\langle p \rangle = \lambda^{-1} F$, where the current described by $\langle p \rangle$ is proportional to the field applied, F, and proportionally inverse to the resistance or friction coefficient.

Kostin [48, 49] derived the so-called Schrödinger–Langevin equation for a Brownian particle interacting with a thermal bath. The random force was assumed to arise from a random potential linearly dependent on the particle position, V_r. In one dimension this equation reads as

$$i\hbar \frac{\partial \Psi(q,t)}{\partial t} = -\frac{\hbar^2}{2m} \frac{\partial^2 \Psi(q,t)}{\partial q^2} + [V(q) + V_r(q,t)]\Psi(q,t) + K(\Psi(q,t)), \quad (5.13)$$

where

$$K(\Psi(q,t)) = \frac{\hbar \gamma}{2im} \ln\left[\frac{\Psi(q,t)}{\Psi^*(q,t)}\right] \quad (5.14)$$

is known as the *energy dissipation operator*. This nonlinear wave equation has been applied to several problems, such as the damped harmonic oscillator and the motion of a charged particle in the presence of damping while it is moving in an external electromagnetic field [6]. Because of the nonlinearity of this equation, the superposition principle does not hold and a general, unique solution cannot be found. Indeed, Hasse [50] showed that Kostin's Hamiltonian is a special case of a more general nonlinear Hamiltonian,

$$H = T + V + \gamma W, \quad (5.15)$$

where T and V are the usual kinetic and potential energy operators, and W satisfies the requirements: (i) $\langle W \rangle = 0$ and (ii) $\langle p \rangle = \partial W/\partial q$ according to Ehrenfest's theorem. The second requirement can also be fulfilled for a number of different W operators. For example, Süssmann's Hamiltonian is a special case, originally found by this author in a completely empirical way when studying the force-free motion of wave packets traveling along classical damped paths. W can also be expressed as

$$W = \int_0^q \dot{q}^n dq, \quad (5.16)$$

which, with the usual quantization rule for the momentum, becomes a differential operator of order n. The corresponding Hamiltonian is, in general, complex and non-Hermitian. For linear damping ($n = 1$), the interaction potential becomes complex with negative imaginary part, i.e., $V(q) - i\lambda\hbar/m$. This potential, called the *optical potential*, is in general nonlocal and has been widely used in atomic and nuclear physics (for example, in Feshbach's theory). Velocity dependent interactions have also been introduced in nuclear scattering and band theory of solids.

One can also follow Schrödinger's procedure to generate a wave equation. As shown in Sect. 3.2.1, from the Hamilton–Jacobi equation for the action S and momentum $p = \partial S/\partial q$,

$$\frac{\partial S}{\partial t} + H(q, \partial S/\partial q, t) = 0, \tag{5.17}$$

and introducing the wave function Ψ through the relation

$$S(q, t) = -i\hbar \ln(\Psi(q, t)), \tag{5.18}$$

the continuity equation for $\rho = \Psi^*\Psi$ reads as

$$\frac{\partial \rho}{\partial t} + \frac{1}{m} \nabla (\rho \, \text{Re}\{p\}) = 0, \tag{5.19}$$

where $\text{Re}\{p\}$ stands for the real part of p, which is in general a complex quantity if Ψ is complex, such that

$$\text{Re}\{p\} = -\frac{i\hbar}{2m} \frac{\partial \ln (\Psi/\Psi^*)}{\partial q},$$
$$\text{Im}\{p\} = -\frac{i\hbar}{2m} \frac{\partial (\ln \rho)}{\partial q}, \tag{5.20}$$

with the mean value of the imaginary part always vanishing. Thus, the expression for p does again fulfill Hasse's second requirement. The Kostin energy dissipation operator can then be expressed as

$$W = -\frac{i\hbar\gamma}{2} \left[\ln\left(\frac{\Psi}{\Psi^*}\right) - \left\langle \ln\left(\frac{\Psi}{\Psi^*}\right)\right\rangle \right]. \tag{5.21}$$

For a damped harmonic oscillator, the solutions contain the undamped frequency ω instead of the damped or reduced frequency Ω. Even more, the density ρ verifies the reversible continuity equation for a system displaying damping which follows an irreversible dynamics. This contradiction was avoided by Schuch et al. [51] by introducing a diffusion term in the continuity equation arriving at the Fokker–Planck equation,

$$\frac{\partial \rho}{\partial t} + \frac{1}{m} \nabla(\rho \, \text{Re}\{p\}) - D\frac{\partial^2 \rho}{\partial q^2} = 0, \tag{5.22}$$

with the additional condition

$$-\frac{D}{\rho} \frac{\partial^2 \rho}{\partial q^2} = \gamma(\ln \rho - \langle \ln \rho \rangle) \tag{5.23}$$

to be satisfied in order to achieve separation of the two equations for the amplitudes Ψ and Ψ^*. The mean value on the right-hand side guarantees the normalization.

The so-called logarithmic nonlinear Schrödinger equation is written as

$$i\hbar \frac{\partial \Psi}{\partial t} = [H - i\hbar\gamma(\ln \rho - \langle \ln \rho \rangle)] \, \Psi, \tag{5.24}$$

where Hasse's second requirement comes from considering the irreversible diffusion term. As is well known, the damping motion described by the Fokker–Planck equation can always be expressed in terms of its corresponding Langevin equation [52].

An extension to the nonlinear Schrödinger equation can be stated by writing [53]

$$i\hbar\frac{\partial\Psi}{\partial t} = H\Psi + i\hbar DG(\Psi), \tag{5.25}$$

where $H = T + V$ and the nonlinear term is given by

$$G(\Psi) = \nabla^2\Psi + \frac{|\nabla\Psi|^2}{|\Psi|^2}\Psi. \tag{5.26}$$

The continuity equation is then modified to a Fokker–Planck equation according to

$$\frac{\partial\rho}{\partial t} + \nabla\cdot\mathbf{J} = D\nabla^2\rho, \tag{5.27}$$

where $\mathbf{J}$ is the usual quantum probability current density.

A phenomenological nonlinear wave equation with complex interaction was also proposed by Gisin [54, 55] to account for decaying states. This wave equation is

$$i\hbar\frac{\partial\Psi}{\partial t} = \left(1 - \frac{i\kappa}{2}\right)H\Psi + \frac{i\kappa}{2}\langle\Psi|H|\Psi\rangle\Psi, \tag{5.28}$$

where H is the usual Hamiltonian for the undamped system and κ is a dimensionless positive and real damping constant. This wave equation presents some advantages, such as: the norm is independent of time, it reduces to Schrödinger's equation when Ψ is an eigenstate of H, the rate of change of the energy expectation value is negative definite and the equation of motion for the damped harmonic oscillator is obtained in terms of $\langle q\rangle$.

Finally, it is worth mentioning that Razavy [56], Wagner [57] and Schuch [58] have shown a connection between the Caldirola–Kanai Hamiltonian and the log-nonlinear Schrödinger equation following Schrödinger's quantization procedure.

5.2.3 The System-plus-Environment Hamiltonian

This approach starts from a total conservative system, formed by a physical system of primary interest coupled to an environment. As it has been shown in Chap. 2, there are several models where the total Hamiltonian is expressed as a sum of Hamiltonians for the physical system of interest, the environment and coupling between them. The quantization procedure is mainly carried out in the Heisenberg picture and, therefore, within the quantum Langevin framework. As previously mentioned, Senitzky [46] and Ford et al. [47] showed that a system of coupled harmonic oscillator can model a heat bath. Different models of heat bath (assuming a linear coupling to the physical system) have been proposed in the literature [59] and probably the most popular ones are: the rotating wave model [8], Ullersma's model [6] and the Caldeira–Leggett (CL) model [60]. Ullersma's model was discussed previously by Magalinskii [3] and, in the path integral formulation of quantum mechanics, also by Feynman and Vernon [61], Dissipative dynamics in tunneling processes have been widely studied in the path-integral context [3, 60, 62]. Using the CL model with Ohmic friction, Yu and Sun [63, 64] also calculated the wave function of the composite system in the Schrödinger picture. In classical mechanics the random force can be neglected at zero temperature, but quantum-mechanically this force is always present due to the zero point motion. They also showed the connection between the CL and the CK model Hamiltonians when the quantum fluctuation is neglected. Even more, the dissipation was to be suppressed the spreading of the free wave packet if the breadth of the initial wave packet was so wide that the effect of the random force could be ignored.

On the other hand, the optical potential can also be derived from a conservative approach. This case has been widely studied in nuclear Feshbach theory [65, 66], where inter-particle forces are assumed to be short-ranged (exponential dependence). Special emphasis has to be given on decaying systems with no classical analogues. The quantum theory of line width using the so-called Wigner–Weisskopf model [8] is the paradigmatic case. Finally, if a chemical or physical problem is not easily describable by a system-plus-environment Hamiltonian or is simply unknown, a phenomenological method can be applied based on the so-called Pauli master equation [12], which contains transition rates between time-dependent occupation probabilities. It describes an incoherent motion of the physical system since the coupling with the environment becomes predominant.

In what follows, the interest will focuss on conservative approaches where theoretical formalisms have been widely developed. Since the main goal of this monograph is to present a trajectory view of quantum mechanics, only those formalisms that can provide a trajectory image of physical processes will be considered.

5.3 Conservative Approach to Dissipative and Stochastic Dynamics

As previously mentioned, the whole system (physical system and bath) is considered as an isolated system. Therefore, the powerful standard quantum-mechanical theoretical formalisms developed so far are at our complete disposal. Dissipation can be described following one of the three standard pictures of quantum mechanics: Schrödinger, Heisenberg and interaction. In the first and third pictures, one always tries to find master equations that account for the time-evolution of the so-called reduced system, where the bath degrees of freedom have been traced out. Among this type of equations of motion, the simplest class is that of Markovian character where one assumes that the bath has no memory and the time-evolution of the reduced density matrix depends only on its present time. An alternative way to the density matrix formalism comes from the path integral formulation of quantum mechanics. Many efforts are being addressed along this way with also great success. If the Heisenberg picture is followed, the quantum (generalized or standard) Langevin equation is reached (needless to say that all formulations are equivalent). Essentially, this procedure consists of replacing the reservoir by damping terms in the Heisenberg equations of motion of a conservative system and then adding random forces as driving terms that give rise to fluctuations over the system. These stochastic quantum formalisms are being addressed to more and more complicated systems. There are recent review works [67, 68] and books [2, 3, 5] which can provide the reader with a more detailed overview of the field.

Hamiltonians describing system-plus-environment (or system-plus-bath) interactions are generally expressed as

$$H = H_S + H_B + H_{SB} = H_0 + H_{SB}, \qquad (5.29)$$

where H_S and H_B describe, respectively, the free evolution of the system and the reservoir or bath. In the literature, it is commonly assumed that the environment consists of a large or infinite collection of independent harmonic oscillators. Different damping mechanisms may require different forms or models for H_B. However, the results should not be in general very sensitive to the particular model chosen. In the analytical treatments presented below, the simplest model is considered: an infinite collection of non-interacting harmonic oscillators in thermal equilibrium at a temperature T. This model is generally used to describe a reservoir consisting of phonon or photon fields [3, 5]. Since H_S and H_B deal with different sets of degrees of freedom (the system and environment subspaces do not overlap), they commute. H_{SB} in (5.29), on the other hand, describes the system-environment coupling. Sometimes, when the environment dynamics is not relevant for the solution of the problem, but only its effects over the system, one can conveniently express the system–reservoir interaction as given by the right-hand side of the second equality in (5.29). Then, within a perturbation scheme, $H_0 \equiv H_S + H_B$ would correspond to the zero order Hamiltonian. In general, the system–reservoir coupling term, H_{SB}, is assumed to be initially "turned off" at $t = 0$. In scattering problems, this condition holds since one can assume that the incoming particle and the target are not interacting at $t \to -\infty$. In many cases the system and reservoir are in continuous contact (e.g., a diatomic solute in a liquid phase solvent) and one can make use of such a hypothesis only in the weak coupling limit (see Appendix B). When this limit does not hold, the trace operation over the bath variables leading to the master equation for the reduced density matrix is questionable since the entanglement prevents it. The role of initial conditions has largely been discussed in many works since it is critical in the time-evolution of open quantum systems [69–71].

5.3.1 The Langevin Formalism

When working in the Heisenberg picture, one readily reaches the *generalized quantum Langevin* equation. The reservoir is then completely eliminated by incorporating suitable quantum noise operators, which act as driving terms in the motion equation leading to fluctuations in the system. Thus, the equation of motion of a general system operator A in the Heisenberg picture reads as

$$\dot{A}(t) = -\frac{i}{\hbar} [A(t), H_S + H_{SB}]. \tag{5.30}$$

After some convenient manipulations, this equation can be transformed into a generalized, non-Markovian Langevin equation [3]. In doing so, (5.30) becomes [8]

$$\dot{A}(t) = -\frac{i}{\hbar}[A(t), H_S] - \int_0^t \gamma_A(t')A(t - t') \, dt' + G_A(t). \tag{5.31}$$

As it can be noticed, in this expression there is a formal separation of the total force into three different components: a driving force, a systematic or dissipative force with retardation effects characterized by the kernel γ_A, and a random force or quantum noise, G_A, also called the *random operator Langevin noise source*. Noise sources are always chosen such that their reservoir averages are zero. Moreover, according to the fluctuation-dissipation theorem, the spectral density of the time autocorrelation function of the random force is related to the kernel in the frequency domain—actually, as is well-known, this theorem states that the dissipation force is actually determined by the random force autocorrelation function. The memory kernel, which gives the dissipation coefficient, represents the resistance or impedance of the system by means of which external work is dissipated into thermal energy. In the Markovian approximation, the relaxation time of A is much larger than reservoir correlations, and (5.31) becomes

$$\dot{A}(t) = -\frac{i}{\hbar}[A(t), H_S] - \gamma_A A(t) + G_A(t), \tag{5.32}$$

where γ_A is the dissipation coefficient. This general formalism, which has been applied to different physical processes, can also be extended to a c-number Langevin formalism known as the quasi-classical version of the Langevin equation due mainly to Senitzky [46] and Schmid [72]. Furthermore, a very interesting and active line of research in this context is Kramers' turnover problem in the quantum domain [73, 74]. In particular, the strong friction limit described by the so-called quantum Smoluchowski equation has also attracted a lot of interest [75].

Now, consider the Caldeira–Leggett Hamiltonian model. The role of initial conditions has also been largely discussed in this model since some problems when solving the Langevin equation can appear [76, 77]. This model will be applied here to describe some simple physical processes, such as atom diffusion on flat surfaces and the vibrational relaxation of a particle adsorbed on a surface or adsorbate, since both problems lead to full analytical descriptions [78]. Within this context, space–time correlation functions play a key role, since they are used to describe the decay of spontaneous thermal fluctuations at surfaces, this being central to the study of transport phenomena. These functions are defined as the thermodynamic average of the product of two dynamical variables, each one expressing the instantaneous deviation from its corresponding equilibrium value at particular points on the surface and time. A complete description of the particle dynamics in a many-body system is then reached when the behavior of the corresponding correlation functions over the entire wavenumber range is studied. This range splits into different characteristic regions, each one associated with a different set of properties of the system. In the case of scattering experiments, since the momentum and energy transfers of the probe particles are the relevant quantities, any correlation-function-based theory has to be developed necessarily in terms of such quantities. Space–time correlation functions can also be used to describe the linear response of a fluid under a weak, external perturbation.

Thus, to start with, consider the so-called *differential reflection coefficient*,

$$\frac{d^2 \mathcal{R}(\Delta \mathbf{K}, \omega)}{d\Omega d\omega} = n_d \mathcal{F} S(\Delta \mathbf{K}, \omega). \tag{5.33}$$

In analogy to scattering of slow neutrons by crystals and liquids [79, 80], this quantity constitutes the observable magnitude in surface diffusion experiments. More specifically, this coefficient gives the probability that the probe particles (usually He atoms) scattered from the interacting adsorbates on the surface reach a certain solid angle Ω with an energy exchange $\hbar\omega = E_f - E_i$ and wave vector transfer parallel to the surface $\Delta \mathbf{K} = \mathbf{K}_f - \mathbf{K}_i$. In (5.33), n_d is the concentration of adparticles; $\mathcal{F}$ is the *atomic form factor*, which depends on the interaction potential between the probe atoms in the beam and the adparticles on the surface; and $S(\Delta \mathbf{K}, \omega)$ is the dynamic structure factor, which provides information about diffusion and low frequency vibrational relaxation. Experimental information about long distance correlations is obtained from the dynamic structure factor when considering small values of $\Delta \mathbf{K}$, while information on long time correlations is provided at small energy transfers, $\hbar\omega$.

Pair distribution functions are usually given in terms of the so-called van Hove or time-dependent pair correlation function $G(\mathbf{R}, t)$. This function is related to the dynamic structure factor by a double Fourier transform—in space and time—, as

$$S(\Delta \mathbf{K}, \omega) = \frac{1}{2\pi \hbar N} \iint G(\mathbf{R}, t) e^{i(\Delta \mathbf{K} \cdot \mathbf{R} - \omega t)} \, d\mathbf{R} \, dt. \tag{5.34}$$

Given an adparticle at the origin at some arbitrary initial time, $G(\mathbf{R}, t)$ represents the average probability to find a particle (the same or another one) at the surface position $\mathbf{R} = (x, y)$ at a time t. This function thus generalizes the well-known pair distribution function $g(\mathbf{R})$ from statistical mechanics, since it provides information about the interacting particle dynamics.

Adsorbate position operators are given, in general, by the respective Heisenberg operators (defined for all $j = 1, \ldots, N$ adparticles and time t),

$$\mathbf{R}_j(t) = e^{iHt/\hbar}\mathbf{R}_j e^{-iHt/\hbar}, \tag{5.35}$$

where H is the Hamiltonian of the total system. The space Fourier transform of the G-function is the intermediate scattering function,

$$I(\Delta\mathbf{K}, t) = N \int \int G(\mathbf{R}, t)e^{i\Delta\mathbf{K}\cdot\mathbf{R}}d\mathbf{R} = \frac{1}{N}\langle\rho_{\Delta\mathbf{K}}(t)\rho^{\dagger}_{\Delta\mathbf{K}}(0)\rangle_\beta, \tag{5.36}$$

where the $\rho_{\Delta\mathbf{K}}$ operator defined as

$$\rho_{\Delta\mathbf{K}}(t) = \sum_{j=1}^{N} e^{-i\Delta\mathbf{K}\cdot\mathbf{R}_j(t)} = \rho^{\dagger}_{-\Delta\mathbf{K}}(t) \tag{5.37}$$

is the Fourier component of the adsorbate number density operator,

$$\rho(\mathbf{R}, t) = \frac{1}{\sqrt{N}} \sum_{j=1}^{N} \delta(\mathbf{R} - \mathbf{R}_j(t)). \tag{5.38}$$

In (5.36) the brackets denote the ensemble average over the trajectories associated with each adsorbate $\mathbf{R}_j(t)$. The intermediate scattering function is the typical observable issued from He and neutron spin-echo experimental techniques. From (5.34)–(5.38), it is seen that the dynamic structure factor can be expressed in terms of a density–density correlation function and determined by the spectrum of the spontaneous fluctuations. Moreover, the *static structure factor*, defined as $S(\Delta\mathbf{K}, t = 0)$, is related to $g(\mathbf{R})$, which describes the instantaneous correlation between adsorbates.

Due to the quantum character of the different operators introduced above, several comments are worth stressing. First, $\rho_{\Delta\mathbf{K}}(t)$ and $\rho^{\dagger}_{\Delta\mathbf{K}}(0)$ commute only at $t = 0$. Second, the system studied here is assumed to be stationary and, therefore, the origin of time is arbitrary for the correlation function associated with the density operators. Third, the complex character of the corresponding correlation function is a signature of the quantum dynamics of the interacting system. Fourth, the G-function is also complex, but the dynamic structure factor is real and positive definite because it represents a cross-section. More properties of the $\rho_{\Delta\mathbf{K}}(t)$ operator, the G-function and the dynamic structure factor can be found in Lovesey's book [80]. And fifth, the so-called *detailed balance principle* can be expressed as

$$S(\Delta\mathbf{K}, \omega) = e^{\hbar\omega\beta} S(-\Delta\mathbf{K}, -\omega), \tag{5.39}$$

with $\beta = 1/k_B T$ and k_B Boltzmann's constant, which expresses that the probability that a He atom loses an energy $\hbar\omega$ is equal to $e^{\hbar\omega\beta}$ times the probability that a He atom gains an energy $\hbar\omega$.

After van Hove [79], if R_0 is the range of the G-function and T_0 its relaxation time, $\hbar/R_0$ and $\hbar/T_0$ determine the orders of magnitude of average momentum and energy transfers in the scattering process of the probe particles, which for light masses display the observable recoil effect. Thus, the time variation of G affects the total scattering and angular distributions only for a particle spending at least a time of order T_0 over a correlation length R_0. Moreover, if the mean de Broglie wavelength, Γ, defined in Chap. 1, is small compared to inter-adparticle distances or the range of adsorbate–adsorbate interaction, no quantum effect will manifest in the G-function, which deals with pairs of adparticles separated by distances of the order of R_0. Nevertheless, for small timescales ($t \ll T_0$ or $t \sim \hbar\beta$), the dynamics entirely concentrates on a region of the order of or less than Γ, and quantum effects are noticeable. This time could be considered as the coherence time, afterwards the diffusion process starts to be decoherent. The imaginary part of the G-function is greater at small values of time.

The dynamic structure factor can also be related to the system linear response function [80],

$$\phi(\Delta\mathbf{K}, t) = \frac{i}{\hbar N}\, \langle[\rho_{\Delta\mathbf{K}}(t), \rho_{\Delta\mathbf{K}}^{\dagger}]\rangle, \tag{5.40}$$

through the fluctuation-dissipation theorem, as

$$S(\Delta\mathbf{K}, \omega) = \frac{1}{2\pi i}\, [1 + n(\omega)] \int_{-\infty}^{\infty} e^{i\omega t}\, \phi(\Delta\mathbf{K}, t)\, dt, \tag{5.41}$$

where $1 + n(\omega) = [1 - \exp(-\hbar\omega\beta)]^{-1}$ is the detailed balance factor, with $n(\omega)$ being the Boltzman factor. Equation (5.41) links the spectrum of spontaneous fluctuations, $S(\Delta\mathbf{K}, \omega)$, to the dissipation part of the response function. The time derivatives of the response function are related to the Heisenberg equation of motion (3.50) of the $\rho_{\Delta\mathbf{K}}$ operator; moments of the dynamic structure factor involve nested commutators to evaluate them. The ϕ-function is a causal function because it cannot be defined before the external perturbation has been switched on. For scattering with He atoms, the perturbation is assumed to start at $-\infty$ and finish at $+\infty$, having typically a bell shape. In (5.41), the time Fourier transform of ϕ defines a *generalized susceptibility function*, $\chi(\Delta\mathbf{K}, \omega)$ and, therefore, can be again expressed as

$$S(\Delta\mathbf{K}, \omega) = -i\, [1 + n(\omega)]\, \chi(\Delta\mathbf{K}, \omega). \tag{5.42}$$

This susceptibility is complex and the real and imaginary parts are related through the well-known Kramers–Kroning or dispersion relations [80].

In order to go a step further into the dynamics, a Hamiltonian has to be specified. In surface diffusion, the full system + bath Hamiltonian is usually written [81] as

$$H = \frac{p_x^2}{2m} + \frac{p_y^2}{2m} + V(x, y)$$

$$+ \sum_{i=1}^{N} \left[\frac{p_{x_i}^2}{2m_i} + \frac{m_i}{2} \left(\omega_{x_i} x_i - \frac{c_{x_i}}{m_i \omega_{x_i}} x \right)^2 \right]$$

$$+ \sum_{i=1}^{N} \left[\frac{p_{y_i}^2}{2m_i} + \frac{m_i}{2} \left(\omega_{y_i} y_i - \frac{c_{y_i}}{m_i \omega_{y_i}} y \right)^2 \right],$$

$$(5.43)$$

where (p_x, p_y) and (x, y) are the adparticle momenta and positions with mass m; and (p_{x_i}, x_i) and (p_{y_i}, y_i) with $i = 1, \cdots, N$ are the momenta and positions of the bath oscillators (phonons), with mass and frequency given by m_i and ω_i, respectively. Phonons with polarization along the z-direction are not considered. The Hamiltonian was originally considered by Magalinskii and Caldeira and Leggett, who used it for weak and strong dissipation (a general discussion about the Hamiltonian (5.43) can be found in [3]). In surface diffusion, $V(x, y)$ is in general a periodic function describing the surface corrugation at zero temperature. The harmonic frequencies of the bath modes and the coupling coefficients are expressed in terms of spectral densities, defined as

$$J_i(\omega) = \frac{\pi}{2} \sum_{j=1}^{N} \frac{c_{i_j}^2}{m_j \omega_{i_j}^2} \left[\delta(\omega - \omega_{i_j}) \right],$$

$$(5.44)$$

with $i = x, y$. These densities enable the passage to a continuum model.

In the Heisenberg picture, the time-evolution of the position operators is given by a generalized Langevin equation for each system coordinate

$$m\ddot{x}(t) + m \int_0^t \gamma_x(t - t') \dot{x}(t') \, dt' + \frac{\partial V(x, y)}{\partial x} = N_x(t),$$

$$(5.45a)$$

$$m\ddot{y}(t) + m \int_0^t \gamma_y(t - t') \dot{y}(t') \, dt' + \frac{\partial V(x, y)}{\partial y} = N_y(t),$$

$$(5.45b)$$

where the associated friction functions are defined through the cosine Fourier transform of the spectral densities,

$$\gamma_i(t) = \frac{2}{\pi m} \int_0^{\infty} \frac{J_i(\omega)}{\omega} \cos \omega t \, d\omega,$$

$$(5.46)$$

with $i = x, y$. The nonhomogeneity of (5.45) represents a fluctuating force which depends on the initial position of the system and initial positions and momenta of the oscillators of each bath (see Eq. (2.78) of Chap. 2) [3]. For each Cartesian component of the noise, it can be easily shown that its equilibrium (canonical ensemble) expectation value with respect to the heat bath including the corresponding bilinear coupling to the system vanishes. On the contrary, the noise autocorrelation function (each Cartesian component) is a complex quantity because in general it does not commute at different times. In the classical limit $\hbar \to 0$, each noise correlation reduces to $mk_BT\,\gamma_i(t)$, with $i = x, y$. For Ohmic friction, $\gamma_i(t) = 2\,\gamma_i\,\delta(t)$, where γ_i is a constant and $\delta(t)$ is Dirac's δ-function. Under the assumption of Ohmic friction, it can be shown that noise in this model is *white*. The paradigm of this type of noise is the *Gaussian white noise*. Dealing with large systems (the surface seen as a thermal bath) where the number of collisions between substrate and adsorbate is very high, one of the fundamental theorems of the theory of probability, namely the central limit theorem, ensures that the fluctuations of the bath will be Gaussian distributed. Diffusion can then be described by a Brownian-type motion involving a continuous Gaussian stochastic process (see Appendix B). In virtue of the fluctuation-dissipation theorem, such fluctuations can be related to the friction coming mainly from surface phonons: the phonon friction. Electronic friction due to low-lying electron–hole pair excitations is usually neglected in most of cases. Moreover, quantum mechanically [3], for Ohmic friction the imaginary part of each noise function is a step function and its real part goes with $\mathrm{csch}^2(\pi t/\hbar\beta)$. Thus, at zero surface temperature, the noise is still correlated even for long time (it decays as t^{-2}) in contrast to the classical case. These facts give rise to important differences with respect to the classical case such as, for example, the noise and the system coordinates are correlated instead of being zero. A detail study of surface diffusion at very low (or even zero) temperatures as well as the role of the measurement process (Zeno and anti-Zeno effects) will be given in Volume 2.

In order to simplify this theoretical treatment, only classical noise will be considered, though keeping in mind that the quantum results will be only valid for not too low surface temperatures. Moreover, the motion of only one adsorbate is considered within the so-called single adsorbate approximation since, for very low coverage, adparticles are considered non-interacting. Thus, if Ohmic friction is assumed, Eqs. (5.45) reduce to two coupled standard Langevin equations[2] (Markovian approximation),

$$m\ddot{\mathbf{R}} = -m\,\gamma\,\dot{\mathbf{R}} - \mathbf{F}(\mathbf{R}) + \delta\mathbf{N}, \qquad (5.47)$$

and where $\delta\mathbf{N}$ is the fluctuation due to the lattice (thermal) vibrational effects which are simulated by a Gaussian white noise acting on the adparticle.

In general, an exact, direct calculation of $I(\Delta\mathbf{K}, t)$ or $S(\Delta\mathbf{K}, \omega)$ is difficult to carry out due to the noncommutativity of the adparticle position operators at different times obeying the Markovian Langevin equation (5.47) and for a nonseparable interaction potential. However, for certain simple cases, closed formulas can be easily obtained [78]. The product of the two exponential operators in (5.36) can be evaluated according to a special case of the Baker–Hausdorff theorem, namely the *disentangling theorem* [8]. If A and B are two noncommuting operators satisfying the condition $[A, [A, B]] = [B, [A, B]] = 0$, then $e^A e^B = e^{A+B}e^{[A,B]/2}$. In particular, this theorem holds when the commutator of A and B is a c-number. Thus, (5.36) can

---

[2] The δ-function counts only one half when the integration is carried out from zero to infinity.

be expressed as a product according to

$$I(\Delta\mathbf{K}, t) = I_1(\Delta\mathbf{K}, t)I_2(\Delta\mathbf{K}, t), \qquad (5.48)$$

which is a product of two quantum intermediate scattering functions $I_j(\Delta\mathbf{K}, t)$, with $j = 1, 2$, associated with the exponentials of the commutator $[A, B]$ and $A + B$, respectively. By identifying the operators A and B with $A = i\,\Delta\mathbf{K} \cdot \mathbf{R}(0)$ and $B = i\,\Delta\mathbf{K} \cdot \mathbf{R}(t)$, the factor I_1 involving their commutator will depend on the character of the dynamics; for classical dynamics, this factor is one. Within the so-called Gaussian approximation, the second factor can also be written as follows

$$I_2(\Delta\mathbf{K}, t) = \langle e^{-i\,\Delta\mathbf{K}\cdot[\hat{\mathbf{R}}(0)-\hat{\mathbf{R}}(t)]}\rangle = \langle e^{-i\,\Delta K \int_0^t \hat{v}_K(t')dt'}\rangle$$
$$\simeq e^{-\Delta K^2 \int_0^t (t-t')C_v(t')dt'}, \qquad (5.49)$$

where $C_v(t) = \langle v_{\Delta K}(t)v_{\Delta K}(0)\rangle$ is the velocity autocorrelation function along the direction given by $\Delta\mathbf{K}$ or the longitudinal direction. Equation (5.49) is exact if the velocity operator gives rise to a Gaussian stochastic process.

In the case of diffusion on flat or very low corrugated surfaces, no particular direction is privileged and the role of the adiabatic adsorbate–substrate interaction potential is negligible (one can assume $V(x, y) \approx 0$). Thus, only the action of the thermal phonons is relevant and the stochastic single-adparticle trajectories $\mathbf{R}(t)$ running on the surface obey the following Markovian Langevin equation (5.47),

$$m\ddot{\mathbf{R}}(t) = -m\,\gamma\,\dot{\mathbf{R}}(t) + \delta\mathbf{N}(t), \qquad (5.50)$$

Usually it is assumed that probe particles do not influence the surface dynamics, i.e., their influence can be considered a perturbation. Therefore, if the adsorbate motion is driven by the external force $F_e(t)$, then from (5.50) an average on a canonical ensemble leads to

$$\langle\ddot{x}(t)\rangle + \gamma\langle\dot{x}(t)\rangle = \frac{1}{m}\,F_e(t). \qquad (5.51)$$

Within the framework of the linear response theory, a particular solution of the corresponding differential equation is written as

$$\langle\tilde{x}(t)\rangle = \int_{-\infty}^{t} \phi(t - s)F_e(s)ds, \qquad (5.52)$$

or, after the time Fourier transform, as

$$\langle\tilde{x}(\omega)\rangle = \chi(\omega)\tilde{F}_e(\omega). \qquad (5.53)$$

The dynamic susceptibility is written as

$$\chi(\omega) = \frac{1}{m}\frac{1}{-\omega^2 - i\,\gamma\,\omega}, \qquad (5.54)$$

its time behavior being given by

$$\chi(t) = \frac{2}{m\gamma} e^{-\gamma t/2} \sinh(\gamma t/2) \Theta(t), \tag{5.55}$$

where $\Theta(t)$ is the step function due to causality. This expression, valid for both the classical and quantum case, is exact whenever an Ohmic friction γ is assumed and any direction given by $\Delta\mathbf{K}$ is considered.

In the Heisenberg representation, (5.50) still holds, its formal solution being

$$\mathbf{R}(t) = \mathbf{R}(0) + \frac{\mathbf{P}(0)}{m\gamma} \Phi(\gamma t) + \frac{1}{m\gamma} \int_0^t \Phi(\gamma t - \gamma t') \delta\mathbf{N}(t') dt', \tag{5.56}$$

where $\mathbf{P}(0)$ is the initial adparticle momentum operator and $\Phi(x) = 1 - e^{-x}$. The commutator between $\mathbf{R}(0)$ and $\mathbf{R}(t)$ is obtained from (5.56), which yields a c-number. Then, assuming a classical noise as previously mentioned (for quantum noise, the corresponding commutator is also a c-number since the noise function only depends on the initial position of the adsorbate, see Eq. (2.78) of Chap. 2 and Ref. [3]), the factor I_1 can be expressed as a time dependent phase

$$I_1(\Delta\mathbf{K}, t) = \exp\left[\frac{i\hbar\Delta\mathbf{K}^2}{2\gamma m}\Phi(\gamma t)\right] = \exp\left[\frac{iE_r}{\hbar}\frac{\Phi(\gamma t)}{\gamma}\right], \tag{5.57}$$

where $E_r = \hbar^2 \Delta\mathbf{K}^2/2m$ is the *adsorbate recoil energy*. As is apparent, the argument of the exponential function becomes less important as the adparticle mass and the total friction increase. The time-dependence only comes from $\Phi(\gamma t)$. At short times ($\lesssim \hbar\beta$), $\Phi(\gamma t) \approx \gamma t$ and the argument of I_1 becomes independent of the total friction, thus increasing linearly with time. On the other hand, in the asymptotic time limit, this argument approaches a constant phase.

The I_2 factor can be evaluated as follows. The fluctuation-dissipation theorem allows us to express the equilibrium position autocorrelation function, $C_x(t) = \langle x(t)x(0) \rangle$, in terms of the imaginary part of the dynamic susceptibility and, after Fourier transforming,

$$C_x(t) = \frac{\hbar}{\pi m} \int_{-\infty}^{+\infty} \frac{\gamma\omega}{\omega^4 + \gamma^2\omega^2} \frac{e^{-i\omega t}}{1 - e^{-\beta\hbar\omega}} d\omega. \tag{5.58}$$

From the relations

$$\frac{1}{1 - e^{-\beta\hbar\omega}} = \frac{1}{2} + \frac{1}{2}\coth(\beta\hbar\omega/2), \tag{5.59a}$$

$$\coth(\beta\hbar\omega/2) = \frac{2}{\beta\hbar\omega}\left(1 + 2\sum_{n=1}^{\infty} \frac{\omega^2}{v_n^2 + \omega^2}\right), \tag{5.59b}$$

where

$$\nu_n = \frac{2\pi n}{\hbar\beta} \tag{5.60}$$

are the so-called *Matsubara frequencies* [82], the correlation function can be split up into its symmetric and antisymmetric parts, $C_x(t) = S_x(t) + iA_x(t)$. For $t > 0$, these functions read as

$$S_x(t) = -\frac{1}{m\beta\gamma}\left(t\ \mathrm{sign}\{t\} + \frac{1}{\gamma}e^{-\gamma t}\right) + \frac{2}{\beta m}\sum_{n=1}^{\infty}\left(\frac{\gamma e^{-\nu_n t} - \nu_n e^{-\gamma t}}{\nu_n(\gamma^2 - \nu_n^2)}\right), \tag{5.61a}$$

$$A_x(t) = -\frac{\hbar}{2\gamma m}\left(1 - e^{-\gamma t}\right), \tag{5.61b}$$

which can be trivially related to (5.55) through the fluctuation–dissipation theorem. In (5.61a), the sign function of the real number t is defined as being $+1$ for $t > 0$ and -1 for $t < 0$. Now, since

$$C_v(t) = -\frac{d^2}{dt^2}C_x(t), \tag{5.62}$$

then

$$C_v(t) = \left(\frac{1}{\beta m} - \frac{i\hbar\gamma}{2m}\right)e^{-\gamma t} - \frac{2\gamma}{\beta m}\sum_{n=1}^{\infty}\frac{\nu_n e^{-\nu_n t} - \gamma e^{-\gamma t}}{\gamma^2 - \nu_n^2}, \tag{5.63}$$

with the real part being identical to the corresponding classical expression except for the infinite sum of Matsubara frequencies. Quantum effects are important at low surface temperatures, the long time behavior being mainly determined by the first term of the Matsubara series. In such cases, relaxation is no longer governed only by the damping constant [3]. Substituting now (5.63) into (5.49), the I_2 factor is finally obtained,

$$I_2(\Delta\mathbf{K}, t) = e^{-\Delta K^2[f(t)+g(t)]}, \tag{5.64}$$

where the time-dependent functions $f(t)$ and $g(t)$ are given by

$$f(t) = \left(\frac{1}{m\beta\gamma^2} - \frac{i\hbar}{2m\gamma}\right)[e^{-\gamma t} + \gamma t - 1], \tag{5.65a}$$

$$g(t) = \frac{2}{m\beta}\sum_{n=1}^{\infty}\frac{\nu_n e^{-\gamma t} - \gamma e^{-\nu_n t} + \gamma - \nu_n}{\nu_n(\gamma^2 - \nu_n^2)}. \tag{5.65b}$$

The total intermediate scattering function (5.48) can then be expressed as

$$I(\Delta\mathbf{K}, t) = e^{-\chi^2[\alpha^*\gamma t - \Phi(\gamma t)]}e^{-\Delta\mathbf{K}^2 g(t)}, \tag{5.66}$$

with $\chi = \Delta\mathbf{K}^2\langle v_0^2\rangle/\gamma^2$ and $\alpha = 1 + i\hbar\beta\gamma/2$, the thermal square velocity being $\langle v_0^2\rangle = 1/m\beta$. The recoil energy is included in the imaginary part of the product $\chi^2\alpha^*$, which disappears when $\hbar \to 0$. Equation (5.66) is the generalization of the intermediate scattering function for the quantum motion of non-interacting adsorbates in a flat surface. The dependence of this function on $\Delta\mathbf{K}^2$ through the shape parameter χ is the same as in the classical theory [78]. No previous information about the velocity autocorrelation function is needed. However, classically, the intermediate scattering function is usually obtained from Doob's theorem, which states that the velocity autocorrelation function for a Gaussian, Markovian stationary process decays exponentially with time. The ballistic or free-diffusion regime and the diffusive regime are apparent from (5.66). The first one is dominant at very low times, $\gamma t \ll 1$, and the second one at very long times, $\gamma t \gg 1$.

The diffusion coefficient can be obtained from the real part of the expression

$$D = \lim_{t\to\infty}\int_0^t C_v(t')dt', \tag{5.67}$$

which renders

$$D = \frac{k_B T}{m\gamma}, \tag{5.68}$$

and coincides with Einstein's law for the classical case (ensuring that the adparticle velocity distribution becomes Maxwellian asymptotically). The same result is reached from the MSD, $\langle[\mathbf{R}(t) - \mathbf{R}(0)]^2\rangle$, which takes into account only the symmetric part of the position autocorrelation function. Quantum fluctuations (in terms of the Matsubara frequencies) do not affect this result at low temperatures except the time limit to which the MSD is linear with time may become very large. At zero temperature, D is also zero and the MSD is no longer linear with time. The infinite sum of Matsubara frequencies determines now the long time limit behavior. As previously mentioned, the limit to zero surface temperature is questionable if the correlation between the noise and the coordinate system is neglected in the commutator. These issues will be discussed in more detail in Volume 2. Diffusion has also been treated quantum-mechanically from the viewpoint of Kramers' theory [74].

The harmonic model is an appropriate working model to understand the bound motion inside the wells of a corrugated surface. This motion comes precisely from the oscillating behavior undergone by the adparticle when the diffusive motion is temporarily frustrated. Now, the dynamic susceptibility will be that of an adparticle subject to a one-dimensional harmonic potential

$$\chi(\omega) = \frac{1}{m}\frac{1}{-\omega^2 - i\gamma\omega + \omega_0^2}, \tag{5.69}$$

where ω_0 is the frequency of the harmonic oscillator. This expression is again exact whenever an Ohmic friction γ is assumed and valid for both the classical and quantum cases. Thus, consider the formal solution of (5.47),

$$\mathbf{R}(t) = \mathbf{R}(0) + \frac{\mathbf{P}(0)}{m\gamma}\,\Phi(\gamma t) + \frac{1}{m\gamma}\int_0^t \Phi(\gamma t - \gamma t')[\mathbf{F}(\mathbf{R}(t')) \mid \delta\mathbf{N}_U(t')]dt', \quad (5.70)$$

where the force $\mathbf{F}$ is given by Hooke's law, $\mathbf{P}(0)$ is the initial adparticle momentum operator. The presence of the adiabatic force introduces an additional commutator, $[\mathbf{R}(0), \mathbf{F}(\mathbf{R}(t))] = i\hbar\partial\mathbf{F}(\mathbf{R}(t))/\partial\mathbf{P}(0)$, where the dependence of the adiabatic force on the initial state $(\mathbf{R}(0), \mathbf{P}(0))$ comes through $\mathbf{R}(t)$, which is negligible in a quantum Markovian framework. The factor I_1 is thus the same as for a flat surface, given by (5.57). On the other hand, in order to obtain the I_2 factor, one needs to start again from (5.47). The dynamic susceptibility is also given by (5.69) and its time behavior by

$$\chi(t) = \frac{1}{m\bar{\omega}}\,e^{-\gamma t/2}\sin\bar{\omega}t\,\Theta(t), \quad (5.71)$$

where $\Theta(t)$ is again the step function due to causality and

$$\bar{\omega} = \sqrt{\omega_0^2 - \frac{\gamma^2}{4}}. \quad (5.72)$$

According to the fluctuation-dissipation theorem, as before, the equilibrium position autocorrelation function can be expressed as

$$C_x(t) = \frac{\hbar}{\pi m}\int_{-\infty}^{+\infty}\frac{\gamma\omega}{(\omega^2 - \omega_0^2)^2 + \gamma^2\omega^2}\frac{e^{-i\omega t}}{1 - e^{-\beta\hbar\omega}}\,d\omega. \quad (5.73)$$

For $t > 0$, the symmetric and antisymmetric parts of (5.73) read as

$$S_x(t) = \frac{e^{-\gamma t/2}}{m\beta\bar{\omega}\omega_0^2}[\bar{\omega}\cos\bar{\omega}t + (\gamma/2)\sin\bar{\omega}t] - \frac{2}{\beta m}\sum_{n=1}^{\infty}\left[\frac{e^{-\nu_n t}}{(\gamma/2 - \nu_n)^2 + \bar{\omega}^2}\right], \quad (5.74a)$$

$$A_x(t) = -\frac{\hbar}{2m\bar{\omega}}\,e^{-\gamma t/2}\sin\bar{\omega}t, \quad (5.74b)$$

respectively, and the velocity autocorrelation function, as

$$C_v(t) = \frac{\omega_0}{m\beta\bar{\omega}}\,e^{-\gamma t/2}\cos(\bar{\omega}t + \delta_1) - \frac{i\hbar\omega_0^2}{2m\bar{\omega}}\,e^{-\gamma t/2}\cos(\bar{\omega}t + \delta_2)$$

$$+ \frac{2}{\beta m}\sum_{n=1}^{\infty}\frac{\nu_n^2 e^{-\nu_n t}}{(\gamma/2 - \nu_n)^2 - \bar{\omega}^2}, \quad (5.75)$$

with $\tan \delta_2 = \gamma \bar{\omega}/\omega_0^2$. Again, the real part is the same as in the classical case except for the presence of the Matsubara series. The same considerations about the surface temperature in the quantum regime can be mentioned as before. Thus, the I_2 factor is again expressed as in (5.64), but where

$$
f(t) = \frac{t}{m\beta\bar{\omega}} e^{-\gamma t/2} \sin \bar{\omega}t + \frac{2}{\beta m} \sum_{n=1}^{\infty} \frac{v_n^2 e^{-v_n t}}{(\gamma/2 - v_n)^2 - \bar{\omega}^2}
$$

$$
+ \frac{i\hbar t}{2m} e^{-\gamma t/2} \left[(1 - \gamma^2/\omega_0^2)(e^{\gamma t/2} - \cos \bar{\omega}t) + \frac{\gamma^3 - 3\gamma\omega_0^2}{2\omega_0\bar{\omega}} \sin \bar{\omega}t \right],
$$

$$\tag{5.76a}$$

$$
g(t) = \frac{1}{m\beta\omega_0^2\bar{\omega}} \left\{ \bar{\omega} - e^{-\gamma t/2}[\bar{\omega} \cos \bar{\omega}t + (\omega_0^2 t + \gamma/2) \sin \bar{\omega}t] \right\}
$$

$$
- \frac{2}{\beta m} \sum_{n=1}^{\infty} \frac{1 - e^{-v_n t}(v_n t + 1)}{(\gamma/2 - v_n)^2 - \bar{\omega}^2} + \frac{i\hbar\omega_0^2}{2m\bar{\omega}} [g_0 + g_1(t) + g_2(t)], \tag{5.76b}
$$

with

$$
g_0 = \frac{\bar{\omega}\gamma}{4\omega_0^7} \left(\gamma^3 + 2\omega_0\gamma^2 - 2\omega_0^2\gamma - 4\omega_0^3 \right), \tag{5.77a}
$$

$$
g_1(t) = \frac{e^{-\gamma t/2}}{\omega_0^6} \left[\bar{\omega}\omega_0^2(\omega_0^2 - \gamma^2)t + 2\omega_0^2\gamma\bar{\omega} - \gamma^3\bar{\omega} \right] \cos \bar{\omega}t, \tag{5.77b}
$$

$$
g_2(t) = \frac{e^{-\gamma t/2}}{\omega_0^6} \left[(\gamma/2)\omega_0^2(3\omega_0^2 - \gamma^2)t + 4\omega_0^2\gamma^2 - \omega_0^4 - \gamma^4/2 \right] \sin \bar{\omega}t. \tag{5.77c}
$$

The total intermediate scattering function will then be the product of the factors I_1 and I_2 given by (5.57) and (5.64), taking into account (5.76a) and (5.76b), respectively. Dephasing can be considered if anharmonic terms are included [78].

When the coverage of the surface is increased, adsorbates can no longer be considered isolated on the surface and they start interacting through multiple collisions while diffusing. At intermediate coverages (around 10%), a two bath model has been recently proposed [78, 83]. This gives rise to two frictions, one due to the phonon motion (γ) and the other to the collisional friction among adsorbates (λ). If the corresponding noise functions are uncorrelated, a total friction $\eta = \gamma + \lambda$ is obtained and the previous theoretical treatment can be straightforwardly generalized to interacting adsorbates by replacing γ by η.

5.3.2 Path Integral Formulation: Propagators

Feynman's path integral approach to quantum mechanics [84–86] deals with quantum fluctuations around classical paths. As seen in Sect. 3.4, this formulation is free of operators and needs the Lagrangian function for the construction of the quantum propagator. The non-uniqueness of Lagrangians leads this approach to provide different results for a given physical problem. In the quantization procedure, one finds additionally the ambiguity arising from the ordering of the dynamical variables that define the studied physical process.

Consider the path integral representation for the propagator (3.117),

$$K[q_f, t; q_0, t_0] = \int_{q_0}^{q_f} e^{i S[q]/\hbar} \mathcal{D}q, \qquad (5.78)$$

going from the initial position q_0 at t_0 to the final position q_f at t_f along the path $q(t)$. In (5.78), the classical action $S[q]$ is a functional of the path $q(t)$ through the classical Lagrangian $L(q, \dot{q}, t)$,

$$S[q] = \int_{t_0}^{t} L(q, \dot{q}, t') dt'. \qquad (5.79)$$

The symbol $\mathcal{D}$ indicates us that the integration is not carried out over an interval but over all paths satisfying the boundary conditions. A general path consists of a classical path plus a fluctuation part which vanishes at the initial and final time. The classical path (one or more) corresponds to a stationary point of the action. This path is the only one which survives in the classical limit, $\hbar \to 0$. The first quantum correction comes from the quadratic term when the action is expanded in the fluctuations. For the free particle and the harmonic oscillator, the corresponding expansion breaks off after the second term, thus resulting that both cases are exact in this formulation. The semiclassical approximation is obtained when stopping in the quadratic term. This approach tries to explain quantum phenomena in terms of classical concepts. Real-time propagators can be split up into two types depending on whether one solves a boundary value problem, such as the Van Vleck–Gutzwiller propagator, or an easier initial value problem, such as the Herman–Kluk or initial value representation propagator [62]. Recently, it has been formulated [87] a correction operator formalism in terms of an asymptotic series, where the first term gives the Herman–Kluk propagator.

When dealing with dissipative systems, Feynman's approach usually starts from the density matrix due to the involvement of mixed states. Here, only the main steps will be accounted for, finding the interested reader further analyses in the related literature [3, 88–90]. Thus, consider the canonical density matrix is written as

$$\rho = \frac{1}{Z} e^{-\beta H}, \qquad (5.80)$$

where $Z = \text{Tr}[e^{-\beta H}]$ is the partition function. Then, the corresponding (imaginary time) path integral representation is [3, 88, 90]

$$\rho(q, q') = \frac{1}{Z} \int_q^{q'} \mathcal{D}\bar{q}\, e^{-S^E[\bar{q}]/\hbar}. \tag{5.81}$$

Here, temperature is interpreted as an imaginary time, $t = -i\hbar\beta$, i.e., a *Wick rotation*. The boundary conditions are given by $\bar{q}(0) = q'$ and $\bar{q}(\hbar\beta) = q$ and S^E is the so-called Euclidean action, where the motion takes place in the inverted potential (see Chap. 2). Furthermore, the partition function in the same representation is written as

$$Z = \oint \mathcal{D}\bar{q}\, e^{-S^E[\bar{q}]/\hbar} \tag{5.82}$$

for all closed paths $q(0) = q(\hbar\beta)$.

The removal of the bath degrees of freedom when, for example, a Caldeira–Leggett Hamiltonian is assumed, leads to the so-called influence functional for the reduced density matrix,

$$\rho(q, q') = \frac{1}{Z} \int_q^{q'} \mathcal{D}\bar{q}\, e^{iS^E[\bar{q}]/\hbar} \mathcal{F}[\bar{q}], \tag{5.83}$$

which contains all the information concerning the influence of the heat bath on the system. As can be shown, such an influence may be taken into account by adding a nonlocal contribution to the action [90].

Tunneling with dissipation has played a very important role in this formulation at a fundamental and an applied level; in particular, at low temperatures and weak frictions where the quantum regime dominates. Caldeira and Leggett developed [60] path integral techniques to adapt a thermodynamical approach following Langer ideas [91] to calculate escape rates. This method is based on the calculation of the imaginary part of the free energy, since it is related to the corresponding rates. The close orbits necessary to compute partition functions are obtained from orbits called bounces which reflect the onset of incoherent quantum tunneling [62]. A similar technique but for coherent tunneling is the so-called instanton orbit technique. For a treatment of real-time dynamics of open quantum systems, see, for example, [92]. Driven tunneling has also been studied with the Markov–Floquet theory, applying the path-integral formulation, by Grifoni and Hänggi [93]. In open quantum systems, it is also worth mentioning that numerical path integral techniques have also been very much developed to deal with more complex physical processes [94]. Simulations based on quantum Monte Carlo methods at very low temperatures and, therefore, long times are prohibitive.

5.3.3 Markovian Master Equations: The Linblad Equation

In problems involving dissipation one always tries to find master equations accounting only for the system dynamics in order to neglect the details of the bath dynamics. These master equations usually describe the time-evolution of the reduced density matrix which is obtained once the trace over the bath coordinates of the full density matrix is carried out. Among this type of equations of motion, the simplest class is the Markovian one in which it is assumed the bath has no memory. These theories have been widely reviewed in articles and books during the last 20 years; a number of them can be found at the end of this chapter. In particular, two approaches are worth mentioning, namely the phase approach (through the Wigner distribution) [11] and the unified semiclassical approach to the density matrix [62] for dissipative systems, since they deal with classical entities.

For any theory of quantum dissipation to be fully satisfactory the reduced density matrix should fulfill the following properties [11]:

1. It should approach an appropriate equilibrium state (the canonical equilibrium state) at long times.
2. It should satisfy the principle of translational invariance (i.e., coordinate-independent frictional forces are required, as happens with classical Brownian motion).
3. It should remain positive semidefinite (no negative eigenvalues) at any time.

However, it turns out that no Markovian theory can fulfill these three criteria at the same time except for very special cases. The three criteria were proved by Lindblad [95, 96] to be numerically exclusive for the specific case of the harmonic oscillator, although it can be conjectured that they are quite generally exclusive for any Markovian dynamics. Hence, quantum Markovian approaches can be classified, in principle, according to which criterion one chooses to sacrifice. In general, whenever a non-Markovian approach must be adopted, it leads to a loss of analytical and conceptual simplicity as well as larger computational times. In such cases, projection operator techniques, such as the well-known Nakajima–Zwanzig and time-convolutionless techniques [2], are widely used. The generalized quantum Langevin formulation is also used in non-Markovian treatments.

Markovian approaches, on the other hand, are much simpler and, according to the differential Chapman–Kolmogorov equation of classical probability theory (see Appendix B), the quantum dynamical semigroup gives rise to a first order differential equation for the reduced density matrix also known as the Linblad master equation [95, 96]. The starting point is the quantum Liouville equation

$$\frac{d\rho(t)}{dt} = \mathcal{L}(t)\rho(t), \qquad (5.84)$$

where $\mathcal{L}$ is the Liouville superoperator (see Chap. 2). If the reduced system (S) density operator $\rho_S = \mathrm{Tr_B}[\rho]$ is obtained by tracing out over the bath (B) degrees of freedom,

(5.84) is replaced by

$$\frac{d\rho_S(t)}{dt} = \mathcal{L}\rho_S(t). \tag{5.85}$$

The Liouville superoperator can be seen as the generator of the semigroup expressed in exponential form, as

$$V(t) = e^{\mathcal{L}t}, \tag{5.86}$$

where the only explicit time-dependence comes through the argument; the generator is time-independent. This dynamical map describes the state change of the reduced system with time ($t \geq 0$),

$$\rho_S(t) = V(t)\rho_S(0), \tag{5.87}$$

which is a convex-linear, completely positive and trace-preserving quantum operation [2]. The one-parameter family $\{V(t)/t \geq 0\}$ displays the semigroup property,

$$V(t_1)V(t_2) = V(t_1 + t_2), \tag{5.88}$$

with $t_1, t_2 \geq 0$. As shown by Linblad, the most general diagonal form for the generator of a quantum dynamical semigroup can be written as

$$\mathcal{L}\rho_S = -i[H, \rho_S] + \sum_{k=1}^{N^2-1} \gamma_k \left(A_k \rho_S A_k^\dagger - \frac{1}{2} A_k^\dagger A_k \rho_S - \frac{1}{2} \rho_S A_k^\dagger A_k \right), \tag{5.89}$$

where the first term describes the unitary part of the dynamics given by the corresponding Hamiltonian, the operators A_k are called the Linblad operators (dimensionless) and γ_k give the nonnegative eigenvalues playing the role of relaxation rates for the different modes of the open system. The generator $\mathcal{L}$ is in general assumed to be bounded, though in physical applications this is not always the case [2]. The sum term in (5.89) is usually known as the dissipator, $\mathcal{D}(\rho_S)$.

If an external time-dependent force is applied to the system, (5.85) can be generalized to

$$\frac{d\rho_S(t)}{dt} = \mathcal{L}(t)\rho_S(t), \tag{5.90}$$

where $\mathcal{L}(t)$ is the generator of a quantum dynamical semigroup for each fixed $t \geq 0$. The corresponding propagator is defined as

$$V(t, t_0) \equiv T_o e^{\int_{t_0}^{t} ds \mathcal{L}(s)}, \tag{5.91}$$

where T_o is the time-ordering operator, which orders products of time-dependent operators such that their arguments increase from right to left [2]. Then, the corresponding semigroup property reads now as

$$V(t, t_1) V(t_1, t_0) = V(t, t_0). \tag{5.92}$$

The time-evolution of the reduced density operator is necessary to calculate multi-time correlation functions of observables or, in general, of operators.

For other non-Markovian master equation models which go beyond the scope of this monograph, such as the Nakajima–Zwanzig equation, the interested reader may consult the related literature [3, 5].

5.3.4 Stochastic Approaches: Quantum Trajectories

If the wave function or state vector is considered as a stochastic process (see Appendix B) in Hilbert space, the corresponding formulation is given in terms of a time-dependent density functional $P[\Psi, t]$. The expectation value of the reduced density operator is then expressed as $\rho_S = E\{|\Psi(t)\rangle\langle\Psi(t)|\}$ and the dynamics is no longer described by a master equation, but a stochastic differential equation. This procedure is called the *unravelling* of the master equation. According to Carmichael [97], the realizations of the underlying stochastic process are called *quantum trajectories*[3] and the transition from a state Ψ to a Ψ' state, a *quantum jump*. The Linblad equation leads to a close connection between the quantum dynamical semigroup and the piecewise deterministic process in Hilbert space. This connection is not unique and the stochastic representation is also related to the continuous measurement process [2]. A path integral procedure can also be carried out in this Hilbert space different from the Feynman–Vernon path integral, which consists of a sum over paths $\Psi(t)$ with their corresponding weights [2].

In the diffusion limit of the Liouville master equation, a Fokker–Planck equation for the probability density functional exists which, in turn, is equivalent to a stochastic Schrödinger equation in Itô form [98, 99] for one Linblad operator,

$$d\Psi(t) = -iK[\Psi(t)]\, dt + \sqrt{\gamma_0} M[\Psi(t)]\, dW(t), \tag{5.93}$$

where K is the nonlinear drift operator and γ_0 the relaxation time. The nonlinear operator M is related to the first-order fluctuation (noise operator) of the Linblad operator, i.e., $A = I + \epsilon C$, with I being the identity operator, ϵ a real small number and C an operator. From this time-evolution, multi-time correlation functions through

---

[3] This type of quantum trajectories must not be confused with Bohmian trajectories (see Chap. 6), which are also regarded as quantum or causal trajectories. Here, the concept refers to the time series or realization associated with a given observable, i.e., it is synonymous of *stochastic trajectory* (see Sect. B.2 of Appendix B).

$\rho_S(t)$ of observables are obtained. This type of stochastic Schrödinger equation has been proposed by many authors and are known as stochastic collapse models (see, for example, the quantum state diffusion model [10, 100, 101]). Some numerical simulation methods to solve (5.93) can be found in [2, 102–104]. For example, Strunz [105] has studied the Brownian dynamics of a parametric oscillator in a Markovian and non-Markovian regime. The non-Markovian dynamics is nowadays developing very fast with potential applications to quantum computation, quantum metrology, etc [106, 107, 108]. It is also worth mentioning that in (5.93) the noise is multiplicative, since the coefficient accompanying $dW(t)$ is not a constant but it depends on the wave function. As mentioned in Chap. 2, this noise can induce transitions, namely *noise induced transitions* [109].

To conclude this Section, it is also worth mentioning Wang's proposal [28] of a new stochastic approach to quantum mechanics in complex space. This approach is based on the concept of weak value of a given observable, as defined above. The real part of the trajectory associated with the corresponding stochastic process (*weak trajectory*) is interpreted as the trajectory of a particle in real configuration space, reducing to the correct classical stochastic trajectory when the de Broglie wavelength vanishes.

References

1. Lieb, E.H., Seiringer, R.: The Stability of Matter in Quantum Mechanics. Cambridge University Press, Cambridge (2009)
2. Breuer, H.P., Petruccione, F.: The Theory of Open Quantum Systems. Oxford University Press, Oxford (2002)
3. Weiss, U.: Quantum Dissipative Systems. World Scientific, Singapore (1999)
4. Accardi, L., Lu, Y.G., Volovich, I.: Quantum Theory and Its Stochastic Limit. Springer, Berlin (2002)
5. Gardiner, C.W., Zoller, P.: Quantum Noise. Springer Complexity, Berlin (2004)
6. Razavy, M.: Classical and Quantum Dissipative Systems. Imperial College Press, London (2005)
7. Yang, C.N., Feldman, D.: The S-matrix in the Heisenberg representation. Phys. Rev. **79**, 972–978 (1950)
8. Louisell, W.H.: Quantum Statistical Properties of Radiation. Wiley, New York (1990)
9. Jung, P.J.: Periodically driven stochastic systems. Phys. Rep. **234**, 175–295 (1993)
10. Percival, I.C.: Quantum State Diffusion. Cambridge University Press, Cambridge (1998)
11. Kohen, D., Tannor, D.J.: Phase space approach to dissipative molecular dynamics. Adv. Chem. Phys. **111**, 219–398 (2000)
12. May, V., Kuhn, O.: Charge and Energy Transfer Dynamics in Molecular Systems. Wiley-VCH Verlag, Weinheim (2003)
13. Kapral, R., Kapral, R.: Mixed quantum-classical dynamics. J. Chem. Phys. **110**, 8919–8929 (1999)
14. Kapral, R.: Quantum-classical dynamics in a classical bath. J. Phys. Chem. A **105**, 2885–2889 (2001)
15. Toutounji, M., Kapral, R.: Subsystem dynamics in mixed quantum-classical systems. Chem. Phys. **268**, 79–89 (2001)

16. Oxtoby, D.W.: Dephasing of molecular vibrations in liquids. Adv. Chem. Phys. **40**, 1–48 (1979)
17. Levine, A.M., Shapiro, M., Pollak, E.: Hamiltonian theory for vibrational dephasing rates of small molecules in liquids. J. Chem. Phys. **88**, 1959–1966 (1988)
18. Bader, J.S., Berne, B.J., Pollak, E., Hänggi, P.: The energy relaxation of a nonlinear oscillator coupled to a linear bath. J. Chem. Phys. **104**, 1111–1119 (1996)
19. Bader, J.S., Berne, B.J.: Quantum and classical relaxation rates from classical simulations. J. Chem. Phys. **100**, 8359–8366 (1994)
20. Egorov, S.A., Berne, B.J.: Vibrational energy relaxation in the condensed phases: Quantum vs classical bath for multiphonon processes. J. Chem. Phys. **107**, 6050–6061 (1997)
21. Ford, G.W., O'Conell, R.F.: There is no quantum regression theorem. Phys. Rev. Lett. **77**, 798–801 (1996)
22. Aharonov, Y., Albert, D.Z., Vaidman, L.: How the result of a measurement of a component of the spin of a spin-1/2 particle can turn out to be 100. Phys. Rev. Lett. **60**, 1351–1354 (1988)
23. Aharonovand, Y., Rohrlich, D.: Quantum Paradoxes. Wiley-VCH, Weinheim (2005)
24. Ritchie, N.W.M., Story, J.G., Hulet, R.G.: Realization of a measurement of a "weak value". Phys. Rev. Lett. **66**, 1107–1110 (1991)
25. Jozsa, R.: Complex weak values in quantum measurement. Phys. Rev. A **76**(1–3), 044103 (2007)
26. Vogel, K., Risken, H.: Determination of quasiprobability distributions in terms of probability distributions for the rotated quadrature phase. Phys. Rev. A **40**, 2847–2849 (1989)
27. Smithey, D.T., Beck, M., Raymer, M.G., Faridani, A.: Measurement of the Wigner distribution and the density matrix of a light mode using optical homodyne tomography: Applications to squeezed states and the vacuum. Phys. Rev. Lett. **70**, 1244–1247 (1993)
28. Wang, M.S.: Stochastic interpretation of quantum mechanics in complex space. Phys. Rev. Lett. **79**, 3319–3322 (1997)
29. Sudarshan, E.C.G., Misra, B.: The Zeno's paradox in quantum theory. J. Math. Phys. **18**, 756–763 (1977)
30. Peres, A.: Zeno paradox in quantum theory. Am. J. Phys. **48**, 931–932 (1980)
31. Itano, W.M., Heinsen, D.J., Bokkinger, J.J., Wineland, D.J.: Quantum Zeno effect. Phys. Rev. A **41**, 2295–2300 (1990)
32. Kofman, A.G., Kurizki, G.: Acceleration of quantum decay processes by frequent observations. Nature **405**, 546–549 (2000)
33. Vineyard, G.H.: Scattering of slow neutrons by a liquid. Phys. Rev. **110**, 999–1010 (1958)
34. Nyquist, H.: Certain topics in telegraph transmission theory. Trans. AIEE **47**, 617–644 (1928)
35. Shannon, C.E.: Communication in the presence of noise. Proc. Inst. Radio Eng. **37**, 10–21 (1949) (see also reprint in: Proc. IEEE **86**, 447–457 (1998))
36. Whittaker, E.T.: On the functions which are represented by the expansions of the interpolation theory. Proc. R. Soc. Edinburgh A **35**, 181–194 (1915)
37. Kotel'nikov, V.A.: On the capacity of the 'ether' and of cables in electrical communication. In: Proceedings of the First All-Union Conference on the Technological Reconstruction of the Communications Sector and Low-Current Engineering—Izd. Red. Upr. Svyazi RKKA (Moscow, 1933). Translated into English in: Modern Sampling Theory, Benedetto, J.J., Ferreira, P.J.S.G. Birkhäuser, Berlin (2001), Chap. 2
38. Kubo, R.: The fluctuation-dissipation theorem. Prog. Theor. Phys. **29**, 255–284 (1966)
39. Caldirola, P.: Forze non conservative nella meccanica quantistica. Nuovo Cimento **18**, 393–400 (1941)
40. Kanai, E.: On the quantization of the dissipative systems. Prog. Theor. Phys. **3**, 440–442 (1948)
41. Kerner, E.H.: Note on the forced and damped oscillator in quantum mechanics. Can. J. Phys. **3**, 371–377 (1958)
42. Jannussis, A.D., Brodimas, G.N., Streclas, A.: Propagator with friction in quantum mechanics. Phys. Lett. A **74**, 6–10 (1979)

43. Jannussis, A., Filipakis, P., Philipakis, Th.: Quantum mechanics in phase space. Physica A **102**, 561–567 (1980)
44. Schuch, D.: A new Lagrangian–Hamiltonian formalism for dissipative systems. Int. J. Quantum Chem. **24**, 767–780 (1990)
45. Dekker, H.: Classical and quantum mechanics of the damped harmonic oscillator. Phys. Rep. **80**, 1–110 (1981)
46. Senitzky, I.R.: Dissipation in quantum mechanics. The harmonic oscillator. Phys. Rev. **119**, 670–679 (1960)
47. Ford, G.W., Kac, M., Mazur, P.: Statistical mechanics of assemblies of coupled oscillators. J. Math. Phys. **6**, 504–515 (1965)
48. Kostin, M.D.: On the Schrödinger–Langevin equation. J. Chem. Phys. **57**, 3589–3591 (1972)
49. Kostin, M.D.: Friction and dissipative phenomena in quantum mechanics. J. Stat. Phys. **12**, 145–151 (1975)
50. Hasse, R.W.: On the quantum mechanical treatment of dissipative systems. J. Math. Phys. **16**, 2005–2011 (1975)
51. Schuch, D.: Nonunitary connection between explicitly time-dependent and nonlinear approaches for the description of dissipative quantum systems. Phys. Rev. A **55**, 935–940 (1997)
52. Risken, H.: The Fokker–Planck Equation. Springer, Berlin (1984)
53. Doebner, H.-D., Goldin, G.A.: On a general nonlinear Schrödinger equation admitting diffusion currents. Phys. Lett. A **162**, 397–401 (1992)
54. Gisin, N.: A simple nonlinear dissipative quantum evolution equation. J. Phys. A **14**, 2259–2267 (1981)
55. Gisin, N.: Microscopic derivation of a class of non-linear dissipative Schrödinger-like equations. Physica A **111**, 364–370 (1982)
56. Razavy, M.: Quantization of dissipative systems. Z. Phys. B **26**, 201–206 (1977)
57. Wagner, H.-J.: Schrödinger quantization and variational principles in dissipative quantum theory. Z. Phys. B **95**, 261–273 (1994)
58. Schuch, D.: Effective description of the dissipative interaction between simple and model-system and their environment. Int. J. Quantum Chem. **72**, 537–547 (1999)
59. Ford, G.W., Lewis, J.T., O'Connell, R.F.: Quantum Langevin equation. Phys. Rev. A **37**, 4419–4428 (1988)
60. Caldeira, A.O., Leggett, A.J.: Path integral approach to quantum Brownian motion. Physica A **121**, 587–616 (1983)
61. Feynman, R.P., Vernon, F.L.: The theory of a general quantum system interacting with a linear system. Ann. Phys. (NY) **24**, 118–173 (1963)
62. Ankerhold, J.: Quantum Tunneling in Complex Systems. Springer Tracts in Modern Physics, vol. 224. Springer, Berlin (2007)
63. Yu, L.H., Sun, C.-P.: Evolution of the wave function in a dissipative system. Phys. Rev. A **49**, 592–595 (1994)
64. Sun, C.-P., Yu, L.H.: Exact dynamics of a quantum dissipative system in a constant external field. Phys. Rev. A **51**, 1845–1853 (1995)
65. Feshbach, H.: Unified theory of nuclear reaction. Ann. Phys. (NY) **5**, 357–390 (1958)
66. Mott, N.F., Massey, H.S.W.: Theory of Atomic Collision. Oxford University Press, London (1965)
67. Yan, Y., Xu, R.: Quantum mechanics of dissipative systems. Annu. Rev. Phys. Chem. **56**, 187–219 (2005)
68. Tanimura, Y.: Stochastic Liouville, Fokker–Planck and master equation approaches to quantum dissipative systems. J. Phys. Soc. Jpn **75**, 082001(1–39) (2006)
69. Hakim, V., Ambegaokar, V.: Quantum theory of a free particle interacting with a linearly dissipative environment. Phys. Rev. A **32**, 423–434 (1985)

70. Hu, B.L., Paz, J.P., Zhang, Y.: Quantum Brownian motion in a general environment: exact master equation with nonlocal dissipation and colored noise. Phys. Rev. D **45**, 2843–2861 (1992)
71. Ford, G.W., O'Connell, R.F.: Exact solution of the Hu-Paz-Zhang master equation. Phys Rev. D **64**, 105020-1,13 (2001)
72. Schmid, A.: On a quasiclassical Langevin equation. J. Low. Temp Phys. **49**, 609–626 (1982)
73. Rips, I., Pollak, E.: Quantum Kramers model: solution of the turnover problem. Phys. Rev. A **41**, 5366–5382 (1990)
74. Hänggi, P., Talkner, P., Borbovec, M.: Reaction-rate theory: fifty years after Kramers. Rev. Mod. Phys. **62**, 251–341 (1990)
75. Ankerhold, J., Pechukas, P., Grabert, H.: Strong friction limit in quantum mechanics. The quantum Smoluchowski equation. Phys. Rev. Lett. **87**, 086802(1–4) (2001)
76. Ford, G.W., Kac, M.: On the quantum Langevin equation. J. Stat. Phys. **46**, 803–810 (1987)
77. Sánchez-Cañizares, J., Sols, F.: Translational symmetry and microscopic preparation in oscillator models of quantum dissipation. Physica A **122**, 181–193 (1994)
78. Martínez-Casado, R., Sanz, A.S., Vega, J.L., Rojas-Lorenzo, G., Miret-Artés, S.: Linear response theory of activated surface diffusion with interacting adsorbates. Chem. Phys. **370**, 180–193 (2010)
79. Van Hove, L.: Correlations in space and time and Born approximation scattering in systems of interacting particles. Phys. Rev. **95**, 249–262 (1954)
80. Lovesey, S.W.: Theory of Neutron Scattering from Condensed Matter. Clarendon, Oxford (1984)
81. Miret-Artés, S., Pollak, E.: The dynamics of activated surface diffusion. J. Phys. Condens. Matter **17**, S4133–S4150 (2005)
82. Matsubara, T.: A new approach to quantum-statistical mechanics. Prog. Theor. Phys. **14**, 351–378 (1955)
83. Martínez-Casado, R., Rojas-Lorenzo, G., Sanz, A.S., Miret-Artés, S.: Two-bath model for activated surface diffusion of interacting adsorbates. J. Chem. Phys. **132**, 054704(1–7) (2010)
84. Feynman, R.P., Hibbs, A.R.: Quantum Mechanics and Path Integrals. McGraw-Hill Book Company, New York (1965)
85. Schulman, L.S.: Techniques and Applications of Path Integrals. Wiley, New York (1981)
86. Kleinert, H.: Path Integrals in Quantum Mechanics Statistics, Polymer Physics and Financial Markets. World Scientific, Singapore (2006)
87. Ankerhold, J., Saltzer, M., Pollak, E.: A study of the semiclassical initial value representation at short times. J. Chem. Phys. **116**, 5925–5932 (2002)
88. Feynman, R.P.: Statistical Mechanics. W A Benjamin, Reading (1972)
89. Grabert, H., Schramm, P., Ingold, G.-L.: Quantum Brownian Motion: The Functional Integral Approach. Phys. Rep. **168**, 115–207 (1988)
90. Ingold, G.-L.: Path Integrals and their Application to Dissipative Quantum Systems. Lecture Notes Physics, vol. 611, pp. 1–53. Springer, Berlin (2002)
91. Langer, J.S.: Theory of the condensation point. Ann. Phys. (NY) **41**, 108–157 (1967)
92. Ankerhold, J., Pollak, E. (eds.): Real-time dynamics in complex quantum systems. Chem. Phys. (Special Issue) **322** (2006)
93. Grifoni, M., Hänggi, P.: Driven quantum tunneling. Phys. Rep. **304**, 229–354 (1988)
94. Makri, N.: Numerical path integral techniques for long time dynamics of quantum dissipative systems. J. Math. Phys. **36**, 2430–2457 (1995)
95. Linblad, G.: On the generators of quantum dynamical semigroups. Comm. Math. Phys. **48**, 119–130 (1976)
96. Linblad, G.: Brownian motion of a quantum harmonic oscillator. Rep. Math. Phys. **10**, 393–406 (1976)
97. Carmichael, H.: An Open Systems Approach to Quantum Optics. Lecture Notes in Physics, vol. 18. Springer, Berlin (1993)

98. Itô, K.: Foundations of stochastic differential equations in infinite dimensional spaces. In: CBMS-NSF Regional Conference Series in Applied Mathematics, vol. 47. SIAM, Philadelphia (1984)

99. Ikeda, N., Watanabe, S.: Stochastic Differential Equations and Diffusion Processes, 2nd edn. North-Holland, Amsterdam (1989)

100. Gisin, N., Percival, I.C.: The quantum-state diffusion model applied to open systems. J. Phys. A **25**, 5677–5691 (1992)

101. Gisin, N., Percival, I.C.: The quantum state diffusion picture of physical processes. J. Phys. A **26**, 2245–2260 (1993)

102. Kloeden, P.E., Platen, E.: Numerical Solution of Stochastic Differential Equations. Springer, Berlin (1992)

103. Kloeden, P.E., Platen, E., Schurz, H.: Numerical Solution of Stochastic Differential Equations Through Computer Experiments. Springer, Berlin (1994)

104. Klauder, J.R., Petersen, W.P.: Numerical integration of multiplicative-noise stochastic differential equations. SIAM J. Numer. Anal. **22**, 1153–1166 (1985)

105. Strunz, W.T.: The Brownian motion stochastic Schrödinger equation. Chem. Phys. **268**, 237–248 (2001) (This paper appears in a Special Issue devoted to Quantum Dynamics of Open Systems, issues 1–3)

106. Breuer, H.-P., Laine, E.-M., Piilo, J.: Measure for the degree of non-Markovian behavior of quantum processes in open quantum systems. Phys. Rev. Lett. **103**, 210401(1–4) (2009)

107. Vacchini, B., Smirne, A., Laine, E.-M., Piilo, J., Breuer, H.-P.: Markovianity and non-Markovianity in quantum and classical systems. New J. Phys. **13**, 093004(1–26) (2011)

108. Liu, B.-H., Li, L., Huang, Y.-F., Li, C.-F., Guo, G.-C., Laine, E.-M., Breuer, H.-P., Piilo, J.: Experimental control of the transition from Markovian to non-Markovian dynamics of open quantum systems. Nat. Phys. **7**, 931–934 (2011)

109. Horsthemke, W., Lefever, W.: Noise-Induced Transitions. Springer Series in Synergetics, vol. 15. Springer, Berlin (2006)

Chapter 6
Quantum Mechanics with Trajectories

6.1 Introduction

According to the conventional view of quantum mechanics, the most complete information about a quantum system is specified by the wave function. As seen in Chap. 3, the wave function provides us with a probabilistic or statistical description [1] of the possible outcomes that can be obtained when a measurement on a property of such a system is carried out. This viewpoint has constituted the center of a longstanding debate around the so-called *completeness* of the wave function and the *quantum theory of measurement* [2]. In order to render some light on this issue, different hidden-variable theories and models have been proposed in the literature [3, 4].

As stated by von Neumann [5], quantum mechanics or any alternative theory cannot be derived by considering a statistical approximation from a classical-like deterministic theory. This result, enunciated in the form of a theorem, constitutes an important benchmark to discern whether a hidden-variable model can be or not considered a serious alternative to the standard quantum mechanics. However, this theorem contains an also important conceptual drawback: it only applies to *local* models, as shown in 1952 by Bohm [6, 7], proven mathematically in the 1964 by Bell [8, 9] and verified experimentally in 1981 by Aspect et al. [10–12]. The model developed by David Bohm [6], nowadays known as *Bohmian mechanics*, was essentially based on assuming that a quantum system consists, at the same time, of a wave and a particle. The wave evolves according to Schrödinger's equation and the particle moves according to a certain guidance condition, which makes the particle motion dependent on the wave evolution, giving rise to quantum trajectories (not to be confused with the same terminology used in quantum stochastic theories seen in Chap. 5). Though Bohmian mechanics is usually regarded as a "reinterpretation" or an alternative picture of standard quantum mechanics, referring to it as a "theory" is also common in order to stress the conceptual difference between the two approaches to the microscopic world—i.e., it emphasizes Bohmian mechanics is a "quantum theory of motion", quoting Holland's book title on the subject [13].

A. S. Sanz and S. Miret-Artés, *A Trajectory Description of Quantum Processes.*
I. Fundamentals, Lecture Notes in Physics 850, DOI: 10.1007/978-3-642-18092-7_6,
© Springer-Verlag Berlin Heidelberg 2012

Bohm's ideas were applied to different prototypical models of quantum mechanics [13] during the late 1970s and particularly the 1980s and early 1990s. However, in the last 10 years, Bohmian mechanics has passed from being a mere way to formulate a quantum mechanics "without observers" [14, 15] to become a well-known (and increasingly accepted) theoretical framework used as a source for new quantum computational methods as well as quantum interpretations [16–18]. These two aspects of Bohmian mechanics are what Wyatt [16] has called the *synthetic* and *analytic* approaches of this theory. The first approach essentially starts with the former numerical schemes developed to obtain quantum information without solving Schrödinger's equation, but its equivalent Bohmian counterparts [19–24]. Since then, this computational branch of Bohmian mechanics has diversified into a myriad of numerical approaches (see Volume 2), which can be summarized by the type of answer they try to give. For example, for wave packet propagation, different Lagrangian, Eulerian and combined Eulerian-Lagrangian algorithms have been developed [19–30]; also semiclassical initial value representation schemes based on Bohmian mechanics have been implemented [31–35]. The purpose of avoiding and, therefore, solving the so-called *nodal problem* has led to schemes such as the bipolar ansatz [36–39], the covering function [40] or a mixed wave function representation [41]. In order to cope with the problem of the coupling between quantum and classical degrees of freedom in multidimensional problems, hybrid Bohmian-based quantum-classical approaches have been developed [42–51]; and also, to deal with systems described by a large number of degrees of freedom, schemes based on linearizing the quantum force have been implemented [52–55]. The starting point for the second approach, on the other hand, based on obtaining the quantum trajectories on the fly from the wave function as a tool to interpret realistic experiments, can be established in the former studies of rare gas atom diffraction by metal surfaces [56, 57]. In Volume 2, different applications will be analyzed within this approach in more detail.

Many different studies have appeared in the literature ever since dealing with one or the other approach, or even both. This has given rise to a rich literature, different from the formerly developed, where Bohmian mechanics was applied to different academic [13], well-known paradigms of quantum mechanics [17, 18]. The reason for this is very simple, arising from a nice appealing feature: Bohmian mechanics allows us to understand and explain quantum systems in terms of the motion displayed (in configuration space) by a swarm of quantum trajectories. Each one of these trajectories represents the evolution in time of a particular initial state specified by a point on the configuration space associated with the system. Thus, unlike standard quantum mechanics, where the wave function determines the state of the system on the *whole* available configuration space, in Bohmian mechanics it is possible to follow one particular point of such a space. The time-evolution of this point is given according to some prescribed quantum laws of motion, as will be seen below, and the evolution of the trajectory ensemble is equivalent to the evolution of a quantum flow—this is precisely the viewpoint of quantum hydrodynamics. The Bohmian view, nevertheless, does not invalidate at all other ways to understand quantum systems; it only allows us to think of them on similar grounds as classical ones, i.e., using a similar intuitive scheme, which differs from a purely classical one precisely in the types of motion one can observe. However, unlike any classical approach to quantum mechanics, Bohmian mechanics is not an approximated theory, but an exact one, as is shown below.

6.2 Bohmian Mechanics

6.2.1 Fundamentals

In Bohmian mechanics the description of quantum systems is carried out in terms of waves and quantum trajectories. The wave function Ψ provides with the dynamical information about the whole available configuration space to quantum particles (whatever these particles are or represent; see Sect. 6.6), which will evolve accordingly as if they are "guided"—in this way, the quantum motion displayed by particles reflects the evolution of the wave function. The fundamental Bohmian equations of motion are usually derived from the standard version of quantum mechanics through the transformation $(\Psi, \Psi^*) \rightarrow (\rho, S)$, where Ψ and Ψ^* are generally complex-valued functions of the position $(\mathbf{r})$ and time (t), and ρ and S are real-valued functions of the same variables. More explicitly, the transformation relation between both types of functions (or fields) for a particle of mass m is given by

$$\Psi(\mathbf{r}, t) = \rho^{1/2}(\mathbf{r}, t)e^{iS(\mathbf{r}, t)/\hbar} \tag{6.1}$$

(and its complex conjugate). After introducing (6.1) into the time-dependent Schrödinger equation, two real coupled partial differential equations are obtained,

$$\frac{\partial \rho}{\partial t} + \frac{1}{m}\nabla(\rho \nabla S) = 0, \tag{6.2a}$$

$$\frac{\partial S}{\partial t} + \frac{(\nabla S)^2}{2m} + V_{\text{eff}} = 0, \tag{6.2b}$$

which come from the imaginary and real parts, respectively, of the resulting equation. The former is the continuity equation, which accounts for the probability conservation, while the latter is the *quantum Hamilton–Jacobi equation*, with

$$V_{\text{eff}}(\mathbf{r}, t) = V(\mathbf{r}) - \frac{\hbar^2}{2m}\frac{\nabla^2 \rho^{1/2}(\mathbf{r}, t)}{\rho^{1/2}(\mathbf{r}, t)} \tag{6.3}$$

being an *effective* total potential. The last term in the right-hand side of (6.3) is the so-called *quantum potential*,

$$Q \equiv -\frac{\hbar^2}{2m}\frac{\nabla^2\rho^{1/2}}{\rho^{1/2}} = \frac{\hbar^2}{4m}\left[\frac{1}{2}\left(\frac{\nabla\rho}{\rho}\right)^2 - \frac{\nabla^2\rho}{\rho}\right], \qquad (6.4)$$

which, as well as ρ, depends on both $\mathbf{r}$ and t. This term is regarded as a potential because, like V, it also rules the quantum particle dynamics. However, its nature is fully quantum-mechanical due to its dependence on the quantum state via ρ. Note that, since, statistically, ρ describes the evolution of a swarm of identical non interacting particles, the dependence of Q on ρ means that the dynamics of a single particle from the swarm is going to be influenced by the behavior of the other. That is, quantum particle dynamics are *nonlocal*. Rather than being a particular feature of Bohmian mechanics, this property is inherent to quantum mechanics in general, which manifests through the kinetic operator, $\hat{\mathcal{K}} = -(\hbar^2/2m)\nabla^2$, in its standard version. It is well-known that, from a computational viewpoint (see Volume 2), in order to evaluate the action of $\hat{\mathcal{K}}$ accurately, one has to consider a very good representation of it[1]. In this other sense, since Q arises from the action of this operator on Ψ after considering (6.1), i.e.,

$$\hat{\mathcal{K}}\Psi = -\frac{\hbar^2}{2m}\nabla^2\Psi = \frac{(\nabla S)^2}{2m} - \frac{\hbar^2}{2m}\frac{\nabla^2\rho^{1/2}}{\rho^{1/2}}, \qquad (6.5)$$

it could also be associated with a sort of nonlocal kinetic energy [13]. Nonlocality only disappears when $Q \equiv 0$. Then, the particle dynamics becomes fully classical.

Just as an illustration, in Fig. 6.1 the quantum potential acting on the quantum trajectories in a typical problem in diffraction by periodic surfaces (see Volume 2) is displayed. In particular, from bottom to top, V_{eff} (left-hand side) and a contour-plot of it (right-hand side) are represented as evolving along the z-coordinate (which is proportional to time since the advance of the wave function is along this direction) at three different distances from the physical surface [56, 58, 59]: the near-field or Fresnel region (bottom panels), the transition region (central panels) and the far-field or Franhofer region (top panels). In the particular case illustrated, the classical interaction potential becomes negligible at about 12 Å from the surface and, therefore in the three regions displayed $V_{\text{eff}} \approx Q$. Taking this into account, notice how from a very intricate structure Q evolves gradually towards a much smoother potential consisting of alternating "canyons" and "plateaus", this giving rise to the well-known Bragg diffraction directions, along which quantum trajectories move asymptotically. Nevertheless, since the outgoing wave function consists of maxima corresponding to the Bragg direction plus a collection of secondary minima among

---

[1] For example, when trying to solve the time-dependent Schrödinger equation by means of standard grid methods, $\mathcal{K}$ has to receive a special consideration. In order to avoid truncations of its nonlocal nature, the action of the kinetic operator is assumed in the momentum space (by means of the fast Fourier transform technique, for example), where this operator is local. Then, after acting on Ψ, the result (which is already affected by the value of Ψ in all points of the grid) is put back in the configuration space.

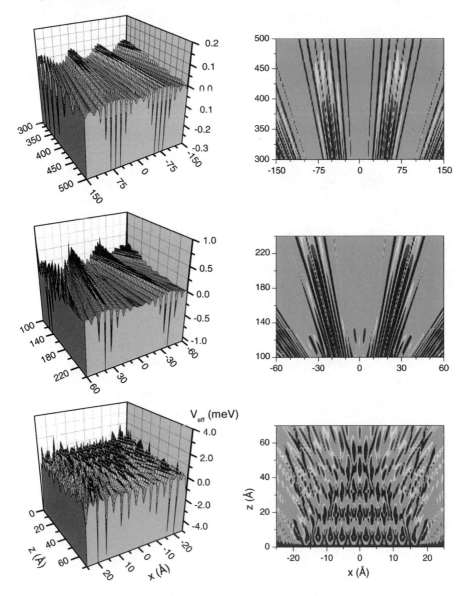

Fig. 6.1 Quantum potential ruling the quantum trajectory dynamics in a typical problem of diffraction by a periodic surface [56, 58, 59]. From *bottom* to *top*, evolution of V_{eff} (*left*) and a contour-plot representation of it (*right*) along the z-coordinate (the surface is parallel to the x-axis) at three distances from the physical surface: the near-field or Fresnel region (*bottom* panels), the transition region (*central* panels) and the far-field or Franhofer region (*top* panels)

them (due to the finiteness of the incident wave), the presence of valleys associated with them is also observable. However, unless the Bragg valleys, the second-order ones are higher in energy on average and the canyons separating two of them are much deeper. In other words, making use again of the "geological" analogy, between two consecutive Bragg valleys there is always a sort of mountain range acting as a barrier (this is specially apparent in the cuts of V_{eff} along the x-axis, for constant z, in the central and top left-hand side panels).

Since (6.2b) is a (quantum) Hamilton–Jacobi equation, paths along which quantum particles travel may be defined according to the *guidance condition*

$$\mathbf{v} = \frac{\nabla S}{m} = \frac{\hbar}{2im}\left[\frac{\nabla\psi}{\psi} - \frac{\nabla\psi^*}{\psi^*}\right], \tag{6.6}$$

where $\dot{\mathbf{r}} = \mathbf{v}$, in analogy to classical mechanics. Equations (6.2) and (6.6) form a closed set of coupled equations, which describes the evolution of a swarm of identical particles under the "guidance" of the quantum state at each time. As will be seen in Volume 2, different numerical techniques have been suggested in the literature [16] to solve this set of equations, which is analogous to those that describe the evolution of classical hydrodynamical flows. Because of this similarity, independently of the particular method considered, two schemes are basically followed. One is the *Eulerian scheme* [60], where the importance relies on the full quantum fluid rather than in the particular trajectory dynamics. Equations (6.2) are then integrated directly with the aid of the velocity field, ∇S, but with no need to obtain any particular trajectory. Alternatively, taking advantage of (6.6), one can also work following a *Lagrangian scheme*, i.e., considering a framework co-moving with the (quantum) fluid [along particular trajectories $x(t)$]. This is the typical scheme considered in order to benefit from the computational advantages of quantum trajectories (either real or complex). In this case, in principle, it is also possible to utilize two equations: (6.2b), which provides information about the quantum flux along a particular trajectory, and (6.6), which will define the particle trajectory. Thus, the first step is to pass from the Eulerian framework to the Lagrangian one by means of the well-known time derivative or Lagrangian operator,

$$\frac{d}{dt} = \frac{\partial}{\partial t} + \mathbf{v}\cdot\nabla, \tag{6.7}$$

where $\mathbf{v}$ is given by (6.6). After using this operator, (6.2b) becomes

$$\frac{dS}{dt} = \frac{1}{2}m\mathbf{v}^2 - V_{\text{eff}}, \tag{6.8}$$

where, as in classical mechanics, the time-derivative of the quantum action S is equal to a generalized quantum Lagrangian,

$$S[\mathbf{r}(t)] - S[\mathbf{r}(0)] = \int_0^t \mathcal{L}_Q\big[\mathbf{r}(t')\big]dt', \tag{6.9}$$

with $\mathcal{L}_Q \equiv m\mathbf{v}^2/2 - V_{\text{eff}}$. Thus, solving this equation together with (6.6) will give us the full dynamics associated with the corresponding quantum system within the Bohmian context. Note that in standard quantum mechanics the system dynamics is only described by Ψ. However, in Bohmian mechanics, one focusses on the particular evolution in time of a given initial system configuration (initial condition), which gives rise to the corresponding Bohmian trajectory. This evolution, though, is strongly determined by the wave function, which acts like a field (apart from any other external field, described by the potential V), prescribing the way how a certain system configuration should evolve. In classical mechanics something similar can be found when going to phase space, although the evolution of the system phase–space configuration is only determined by the external fields.

As it can be inferred from (6.8), in the Lagrangian scheme it is also necessary the evaluation of the quantum potential along the trajectory. However, instead of integrating the continuity equation to obtain $\rho(t)$, one can alternatively proceed as in semiclassical mechanics [61] assuming that the solution of (6.2a) along the Bohmian trajectory $\mathbf{r}(t)$ is given by

$$\rho[\mathbf{r}(t)] = \left| \frac{\partial \mathbf{r}(0)}{\partial \mathbf{r}(t)} \right| \rho[\mathbf{r}(0)]. \tag{6.10}$$

This procedure has been used with practical computational purposes in the literature [31–35], with the aim to apply Bohmian mechanics as a sort of quantum initial value representation numerical approach. Substituting (6.10) into (6.4), one finds

$$Q[\mathbf{r}(t)] = \frac{\hbar^2}{4m} \left[\frac{\partial \mathbf{r}(0)}{\partial \mathbf{r}(t)} \right]^2 \left\{ \frac{1}{2} \left[\frac{\nabla_{\mathbf{r}(0)} \rho[\mathbf{r}(0)]}{\rho[\mathbf{r}(0)]} \right]^2 - \frac{\nabla^2_{\mathbf{r}(0)} \rho[\mathbf{r}(0)]}{\rho[\mathbf{r}(0)]} \right\}, \tag{6.11}$$

where $\nabla_{\mathbf{r}(0)} \equiv \partial/\partial\mathbf{r}(0)$ is the action of the ∇-operator evaluated at $\mathbf{r}(0)$. In this way, $\rho[\mathbf{r}(t)]$ and $Q[\mathbf{r}(0)]$ can be both determined from $\rho[\mathbf{r}(0)]$. On the other hand, $\rho(\mathbf{r}, t)$ can be obtained by sampling the initial probability density with a sufficient number of initial conditions (trajectories). Of course, the numerical evaluation of the derivatives of ρ within this Lagrangian framework can lead to instabilities that will destroy the stability of the method, as also happens in classical hydrodynamics, due to the particle approaching nodal (vortical) regions. Some alternative methods have been proposed in the literature to solve these drawbacks [36, 40, 41].

Within the Lagrangian scheme, the trajectories are calculated one by one, obtaining ρ and S (and, therefore, Ψ) along them. Hence, from a computational viewpoint, this can be regarded as a "local" calculation. However, it is important to stress that this locality has nothing to do with the nonlocal dynamical behavior mentioned above. Note that the information related to the whole ensemble (though evaluated along one particular path) appears in a very precise and unambiguous manner in V_{eff}, as seen in (6.8), and influences the field S. Thus, it is always very important to distinguish between the locality of the calculations and the nonlocality inherent to the dynamical behavior of quantum particles [62].

As seen in Chap. 3, the wave formulation of quantum mechanics can be derived from the (quantum) Lagrangian density [13, 63],

$$\mathcal{L}_q = \frac{i\hbar}{2}\left(\Psi^*\frac{\partial\Psi}{\partial t} - \frac{\partial\Psi^*}{\partial t}\Psi\right) - \frac{\hbar^2}{2}\nabla\Psi\cdot\nabla\Psi^* - V|\Psi|^2, \qquad (6.12)$$

when the corresponding integral is required to be stationary with respect to variations in the complex-valued field variables Ψ and Ψ^*. Then, when variations are taken with respect to Ψ^*, the Euler–Lagrange equations yield the time-dependent Schrödinger equation (as well as its complex conjugate when variations are considered with respect to Ψ). Similarly, one can also proceed taking into account the polar form (6.1), which gives rise to the Lagrangian density [64, 65]

$$\mathcal{L}_q = -\left[\frac{\partial S}{\partial t} + \frac{1}{2}(\nabla S)^2 + V\right]\rho - \frac{\hbar^2}{8}\left(\frac{\nabla\rho}{\rho}\right)^2\rho$$
$$= -\left[\frac{\partial S}{\partial t} + \frac{1}{2}(\nabla S)^2 + \frac{1}{2}(\nabla K)^2 + V\right]\rho, \qquad (6.13)$$

where

$$K \equiv \frac{\hbar}{2}\ln\rho \qquad (6.14)$$

is a term from which the quantum potential emerges. In this regard, note that it would be more appropriate to associate this term (and therefore the quantum potential) with a sort of inner kinetic energy, since Q does not appear explicitly in the Lagrangian density as the external potential V does. This is in correspondence, for example, with the fact that the evolution of a wave packet is ruled by two types of motions [66], one associated with its translation (and, therefore, with ∇S) and another with its spreading (i.e., with ∇K), as will be seen in Volume 2. Furthermore, it is the presence of this term what makes quantum motion so different from the classical one, as can be readily seen when (6.13) is compared with its classical counterpart. This analogy can be better appreciated in Table 6.1.

6.2.2 Expectation Values and Ensemble Averages

Though Bohmian mechanics allows us to describe the evolution of quantum processes and phenomena in terms of individual trajectories, it is clear that any observable will require of a statistical treatment of the corresponding quantum trajectories. That is, any observable will arise as a consequence of the counting of trajectory arrivals at a certain region, which is much connected with the way how experiments occur. For example, in a typical diffraction experiment, the diffraction pattern arises by counting (detecting) individual arrivals [67–71]. Similarly, with Bohmian mechanics one can reproduce the experiment by counting the arrivals of quantum trajectories [56, 59]. Thus, taking this into account, the statistical nature of quantum mechanics arises in

Table 6.1 Comparative scheme of the main elements involved in classical and quantum-mechanical (Bohmian) dynamics

CONCEPT	TRAJECTORY	
	CLASSICAL	BOHMIAN
Associated Lagrangian density	$-\left[\dfrac{\partial S_{cl}}{\partial t}+\dfrac{1}{2}(\nabla S_{cl})^{2}+V\right]\rho_{cl}$	$-\left[\dfrac{\partial S}{\partial t}+\dfrac{1}{2}(\nabla S)^{2}+\dfrac{1}{2}(\nabla K)^{2}+V\right]\rho$
Associated HJ Equation	$\dfrac{1}{2}(\nabla S_{cl})^{2}-\left(-\dfrac{\partial S_{cl}}{\partial t}-V\right)=0$ $\dfrac{\partial \rho_{cl}}{\partial t}+\nabla\cdot(\rho_{cl}\nabla S_{cl})=0$	$\dfrac{1}{2}(\nabla S)^{2}-\left(-\dfrac{\partial S}{\partial t}-V_{\mathit{eff}}\right)=0$ $\dfrac{\partial \rho}{\partial t}+\nabla\cdot(\rho\,\nabla S)=0$
Characteristic Set of **Differential Equations**	$\dfrac{d\mathbf{r}}{dt}=\mathbf{p}=\nabla S_{cl}$ $\dfrac{d\mathbf{p}}{dt}=-\nabla V$ $\dfrac{dS_{cl}}{dt}=\dfrac{1}{2}(\nabla S_{cl})^{2}-V$ $\dfrac{d\rho_{cl}}{dt}=-\rho_{cl}\nabla^{2}S_{cl}$	$\dfrac{d\mathbf{r}}{dt}=\mathbf{p}=\nabla S$ $\dfrac{d\mathbf{p}}{dt}=-\nabla V_{\mathit{eff}}$ $\dfrac{dS}{dt}=\dfrac{1}{2}(\nabla S)^{2}-V_{\mathit{eff}}$ $\dfrac{d\rho}{dt}=-\rho\,\nabla^{2}S$

a very natural way, where expectation values are directly associated with average (ensemble) values.

In order to understand the relationship between the expectation value of a quantum operator and the statistical Bohmian description, consider $\hat{A}$ is a Hermitian operator, which can be a function of the position and momentum operators, $\hat{\mathbf{r}}$ and $\hat{\mathbf{p}} = -i\hbar\nabla$, i.e., $\hat{A} = \hat{A}(\hat{\mathbf{r}}, -i\hbar\nabla)$. The expectation value of this operator is defined as

$$\langle \hat{A}\rangle = \langle\Psi|\hat{A}|\Psi\rangle = \frac{\int \Psi^{*}(\hat{A}\Psi)d\mathbf{r}}{\int \Psi^{*}\Psi\,d\mathbf{r}}, \tag{6.15}$$

where $\Psi(\mathbf{r}, t) = \langle\mathbf{r}|\Psi(t)\rangle$ is the wave function in the system configuration representation and

$$[\hat{A}\Psi](\mathbf{r}, t) \equiv \langle\mathbf{r}|\hat{A}\left(\int |\mathbf{r}'\rangle\langle\mathbf{r}'|d\mathbf{r}'\right)|\Psi(t)\rangle = \int \hat{A}(\hat{\mathbf{r}}, \hat{\mathbf{r}}')\Psi(\mathbf{r}', t)d\mathbf{r}', \tag{6.16}$$

with $\hat{A}(\hat{\mathbf{r}}, \hat{\mathbf{r}}') \equiv \langle\mathbf{r}|\hat{A}|\mathbf{r}'\rangle$. For Hermitian operators, $\hat{A}$, only its real part has to be taken into account in the calculation of its expectation value. Hence, (6.15) can be expressed as

$$\langle \hat{A}\rangle = \frac{\mathrm{Re}\left\{\int \Psi^{*}(\hat{A}\Psi)d\mathbf{r}\right\}}{\int \Psi^{*}\Psi\,d\mathbf{r}} = \frac{\int \mathrm{Re}\left\{\Psi^{*}(\hat{A}\Psi)\right\}d\mathbf{r}}{\int \Psi^{*}\Psi\,d\mathbf{r}}. \tag{6.17}$$

Moreover, one can also consider the quantity

$$A \equiv \frac{\mathrm{Re}\left\{\Psi^*(\hat{A}\Psi)\right\}}{\Psi^*\Psi} \tag{6.18}$$

to represent the *local* value of the operator $\hat{A}$, given in terms of the associated field function $A(\mathbf{r}, t)$. In other words, the quantity (6.18) can be interpreted as the property A for a given particle.

For example, consider the position, momentum and energy operators in the configuration representation,

$$\hat{\mathbf{r}}(\mathbf{r}, \mathbf{r}') = \mathbf{r}\delta(\mathbf{r} - \mathbf{r}'), \tag{6.19a}$$

$$\hat{\mathbf{p}}(\mathbf{r}, \mathbf{r}') = -i\hbar\,\delta(\mathbf{r} - \mathbf{r}')\nabla, \tag{6.19b}$$

$$\hat{H}(\mathbf{r}, \mathbf{r}') = \delta(\mathbf{r} - \mathbf{r}')\left[-\frac{\hbar^2}{2m}\nabla^2 + \hat{V}(\hat{\mathbf{r}})\right], \tag{6.19c}$$

respectively. The associated field functions are

$$\mathbf{r}(\mathbf{r}, t) = \frac{\mathrm{Re}\left\{\Psi^*\hat{\mathbf{r}}\Psi\right\}}{\Psi^*\Psi} = \mathbf{r}(t), \tag{6.20a}$$

$$\mathbf{p}(\mathbf{r}, t) = \frac{\mathrm{Re}\left\{\Psi^*(-i\hbar\nabla)\Psi\right\}}{\Psi^*\Psi} = \nabla S, \tag{6.20b}$$

$$E(\mathbf{r}, t) = \frac{\mathrm{Re}\left\{\Psi^*\left(-\frac{\hbar^2}{2m}\nabla^2 + \hat{V}\right)\Psi\right\}}{\Psi^*\Psi} = \frac{(\nabla S)^2}{2m} + V_{\mathrm{eff}}, \tag{6.20c}$$

which will provide us with the position, momentum and energy of a Bohmian particle when they are evaluated along its trajectory. Indeed, in this case, note that (6.20a) corresponds precisely to the equation of motion (6.6), thus being a solution of (6.20b).

If instead of a particle there is a statistical ensemble of them (or, equivalently, some set of initial conditions has to be sampled) distributed according to $\rho(\mathbf{r}, t)$, the average value of A can be computed as in classical mechanics (see Sect. 1.4.1),

$$\langle A(t)\rangle = \int \rho(\mathbf{r}, t)A(\mathbf{r}, t)d\mathbf{r}. \tag{6.21}$$

Thus, sampling (6.19) over ρ,

$$\bar{\mathbf{r}} = \int \rho\mathbf{r}d\mathbf{r} = \int \Psi^*\hat{\mathbf{r}}\Psi\,d\mathbf{r} = \langle\hat{\mathbf{r}}\rangle, \tag{6.22a}$$

$$\bar{\mathbf{p}} = \int \rho\nabla S d\mathbf{r} = \int \Psi^*(-i\hbar\nabla)\Psi\,d\mathbf{r} = \langle\hat{\mathbf{p}}\rangle, \tag{6.22b}$$

$$\bar{E} = \int \rho\left[\frac{(\nabla S)^2}{2m} + V_{\mathrm{eff}}\right]d\mathbf{r} = \int \Psi^*\left[-\frac{\hbar^2}{2m}\nabla^2 + \hat{V}\right]\Psi\,d\mathbf{r} = \langle\hat{H}\rangle. \tag{6.22c}$$

which coincide with the corresponding expectation values obtained from standard quantum mechanics, this showing the equivalence at a predictive level of both approaches. Obviously, from a trajectory viewpoint, i.e., when the associated local field functions are evaluated along trajectories, (6.22) read as

$$\bar{\mathbf{r}}_B = \frac{1}{N} \sum_{i=1}^{N} w_i \mathbf{r}_i(t), \tag{6.23a}$$

$$\bar{\mathbf{p}}_B = \frac{1}{N} \sum_{i=1}^{N} w_i \nabla S(\mathbf{r}_i(t)), \tag{6.23b}$$

$$\bar{E}_B = \frac{1}{N} \sum_{i=1}^{N} w_i \left\{ \frac{[\nabla S(\mathbf{r}_i(t))]^2}{2m} + V_{\text{eff}}(\mathbf{r}_i(t)) \right\}, \tag{6.23c}$$

where N is the total number of trajectories considered, w_i is the associated weight—if the trajectories are initially sampled according to ρ_0, then $w_i = 1$ for all trajectories, otherwise $w_i \approx \rho(\mathbf{r}(t_0))$ —and the subscript B means that these average values are computed from a sampling of Bohmian trajectories. As in classical statistical treatments, provided the sampling of initial conditions is properly carried out according to some initial distribution function (this role is played here by $\rho(0)$), in the limit $N \rightarrow \infty$, the quantities (6.23) will correspond with their quantum homologous (6.22). Taking this into account, one readily notes that the uncertainty principle can be directly related to a statistical result instead of to an inherent impossibility to measure positions or momenta—the source for this impossibility would be rather associated with the way how things happen (interact) at quantum scales. In this sense, the inequality

$$\Delta r_i \Delta p_i \geq \frac{\hbar}{2} \tag{6.24}$$

expresses the relationship between two statistical quantities (in this case, position and momentum) in quantum mechanics, where

$$(\Delta r_i)^2 = \overline{r_i^2} - \overline{r_i}^2 \approx \overline{r_{B,i}^2} - \overline{r_{B,i}}^2, \tag{6.25a}$$

$$(\Delta p_i)^2 = \overline{p_i^2} - \overline{p_i}^2 \approx \overline{p_{B,i}^2} - \overline{p_{B,i}}^2, \tag{6.25b}$$

and $i = 1, 2, 3$.

It is clear from the above discussion that the role played by time in Bohmian mechanics and in quantum mechanics differs. Time in the quantum theory is not an observable, but a parameter, i.e., there is no time operator such that its eigenvalues provide us with some information, for example, about the time a particle needs to cross a barrier by tunnel effect, the time of a scattering process or the lifetime of a resonance phenomenon. Actually, when the time calculated through a given expression is compared to some experimental data value, things become more troublesome.

This issue constitutes an important question which has been considered in length in the literature [72–74], where several definitions of time can be found, such as dwell time, tunneling time, interaction time, arrival time, etc. This situation changes within the context of Bohmian mechanics, where the concept of arrival time is unambiguously defined because it is based on the concept of well-defined trajectories. Indeed, from the guidance condition or the trajectory, just by integration or by inspecting a graph [13], information about arrival times can be readily obtained with no need for a time operator. This crucial issue will be treated again in Volume 2, when discussing quantum processes, such as diffraction, resonance or reactive scattering.

Another very important issue, which also merits special attention, is that of scattering processes within the context of Bohmian mechanics. In particular, scattering singularities [61, 75], such as resonances, rainbows, glory effect or skipping orbits, mentioned in Chap. 1, will be discussed in more detail in Volume 2.

6.2.3 Quantum Hydrodynamics

In order to provide an interpretation to quantum mechanics, in 1926 Madelung [76] formulated what is known nowadays as *quantum hydrodynamics*, closely related to Bohmian mechanics. This interpretation is directly connected to some relevant phenomena in quantum mechanics, such as superconductivity [77] or Bose–Einstein condensation [78], for example. Furthermore, within chemical physics, it has accommodated very advantageously, for it provides an ideal framework to understand and interpret quantum processes there, from the chemical reactivity in collinear reactions [79–82] to the understanding of molecular magnetic properties within a framework encompassing both electronic structure and topology [83–93].

Here, one also starts by considering the wave function in polar form, (6.1), but focussing on

$$\rho = R^2 = \Psi^* \Psi, \tag{6.26a}$$

$$\mathbf{J} = \rho \mathbf{v} = R^2 \frac{\nabla S}{m}, \tag{6.26b}$$

where $\rho(\mathbf{r}, t)$ is the probability density, $\mathbf{J}(\mathbf{r}, t)$ is the quantum probability current density and $\mathbf{v}$ is the velocity fiel (6.6), which describes the flow of the latter. Taking this into account, (6.2) can be again expressed as

$$\frac{\partial \rho}{\partial t} + \nabla \cdot \mathbf{J} = 0, \tag{6.27a}$$

$$\frac{d\mathbf{v}}{dt} = \frac{\partial \mathbf{v}}{\partial t} + (\mathbf{v} \cdot \nabla)\mathbf{v} = -\frac{1}{m} \nabla(V + Q), \tag{6.27b}$$

which constitute the formal basis of quantum hydrodynamics and have a direct correspondence with those of classical fluid mechanics if m is identified with the mass of a piece of fluid separated from the rest by a closed surface, $m\rho$ is the fluid density and $\mathbf{v}$ is the velocity field of the flow [94]. However, unlike classical fluids, quantum fluids correspond to probability flows, with no material structure [95] That is, they only characterize statistical events at each point in space and time, in spite of the fact that the time evolution of these events can be better understood when compared with the motion of ordinary fluids. Moreover, whereas the classical concept of fluid can be applied to describe the statistical behavior of a macroscopic ensemble of particles, in quantum mechanics it is applied to single particles.

According to the preceding statements, (6.27a) can be interpreted as the continuity equation for the quantum flow, while (6.27b) represents the quantum Euler equation, analogous to the classical one for an ideal classical fluid (incompressible and non-viscous flow) when thermal effects are not taken into account. Indeed, considering the classical Euler equation for the component v_i of $\mathbf{v}$ (see (1.126) from Chap. 1),

$$\rho\left[\frac{\partial v_i}{\partial t} + (\mathbf{v}\cdot\nabla)v_i\right] = \rho f_i + \frac{\partial}{\partial x_j}(-p\delta_{ij}), \tag{6.28}$$

where f_i is the external force acting on the fluid along the i-direction and p the fluid pressure. This expression shows how the flow dynamics is determined by the influence of both an external force and other internal one, given by $\rho^{-1}\partial(-p\delta_{ij})/\partial x_j$ and that depends on the fluid properties. Now, ((6.27b) can be rewritten in the form of (6.28) as

$$\rho\left[\frac{\partial v_i}{\partial t} + (\mathbf{v}\cdot\nabla)v_i\right] = \rho f_i + \frac{\partial T_{ij}}{\partial x_j}, \tag{6.29}$$

by defining the *quantum stress tensor* as

$$T_{ij} = \frac{\hbar^2}{4m^2}\rho\frac{\partial\ln\rho}{\partial x_{ij}}, \tag{6.30}$$

which is the quantum counterpart of the classical stress tensor $-p\delta_{ij}$, and whose explicit dependence on ρ can be easily obtained by expressing Q as

$$Q = -\frac{\hbar^2}{2m}\left[\frac{1}{2}(\nabla\ln\rho)^2 + \nabla^2\ln\rho\right]. \tag{6.31}$$

An important topic in quantum hydrodynamics is the so-called quantum Navier–Stokes equation. In classical fluid dynamics, this equation expresses the rate of change in the momentum density, which is defined as the linear momentum times the fluid density (see Chap. 1). In the quantum version, apart from the classical contributions, quantum contributions are clearly identified and are related to quantum stress and pressure (see, for example, [16]).

As also happens with classical fluids, in quantum hydrodynamics one can also observe the presence of vortices and the corresponding associated (vortical) dynamics. The first theory on quantum vortices was first formulated by Dirac [96] in connection with the existence of magnetic monopoles. This theory has been described in detail in the literature [97, 98], finding some interesting applications [99]. More recent and explicit developments of the quantum theory of magnetic monopoles had led to a generalization of the concept of quantum vortex [100], the well-known Aharonov–Bohm effect [101] being related to this generalization. On the other hand, within the field of surface physics, the presence of this vortical motion has been detected in atom-surface scattering process with presence of impurities, where atoms may undergo a series of loops before they scape from the surface [102, 103].

The conditions leading to the formation of quantum vortices can be obtained from the fact that the complex character of the wave function implies the multi-valuedness of its phase

$$S'(\mathbf{r}, t_0) = S(\mathbf{r}, t_0) + 2\pi n\hbar, \quad n = 0, \pm 1, \pm 2, \ldots \tag{6.32}$$

This multivaluedness can only take place at those points where $\rho = 0$ (nodal regions of Ψ), where the smoothness of the wave function disappears and the value of S may undergo discrete jumps. According to (6.32), under these conditions $\mathbf{J}$ vanishes, but not the velocity field $\mathbf{v}$. By inspecting the circulation of $\mathbf{v}$ along a closed path, C, one finds that this magnitude is quantized,

$$\oint_C d\mathbf{l} \cdot \mathbf{v} = \oint_C d\mathbf{l} \cdot \frac{\nabla S}{m} = \frac{1}{m} \oint_C dS = \frac{2\pi n\hbar}{m}. \tag{6.33}$$

Applying Stoke's theorem to this result, it can be alternatively expressed as

$$\int_{\Sigma} d\mathbf{r} \cdot (\nabla \times \mathbf{v}) = \frac{2\pi n\hbar}{m}, \tag{6.34}$$

where Σ is the region enclosed by C. This result indicates the appearance of vortices when $n \neq 0$, which happens only at those points where the wave function presents nodes. At these points, the streamlines will be closed paths around the nodes, which is consistent with the fact that the quantum current density vanishes at those points and the impossibility of passing through the regions where $\rho = 0$. Conversely, the velocity field $\mathbf{v}$ will be irrotational in those regions free of quantum vortices.

6.2.4 Bohmian Mechanics in Complex Space

In the previous Section, it was shown how Bohmian mechanics arises when a transformation from the complex field variables (Ψ, Ψ^*) to the real variable fields (ρ, S) is considered. This transformation is necessary when looking for a theory based on real-valued fields; since Ψ is complex-valued, one real field will carry the information about the modulus of Ψ and the other one about its phase. Alternatively, the wave function can be expressed in terms of a complex phase,

$$\Psi(\mathbf{r}, t) = e^{i\bar{S}(\mathbf{r},t)/\hbar}, \tag{6.35}$$

and therefore there will be a one-to-one correspondence between Ψ and $\bar{S}$ —except for a constant $2\pi n\hbar$, as in (6.32). Of course, in the particular case

$$\bar{S}(\mathbf{r}, t) = S(\mathbf{r}, t) - i\hbar \ln R(\mathbf{r}, t) = S(\mathbf{r}, t) - \frac{i\hbar}{2} \ln \rho(\mathbf{r}, t), \tag{6.36}$$

standard Bohmian mechanics is recovered. After substitution of (6.35) into the time-dependent Schrödinger equation, we reach

$$\frac{\partial \bar{S}}{\partial t} + \frac{(\nabla \bar{S})^2}{2m} + \bar{V}_{\text{eff}} = 0, \tag{6.37}$$

which is a complex quantum Hamilton–Jacobi equation that can be regarded as the time-dependent Schrödinger equation associated with a logarithmic wave function. In this equation,

$$\bar{V}_{\text{eff}} \equiv V - \frac{i\hbar}{2m} \nabla^2 \bar{S} \tag{6.38}$$

is now an effective total complex potential, whose second component is a *complex quantum potential*,

$$\bar{Q} = -\frac{i\hbar}{2m} \nabla^2 S = \frac{i\hbar}{2} \left[\left(\frac{\nabla \Psi}{\Psi} \right)^2 - \frac{\nabla^2 \Psi}{\Psi} \right]. \tag{6.39}$$

Also as before, this (complex) quantum potential can be referred to as a sort of (complex) quantum kinetic energy, since

$$\hat{\mathcal{K}}\Psi = \frac{(\nabla \bar{S})^2}{2m} - \frac{i\hbar}{2m} \nabla^2 \bar{S}, \tag{6.40}$$

which is the complex analog of (6.5). Actually, if (6.36) is substituted into the right-hand side of (6.40),

$$\begin{aligned}
\hat{\mathcal{K}}\Psi &= \frac{1}{2m} \left\{ \left[(\nabla S)^2 - \frac{\hbar^2}{4} \left(\frac{\nabla \rho}{\rho} \right)^2 - \frac{i\hbar}{\rho} \nabla \rho \nabla S \right] \right. \\
&\quad \left. -i\hbar \left[\nabla^2 S - \frac{i\hbar}{2} \frac{\nabla^2 \rho}{\rho} + \frac{i\hbar}{2} \left(\frac{\nabla \rho}{\rho} \right)^2 \right] \right\} \\
&= \frac{(\nabla S)^2}{2m} + Q - \frac{i\hbar}{2m\rho} \nabla(\rho \nabla S). \tag{6.41}
\end{aligned}$$

From the second equality it is very apparent that the complex kinetic energy not only contains the real quantum potential, but its imaginary part provides us with information about the quantum flux conservation (i.e., about the rate of change $\partial \rho / \partial t$). In other words, it contains some extra information about the quantum flow, in such a way that even in those cases where $\bar{Q}$ may seem to be negligible (in comparison with its real counterpart), this does not mean that the nonlocal information ruling the quantum dynamics disappears and the motion will be classical. In general, as it is inferred from (6.41), due to the complex nature of $\bar{S}$, part of such an information will be contained in the first term of the kinetic energy.

In analogy to standard Bohmian mechanics, complex quantum trajectories can also be defined by analytic continuation of (6.37) to the complex plane, which is necessary for the corresponding equation of motion,[2]

$$\bar{\mathbf{v}} = \frac{\nabla \bar{S}}{m} = \frac{\hbar}{im} \frac{\nabla \bar{\Psi}}{\bar{\Psi}},$$ (6.42)

to be self-consistent: if $\bar{S}$ is complex, $\mathbf{v}$ has also to be complex-valued and, therefore, the integrated trajectory. This also implies that $\bar{S}$ (and eventually Ψ) will be evaluated along a complex trajectory $\mathbf{z}(t)$ rather than along a real coordinate $\mathbf{r}$ (this is the reason why $\bar{\Psi}$ and $\bar{\nabla}$ are used in (6.42), instead of Ψ and ∇), though its value will be (physically) meaningful only along the real axis. The relationship between the real Bohmian velocity and its complex counterpart,

$$\bar{\mathbf{v}} = \mathbf{v} - \frac{i\hbar}{2m} \frac{\nabla \rho}{\rho},$$ (6.43)

can be followed from (6.36). Formerly, Rosen [104–106] considered this expression a sort of total quantum mechanical momentum (or velocity) field which explains why it is possible to observe nonvanishing momenta in cases where the momentum ∇S vanishes. This effect would arise from the second term on the right-hand side of (6.43), which is assumed to be a "local" momentum assigned to the quantum mechanical field with which the particle interacts. In this way, (6.43) becomes the momentum that matches the momentum distributions provided by standard quantum mechanics, rather than the momentum ∇S from Bohmian mechanics, which is quite different [3, 107, 108]. This is consistent with the fact that Bohmian trajectories only carry information about the dynamics of the quantum flow, while complex quantum trajectories will also include information about the probability (as inferred from the analytic continuation of $\bar{S}$ from (6.36)). The dynamics in the complex configuration space thus explains in a natural way how to get the correct momentum distribution.

---

[2] Note that, unlike (6.6), the expression for $\bar{v}$ is not symmetric with respect to $\bar{\Psi}^*$. This is because, as previously mentioned, in this case the transformation is one to one, and therefore the complex conjugate wave field is not needed.

This also explains why algorithms based on complex trajectories are quite stable and accurate.

It is clear that if $\bar{v}$ is assumed to be complex, depending on the z-variable, while v and ρ depend on the real variable x, then (6.43) becomes an inequality. In the literature [109–121], (6.43) has been considered as an identity, particularly within the so-called *stochastic Bohmian mechanics* [120, 121], where the second term on the right hand side is interpreted as a stochastic diffusive term. However, not only this is not mathematical consistent, but the inequality forbids to associate the real part of the complex quantum trajectories with the (real, standard) Bohmian trajectories, as assumed in the literature [109–117, 122, 123]. That is, assuming $z(t) = z_r(t) + i z_i(t)$ describes the complex trajectory, with z_r and z_i being its real and imaginary parts, respectively, both being real functions, the equality $z_r(t) = x(t)$, where $x(t)$ would represent the standard Bohmian trajectory, does not hold as t goes on even if $z_0 = x_0$ at $t = 0$. Note that, from a strict Bohmian viewpoint, the equality $z_r(t) = x(t)$ means that the same (projected) z_r may present different (quantum) velocities, something contrary to what happens in standard Bohmian mechanics, where only one velocity can be associated with a space coordinate. Of course, only when moving to the complex plane one realizes that, effectively, that univaluedness still continues, except for those cases where a node of the wave function appears. This is precisely what can be appreciated when looking at the behavior of the complex counterparts of the Bohmian trajectories associated with an interference process [124], as displayed in the top panel of Fig. 6.2. The four groups of complex trajectories below, from (a) to (d), have been chosen such that the trajectories represented pass through the real axis (i.e., their imaginary part becomes zero) at the times $t = 0, 2, 4$ and 8, respectively. Therefore, if one of the Bohmian trajectories is considered, at such times it coincides with the complex quantum trajectories that cross the real axis at each one of those instants.

Once it is assumed that both terms in (6.43) depend on the complex z-variable, and then v and ρ are some generalized complex Bohmian functions ($\tilde{v}$ and $\tilde{\rho}$, respectively) depending on this variable instead of $\mathbf{r}$, it is instructive to apply $\bar{\nabla}$ on both sides, which yields

$$\bar{\nabla}\bar{v} = \bar{\nabla}\tilde{v} + \frac{i\hbar}{2m}\left[\left(\frac{\bar{\nabla}\tilde{\rho}}{\tilde{\rho}}\right)^2 - \frac{\bar{\nabla}^2\tilde{\rho}}{\tilde{\rho}}\right]. \tag{6.44}$$

As seen, the second term in the right-hand side reminds the functional dependence of Q on ρ in (6.4), except for a 1/2 factor inside the square bracket. In this sense, the effects of the (real) quantum potential (i.e., the nonlocality) are still present in the complex dynamics although the corresponding complex quantum potential, which is proportional to $\bar{\nabla}\tilde{v}$, might be relatively small (or even zero).

Equation (6.37) was formerly derived by Pauli during his studies on the quantum WKB approximation [125, 126]. However, the formalism based on the complex version of Bohmian mechanics is relatively recent, receiving much attention in these last years. The work developed here can also be framed within the framework of

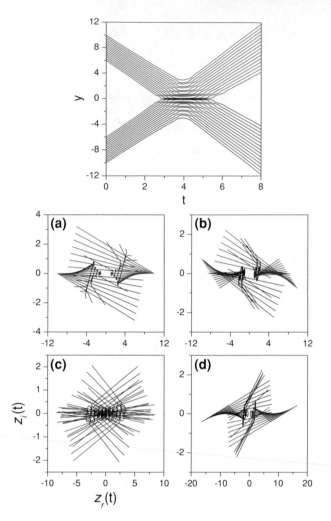

Fig. 6.2 *Top*: Bohmian trajectories associated with an interference of two incoming wave packets. *Bottom*: From (**a**) to (**d**), groups of complex quantum trajectories which cross the real axis at $t = 0, 2$, 4 and 8, respectively. To distinguish the contribution from each wave packet, both real and complex trajectories associated with each one are represented with different color. As their real counterparts, all complex trajectories start at $t = 0$ and end at $t = 8$. However, their initial conditions are chosen such that at the particular time indicated before, all of them (within the same frame) cross the real axis (i.e., their imaginary part vanishes)

the analytic versus synthetic approaches mentioned in the previous Section. From the analytic viewpoint, with interpretational purposes, one of the trends followed is the one aimed at studying stationary states. As can be easily shown, the velocity field **v** vanishes when the wave function is described by energy eigenfunctions associated with zero angular momentum states (i.e., *s*-waves) and, therefore, the corresponding Bohmian particles will remain standing at their initial positions at any time. In order to overcome this problem, different time-independent quantum Hamilton–Jacobi formulations have been formulated in the literature. For example, Floyd [127–134] and Faraggi and Matone [135–140] developed time-independent quantum Hamilton–Jacobi-like formulations starting from real bipolar ansatz, though not fully equivalence to standard quantum mechanics at a predictive level. However, later on, John [122, 123] proposed a time-dependent complex quantum trajectory formalism, namely the "modified de Broglie-Bohm approach to quantum mechanics", which has also been used and further developed by other authors [109–117, 141] to understand the problem of Bohmian stationarity in other kind of problems. On the other hand, also with interpretational purposes, (6.37) was found [58, 59, 61, 62] when trying to discriminate the amount of "quantumness" implicit in Bohmian trajectories (i.e., how different they are with respect to their classical counterparts) within the framework of the semiclassical WKB approximation. Finally, a series of fundamental works can also be found in the literature dealing with the dynamics of problems in the continuum, such as interference [124, 142, 143], entanglement [110] or stochastic complex Bohmian mechanics [109, 117].

On the other hand, from the analytic viewpoint, Leacock and Padgett used [144, 145] the connection formula (6.35)—the same considered later by John—as an alternative way to tackle the problem of the calculation of stationary or bound states. This is formally equivalent to the more recent "Bohmian mechanics with complex action" developed by Tannor and coworkers [146–153] and the methodology developed by Wyatt and coworkers [154–159] for computational purposes without the need to solve the time-dependent Schrödinger equation. In principle, using analytical continuation it is possible to implement numerical codes that benefit from working in a complex configuration space—these advantages are similar to those that in electromagnetism, for example, lead to consider complex fields instead of real ones. More specifically, the main idea behind the development of computational tools based on the complex trajectory methodology is that the wave function on the whole real axis can be synthesized using the information transported by those particles crossing the real axis simultaneously. This allows us to define a curve, namely the *isochrone* [148, 157, 158], which joins the specific initial positions of trajectories such that their crossing with the real axis occurs at the same time. Computationally, the main problem in dealing with the complex dynamics is locating isochrones, which is similar to the root search problem in semiclassical mechanics [75]. In this regard, methods devised to solve the latter might prove useful for the isochrone problem [158, 159]. Actually, within this picture, three points are worth stressing. First, all the complex trajectories associated with an isochrone will reach the real axis simultaneously [156]. Second, the uniqueness in complex Bohmian mechanics arises from the bonds established by the initial real wave function, which is in the end the observable magnitude (through ρ). Therefore, though there might be many initial conditions leading to the same point on the real axis, only the isochrones connect the complex problem with the real one, thus establishing the same uniqueness observed in standard Bohmian mechanics. Third, a Bohmian trajectory does not correspond, therefore, to a given complex trajectory or to the family of complex trajectories associated with a given isochrone, but to a family of complex trajectories such that their crossings with the real axes take place consecutively, one after the other. In other words, a Bohmian trajectory itself defines a family or set of complex quantum trajectories [124].

6.2.5 Feynman's Paths and Bohmian Trajectories

As seen in Sect. 3.4, the essential feature of Feynman's formulation of quantum mechanics consists of assigning a phase or probability amplitude to each classical path from an ensemble, though their associated probabilities can be the same for all of them. In this way, the probability for an event to happen arises as the combined effect—*interference*—of all the classical paths considered. In order to better establish a connection with Bohmian mechanics, consider two causal events, a and b; the first takes place on $\mathbf{r}_a$ at t_a and the later on $\mathbf{r}_b$ at t_b. According to Sect. 1.2.1, the path joining both events is described by a trajectory $\bar{\mathbf{r}}(t)$ determined through the least action principle. That is, the corresponding classical action is an extremum when evaluated along it. In this way, classical mechanics provides us with a criterion to distinguish between physical and nonphysical trajectories. In quantum mechanics, though, according to Feynman such a distinction cannot be considered *a priori*, for any thought trajectory joining those two events is equally valid. Therefore, unlike classical mechanics, in Feynman's approach both the functional form of the action S_{cl} as well as its value when evaluated along any path are relevant.

The combined effect of all possible paths was defined by means of a propagator, which acts according to *Huygens' principle*. That is, following an optical analogy, the (quantum) wave function at the time t_b is not other thing that the interference of a series of secondary wavelets starting from a wave function at a previous time t_a, i.e.,

$$\Psi(\mathbf{r}_b, t_b) = \int K[\mathbf{r}_b, t_b; \mathbf{r}_a, t_a]\Psi(\mathbf{r}_a, t_a)d\mathbf{r}_a, \qquad (6.45)$$

with

$$K[\mathbf{r}_b, t_b; \mathbf{r}_a, t_a] = \int_a^b e^{iS_{cl}[b,a]/\hbar}\mathcal{D}x(t), \qquad (6.46)$$

as seen in Sect. 3.4. The dependence of this sum with each one of the paths considered is contained in the action $S_{cl}[b, a]$ when evaluated along them, thus taking all of them into account—though the "classical" ones are only those which make $S_{cl}[b, a]$ to be an extremum. On the other hand, Huygens' secondary wavelets are described by the term $e^{iS_{cl}[b,a]/\hbar}$, which can be interpreted as dressing each trajectory with a wave.

In order to find a relationship between Feyman's formulation and Bohmian mechanics, it is important to note that while Feynman's paths constitute an ensemble of virtual paths joining two given points, only one Bohmian trajectory will join such points [13]—classically, there would be a set of them in configuration space and only one in phase space. This trajectory is precisely the one that satisfies the guiding condi-tion $\dot{\mathbf{r}} = \nabla S/m$. In particular, S can be associated with the phase of the kernel (6.46) when it is expressed in the form $K = Re^{iS/\hbar}$. In analogy to the way how a trajectory is defined in classical mechanics (see Sect. 1.2.1), here a quantum trajectory can be defined as

$$S(\mathbf{r}, t; \mathbf{r}_0, t_0) = \int_{(\mathbf{r}_0,t_0)}^{(\mathbf{r},t)} \left(\frac{1}{2}m\dot{\mathbf{r}}^2 - V - Q \right) dt, \qquad (6.47)$$

which should be an extremal along such a trajectory. Keeping this in mind, Feynman's formulation could be interpreted as a way that allows us to obtain the quantum action by means of a superposition of all possible classical-like actions. Similarly, a quantum trajectory can be considered as a sort of superposition of classical trajectories, each one of them contributing to the interference with a different phase. This allows us to assume the quantum momentum of a particle at a certain point $(\mathbf{r}, t)$ as given by a function of the "virtual" momenta associated with each one of the Feynman paths which pass through such a point. In order to see this more apparently, if (6.46) is written as a discrete sum $K = N \sum_j e^{iS_j/\hbar}$, where j labels the corresponding classical trajectory, the quantum momentum at the point $(\mathbf{r}, t)$ will be

$$\nabla S = N^2 R^{-2} \left[\sum_j \nabla S_j + \sum_{j \neq k} \nabla S_j \cos[(S_j - S_k)/\hbar] \right]. \qquad (6.48)$$

As can be seen, this expression gives the quantum momentum as an average over clas-sical momenta, ∇S_j, as well as an additional term which accounts for the interference between the different secondary wavelets dressing the classical trajectories—note that, in general, the momentum ∇S_j corresponding to evaluate the classical action along a quantum trajectory is not the same as ∇S.

6.3 Towards the Classical Limit in Bohmian Mechanics

According to the *correspondence principle* stated by Bohr [160] in 1923, quantum systems may behave classical-like under certain conditions, thus being describable by means of the classical mechanical laws. This is usually regarded as the *classical limit* of quantum mechanics. Usually, this limit relies on assuming that the value of a certain magnitude of interest becomes meaningless (e.g., $\hbar$) or, on the contrary, very large (e.g., the principal quantum number)—though not always different clas-sicality criteria lead to the same limit [161, 162]. In the case of Bohmian mechanics, for example, it is typical to regard the classical limit (at least, from a formal view-point) as the regime where Q becomes negligible and, therefore, (6.2b) becomes the classical Hamilton–Jacobi equation. This condition can be satisfied, for example, by increasing the mass m of the particles described. However, though this may seem the correct way to operate, what one really observes is that this condition does not ensure the appearance of classical trajectories, as can be seen in atom-surface scattering [57], for example. In this case, though the corresponding diffraction pattern (i.e., the observable quantity in this type of quantum process) behaves on average like the corresponding classical pattern, the "fine grain" shows very strong oscillations. Similarly, the quantum trajectories do not behave at all as their classical counterparts, for they still contain some nonlocal information (i.e., *coherence*) through ρ. Remember that regarding dynamical effects the "shape" of ρ is more relevant than its intensity. Thus, very tinny values of ρ may lead to very dramatic dynamical effects. This is in sharp contrast with other limits in physics, like the passage from relativistic to Newtonian mechanics, where a gradual, smooth transition is observed as particle velocities become much smaller than the speed of light.

6.3.1 The JWKB Approximation

As seen in Sect. 3.5.1, Ehrenfest's theorem helps us to establish certain criteria of classicality, e.g., obtaining the conditions for the center of a wave packet to move like a classical particle [163]. Now, as also seen in Sect. 6.2.2, such a wave packet can be interpreted as a swarm of non-interacting particles moving according to the motion laws of Bohmian mechanics, which in Newtonian terms read as

$$\frac{d\mathbf{r}}{dt} = \frac{\mathbf{p}}{m}, \tag{6.49a}$$

$$\frac{d\mathbf{p}}{dt} = -\nabla V_{\text{eff}}, \tag{6.49b}$$

and distribute as $\rho(\mathbf{r}, t)$. Therefore, the corresponding average values will evolve as

$$\frac{d\bar{\mathbf{r}}}{dt} = \frac{\langle \mathbf{p} \rangle}{m}, \tag{6.50a}$$

$$\frac{d\bar{\mathbf{p}}}{dt} = -\overline{\nabla V}_{\text{eff}}, \tag{6.50b}$$

under conditions of (Ehrenfest) classicality. This does not mean necessarily that Bohmian particles move like classical ones, but only on average (i.e., their distribution). A good example illustrating this fact is the one mentioned above on atom-surface scattering, where the mass of the incident particles is gradually increased [57]: the average distributions reproduce classical-like results, but quantum trajectories behave very differently with respect to their classical counterparts.

Ehrenfest's theorem may constitute a first step when trying to render some light on the transition to the classical limit. However, the JWKB approximation (see also Sect. 7.3) results more insightful when dealing with Bohmian mechanics. Due to the explicit series developments in terms of $\hbar$—though not always this is a good criterion—, it yields a more direct correspondence between Bohmian and classical mechanics, i.e., to establish a closer connection between two trajectory-based formulations. Thus, as in optics [164], here one also proceeds with the ansatz

$$\Psi(\mathbf{r}, t) = e^{i\bar{S}(\mathbf{r},t)/\hbar}, \qquad (6.51)$$

with $\bar{S}$ being a complex function that varies slowly in space. When (6.51) is substituted into the time-dependent Schrödinger equation,

$$\frac{\partial \bar{S}}{\partial t} + \frac{(\nabla \bar{S})^2}{2m} + V + \frac{\hbar}{2mi} \nabla^2 \bar{S} = 0, \qquad (6.52)$$

which corresponds to (6.37). In this regard, notice that in spite of the assumptions on $\bar{S}$, it is a general equation. Nevertheless, in the classical limit $\hbar \to 0$, it can be assumed that $\bar{S}$ can be expanded as a series of $\hbar/i$,

$$\bar{S} = \sum_{n=0}^{\infty} \left(\frac{\hbar}{i}\right)^n \bar{S}^{(n)}, \qquad (6.53)$$

where the functions $\bar{S}^{(n)}$ are real. Inserting this series into (6.52), one obtains

$$\sum_{n=0}^{\infty} \left(\frac{\hbar}{i}\right)^n \frac{\partial \bar{S}^{(n)}}{\partial t} + \frac{1}{2m} \sum_{n=0}^{\infty} \left(\frac{\hbar}{i}\right)^n \sum_{k=0}^{n} \nabla \bar{S}^{(k)} \cdot \nabla \bar{S}^{(n-k)}$$
$$+ V + \frac{1}{2m} \sum_{n=0}^{\infty} \left(\frac{\hbar}{i}\right)^{n+1} \nabla^2 \bar{S}^{(n)} = 0. \qquad (6.54)$$

The JWKB approximation consists of solving order by order, in powers of $\hbar/i$, the coupled equations involved in (6.54). Thus, at zeroth order,

$$\frac{\partial \bar{S}^{(0)}}{\partial t} + \frac{(\nabla \bar{S}^{(0)})^2}{2m} + V = 0, \qquad (6.55)$$

which is the classical Hamilton–Jacobi equation, with $\bar{S}^{(0)}(\mathbf{r}, t)$ being the classical action, S_{cl}. Note that although $\bar{S}$ is not a real function in general, $\bar{S}^{(0)}$ is real because (6.55) is real.

Regarding the remaining terms of (6.54), i.e.,

$$\sum_{n=1}^{\infty} \left(\frac{\hbar}{i}\right)^n \frac{\partial \bar{S}^{(n)}}{\partial t} + \frac{1}{2m} \sum_{n=1}^{\infty} \left(\frac{\hbar}{i}\right)^n \sum_{k=0}^{n} \nabla \bar{S}^{(k)} \cdot \nabla \bar{S}^{(n-k)}$$
$$+ \frac{1}{2m} \sum_{n=1}^{\infty} \left(\frac{\hbar}{i}\right)^n \nabla^2 \bar{S}^{(n-1)} = 0, \qquad (6.56)$$

they lead us to a hierarchy of equations which couple the different higher orders of $\bar{S}$. These equations can be expressed in a general form as

$$\frac{\partial \bar{S}^{(n)}}{\partial t} + \frac{1}{2m} \sum_{k=0}^{n} \nabla \bar{S}^{(k)} \cdot \nabla \bar{S}^{(n-k)} + \frac{1}{2m} \nabla^2 \bar{S}^{(n-1)} = 0, \tag{6.57}$$

which couple the nth order with the remaining lower ones. Here, it is interesting to note that, since (6.57) is real as well as $\bar{S}^{(0)}$, all the remaining n orders will also be real.

Usually, the *semiclassical* wave function is defined as

$$\overline{\Psi}(\mathbf{r}, t) = e^{\bar{S}^{(1)}(\mathbf{r},t)+i\bar{S}^{(0)}(\mathbf{r},t)/\hbar}, \tag{6.58}$$

i.e., taking into account only the zeroth and first orders of $\bar{S}$—in particular, the first exponential is related to the van Vleck determinant seen in Sect. 3.5.2—, where the latter reads as

$$\frac{\partial \bar{S}^{(1)}}{\partial t} + \frac{1}{m} \nabla \bar{S}^{(0)} \cdot \nabla \bar{S}^{(1)} + \frac{1}{2m} \nabla^2 \bar{S}^{(0)} = 0 \tag{6.59}$$

and comes from (6.57) when $n = 1$. The reason why semiclassically only the first two terms of $\bar{S}$ in (6.58) are considered comes from the fact that they are related to the probability density and the quantum probability current density,

$$\rho(\mathbf{r}, t) \simeq e^{2\bar{S}^{(1)}(\mathbf{r},t)}, \tag{6.60a}$$

$$\mathbf{J}(\mathbf{r}, t) = \frac{\hbar}{m} \text{Im}[\Psi^* \nabla \Psi] \simeq \frac{1}{m} e^{2\bar{S}^{(1)}} \nabla \bar{S}^{(0)}, \tag{6.60b}$$

respectively, which (at this order of approximation) will be independent (at least explicitly) of $\hbar$. Furthermore, it can be noticed that (6.60b) can be expressed as

$$\mathbf{J}(\mathbf{r}, t) \simeq \rho(\mathbf{r}, t)\mathbf{v}_{cl}(\mathbf{r}, t), \tag{6.61}$$

where $\mathbf{v}_0 = \nabla \bar{S}_0/m$ describes the classical velocity (actually, a generalized velocity, $\mathbf{v}_{cl} \equiv \mathbf{p}/m$) within the Hamilton–Jacobi formulation of classical mechanics. That is, taking into account this picture, the classical limit is understood as a motion regime characterized by a continuous mass density, $m\rho(\mathbf{r}, t)$, under the action of a (classical) velocity field $\mathbf{v}_{cl}(\mathbf{r}, t)$.

If the wave function (6.51) is expressed in polar form, taking (6.53) into account,

$$R = \exp\left[\sum_{n=0}^{\infty}(-1)^n \hbar^{2n} \bar{S}^{(2n+1)}\right], \tag{6.62a}$$

$$S = \sum_{n=0}^{\infty}(-1)^n \hbar^{2n} \bar{S}^{(2n)}. \tag{6.62b}$$

Substituting (6.62b) into (6.6), one obtains the expression of the quantum trajectories within the JWKB approach, i.e., in terms of $\hbar$,

$$\dot{\mathbf{r}} = \frac{1}{m}\sum_{n=0}^{\infty}(-1)^n\hbar^{2n}\nabla\bar{S}^{(2n)} = \dot{\mathbf{r}}_{\mathrm{cl}} + \frac{1}{m}\sum_{n=1}^{\infty}(-1)^n\hbar^{2n}\nabla\bar{S}^{(2n)}, \qquad (6.63)$$

where $\dot{\mathbf{r}}_{\mathrm{cl}} = \nabla\bar{S}_t^{(0)}/m$ is the classical law of motion. Therefore, from (6.63) one can interpret quantum trajectories as classical trajectories "dressed" with a series of terms coming from quantum interference, showing the capital difference between both types of trajectories. Moreover, also from (6.63) it is very apparent how classical mechanics underlies quantum mechanics and, therefore, how quantum phenomena will keep a reminiscence of a classical-like feature—which will be stronger as the classical limit is approached. This can also be seen by reexpressing (6.63) in a Newtonian-like way,

$$\ddot{\mathbf{r}} = \frac{\partial}{\partial t}\left(\frac{\nabla S}{m}\right) = -\frac{\nabla}{m}\sum_{n=0}^{\infty}(-1)^n\,\hbar^{2n}\frac{\partial\bar{S}^{(2n)}}{\partial t}$$

$$= -\frac{\nabla}{m}\sum_{n=0}^{\infty}(-1)^n\frac{\hbar^{2n}}{2m}\left[\sum_{k=0}^{2n}\nabla\bar{S}^{(k)}\cdot\nabla\bar{S}^{(2n-k)} + \nabla^2\bar{S}^{(2n-1)}\right]. \qquad (6.64)$$

As can be noticed, from this expression the classical potential can be straightforwardly obtained as a function of $\bar{S}^{(0)}$,

$$V = -\frac{\partial\bar{S}^{(0)}}{\partial t} - \frac{(\nabla\bar{S}^{(0)})^2}{2m}, \qquad (6.65)$$

i.e., in terms of the corresponding classical Hamilton–Jacobi equation. On the other hand, the quantum potential will consist of the remaining terms,

$$Q = -\frac{\hbar^2}{2m}\left[\sum_{n=0}^{\infty}(-1)^n\hbar^{2n}\nabla^2\bar{S}^{(2n+1)} + \sum_{n=0}^{\infty}(-1)^n\hbar^{2n}\sum_{k=0}^{n}\bar{S}^{(2k+1)}\bar{S}^{(2(n-k)+1)}\right],$$

$$\qquad (6.66)$$

which, as is apparent, is a much more complicated function of the $\bar{S}^{(n)}$.

6.3.2 Interaction and Entanglement

Consider two isolated quantum objects regardless of their size or properties and that the total wave function describing them is a product state (factorizable) of the wave functions associated with each object. As already mentioned by Schrödinger [165, 166], as soon as these objects enter into contact one with another, i.e., as they interact, the total wave function becomes no longer factorizable and they cannot be described as independent entities. The new quantum state becomes *entangled*, with the property that any quantum operation performed on one of the objects will have important implications on the other one independently of how far apart they are [10–12, 167]. This property has led to the well-known quantum information theory [168] as well as mechanisms, such as decoherence, which are used to explain the appearance of the classical world from quantum mechanics [169–171].

Before going to the implications of entanglement and, in particular, its implications in the classical limit, let us consider a set of N bodies, each one represented at a given time by a solution of the corresponding separated Schrödinger equation. The total N-body wave function can be expressed, as said above, as a product of N single wave functions,

$$\Psi(\mathbf{r}_1, \mathbf{r}_2, \ldots, \mathbf{r}_N, t) = \psi_1(\mathbf{r}_1, t)\psi_2(\mathbf{r}_2, t)\cdots\psi_N(\mathbf{r}_N, t)$$
$$= \Pi_{i=1}^N \rho_i^{1/2}(\mathbf{r}_i, t) \, e^{i S_i(\mathbf{r}_i, t)/\hbar}. \tag{6.67}$$

In this case, it is easy to show that from this total wave function, individual solutions of the Bohmian equations of motion are obtained according to

$$\mathbf{v}_i = \dot{\mathbf{r}}_i = \frac{\nabla_{\mathbf{r}_i} S_i(\mathbf{r}_i, t)}{m}, \tag{6.68}$$

since the quantum potential is also separable as a sum of partial quantum potentials,

$$Q = \sum_{i=1}^N Q_i = \sum_{i=1}^N -\frac{\hbar^2}{2m} \frac{\nabla^2 \rho_i^{1/2}}{\rho_i^{1/2}}. \tag{6.69}$$

This implies that the Bohmian trajectories described by the different particles are independent and no information is transmitted among them. Of course, this also happens if instead of N different bodies, N different degrees of freedom are considered, for the treatment is exactly the same. In order to illustrate this, consider a wave function which depends on two coordinates, x and y—which may represent two different bodies or two different degrees of freedom—, is factorizable in terms of single–particle partial wave functions,

$$\Psi(x, y, t) = \psi_1(x, t)\psi_2(y, t) = \rho_1^{1/2}(x, t)\rho_2^{1/2}(y, t) \, e^{i\left[S_1(x,t)+S_2(y,t)\right]/\hbar}. \tag{6.70}$$

Thus, each coordinate follows an independent equation of motion,

$$v_1 = \frac{1}{m}\frac{\partial S_1(x, t)}{\partial x}, \quad v_2 = \frac{1}{m}\frac{\partial S_2(y, t)}{\partial y}, \tag{6.71}$$

which are uncoupled because $S(x, y, t) = S_1(x, t) + S_2(y, t)$. This is also apparent from the corresponding total quantum potential,

$$Q(x, y, t) = -\frac{\hbar^2}{2m} \frac{1}{\rho_1^{1/2}(x, t)} \frac{\partial^2 \rho_1^{1/2}(x, t)}{\partial x^2} - \frac{\hbar^2}{2m} \frac{1}{\rho_2^{1/2}(y, t)} \frac{\partial^2 \rho_2^{1/2}(y, t)}{\partial y^2}$$

$$= Q_1(x, t) + Q_2(y, t), \tag{6.72}$$

which is fully separable because $\rho(x, y, t) = \rho_1(x, t)\rho_2(y, t)$.

Typically, factorizability is closely connected to *distinguishability*, i.e., with the idea that both objects are distinguishable and can be monitored independently. For example, this is the case of particles describable by means of a Maxwell–Boltzmann statistics. That is, for two of these particles their total wave function can be expressed as

$$\Psi(\mathbf{r}_1, \mathbf{r}_2, t) = \Phi_A(\mathbf{r}_1, t)\Phi_B(\mathbf{r}_2, t). \tag{6.73}$$

Usually, the particles constituting this kind of ensembles are non-interacting and they can be accounted for by only taking into account how they distribute. When the particles forming an ensemble do interact among themselves and become indistinguishable (i.e., truly quantum-mechanical), apart from their distribution it is also very important how they interact. In this case, even if they are very far apart, their quantum state cannot be described in a simple manner, but it is an entangled state. This is the case of the Fermi–Dirac or Bose–Einstein statistics, which describe fermions or bosons, respectively. That is, if two of these particles are considered, their total wave function will be

$$\Psi(\mathbf{r}_1, \mathbf{r}_2, t) = N_{\pm}[\Phi_A(\mathbf{r}_1, t)\Phi_B(\mathbf{r}_2, t) \pm \Phi_A(\mathbf{r}_2, t)\Phi_B(\mathbf{r}_1, t)], \tag{6.74}$$

where the minus sign stands for fermions (wave function with odd parity under exchange of the positions of the particles) and the plus sign for bosons (even parity).

As a simple illustration of these ideas, consider the case of two particles which interact through a potential V at some time. The two-particle wave function describing this system will be

$$\Psi(x, y, t) = \rho(x, y, t)\, e^{iS(x, y, t)/\hbar}. \tag{6.75}$$

The trajectories for these particles are obtained from

$$\mathbf{v}_1 = \frac{1}{m} \frac{\partial S(x, y, t)}{\partial x}, \quad \mathbf{v}_2 = \frac{1}{m} \frac{\partial S(x, y, t)}{\partial y}, \tag{6.76}$$

which are influenced by the quantum potential

$$Q(x, y, t) = -\frac{\hbar^2}{2m} \frac{1}{R(x, y, t)} \left[\frac{\partial^2 R(x, y, t)}{\partial x^2} + \frac{\partial^2 R(x, y, t)}{\partial y^2} \right]. \tag{6.77}$$

According to Schrödinger [165, 166], after the interaction, even if the two particles are very far apart one from another, the wave function (6.75) becomes non-factorizable, i.e.,

$$\Psi(x, y, t) \neq \psi(x, t)\psi(y, t), \tag{6.78}$$

and the evolution of both particles will remain entangled.

In general, given an entangled state described by the wave function

$$\Psi(\mathbf{r}_1, \mathbf{r}_2, \ldots, \mathbf{r}_N, t) = R(\mathbf{r}_1, \mathbf{r}_2, \ldots, \mathbf{r}_N, t) \, e^{i S(\mathbf{r}_1, \mathbf{r}_2, \ldots, \mathbf{r}_N, t)/\hbar}, \tag{6.79}$$

the corresponding quantum trajectories will be obtained by integrating the equation of motion

$$\mathbf{v}_i = \frac{\nabla_i S}{m}, \tag{6.80}$$

where the right-hand side will be dependent in general on the degrees of freedom involved in the problem, though ∇ is computed with respect to $\mathbf{r}_i$. This scheme gives rise to a set of N equations of motion coupled through the total phase S, where the evolution of a particle will be strongly nonlocally influenced by the other (apart from other classical like interactions through V). This entanglement [165, 166] becomes more apparent through the quantum potential,

$$Q = -\frac{\hbar^2}{2m} \sum_{i=1}^{N} \frac{\nabla_i^2 R}{R}, \tag{6.81}$$

where $Q = Q(\mathbf{r}_1, \mathbf{r}_2, \ldots, \mathbf{r}_N, t)$, which is nonseparable and, therefore, strongly nonlocal. Different works in the literature analyze the trajectory correlation among entangled particles [172–179], the most recent one within the many-body context of transport phenomena in mesoscopic systems [179].

6.3.3 Mixed Bohmian-Classical Mechanics

The reason why quantum systems behave very differently from classical ones is because of the property of coherence, i.e., their ability to interfere (as it also happens in optics). Nonetheless, this phenomenon cannot be gradually suppressed under any limit. In order to eliminate totally interferences and, therefore, achieve an appropriate classical limit, it is necessary to introduce an additional mechanism, though still within the framework of quantum mechanics. This mechanism is *decoherence* [1, 169–171], which is the gradual loss of coherence by an interaction with an environment. This interaction makes the wave function to become entangled. Consequently, if one only "looks" at the system of interest by tracing out over the remaining degrees of freedom, a loss of coherence is observed (i.e., its capability to interfere).

As mentioned in Chap. 3, there are different hybrid approaches to deal with many degree-of-freedom systems, where the system degrees of freedom are described quantum-mechanically while the bath ones are accounted for classically. Among them, there are methods such as the mean-field approximation [180] or the surface hopping trajectories [181, 182]. In all these methods, the key point is the implementation of the so-called *backreaction* [183], i.e., the action of the system over the bath, since the contrary is simple and it is usually done in terms of a time-dependent potential which is function of the bath coordinates (following the BOA scheme). Here, this problem will be analyzed within the framework of the mixed quantum-Bohmian approach [42–46].

Thus, consider a two dimensional total system, where x and y denote the coordinates of the two subsystems, with masses m_x and m_y, respectively. As before, an ansatz wave function with polar form is considered, but taking into account that now it represents a solution of a two-dimensional Schrödinger equation. In this case, the corresponding continuity and quantum Hamilton–Jacobi equations will read as

$$\frac{\partial R^2}{\partial t} + \frac{1}{m_x}\frac{\partial S}{\partial x}\frac{\partial R^2}{\partial x} + \frac{1}{m_y}\frac{\partial S}{\partial y}\frac{\partial R^2}{\partial y} = -R^2\left(\frac{1}{m_x}\frac{\partial^2 S}{\partial x^2} + \frac{1}{m_y}\frac{\partial^2 S}{\partial y^2}\right), \tag{6.82a}$$

$$\frac{\partial S}{\partial t} + \frac{1}{2m_x}\left(\frac{\partial S}{\partial x}\right)^2 + \frac{1}{2m_y}\left(\frac{\partial S}{\partial y}\right)^2 = -V_{\text{eff}}, \tag{6.82b}$$

respectively, where the quantum potential has the form

$$Q(x, y, t) = -\frac{\hbar^2}{2m_x}\frac{1}{R}\frac{\partial^2 R}{\partial x^2} - \frac{\hbar^2}{2m_y}\frac{1}{R}\frac{\partial^2 R}{\partial y^2}. \tag{6.83}$$

Focussing now on the quantum-classical $(x-y)$ coupling, (6.82b) is recast in its Eulerian form. In order to do so, the operators $\partial/\partial x$ and $\partial/\partial y$ are applied to (6.82b). This gives rise to the coupled equations

$$\frac{\partial}{\partial t}\left(\frac{\partial S}{\partial x}\right) + \frac{1}{m_x}\frac{\partial S}{\partial x}\frac{\partial^2 S}{\partial x^2} + \frac{1}{m_y}\frac{\partial S}{\partial y}\frac{\partial^2 S}{\partial y\partial x} = -\frac{\partial V_{\text{eff}}}{\partial x}, \tag{6.84a}$$

$$\frac{\partial}{\partial t}\left(\frac{\partial S}{\partial y}\right) + \frac{1}{m_y}\frac{\partial S}{\partial y}\frac{\partial^2 S}{\partial y^2} + \frac{1}{m_x}\frac{\partial S}{\partial x}\frac{\partial^2 S}{\partial x\partial y} = -\frac{\partial V_{\text{eff}}}{\partial y}, \tag{6.84b}$$

respectively, which are expressed in terms of the Bohmian velocities $p_x = \partial S/\partial x$ and $p_y = \partial S/\partial y$. Note that, taking into account the definition of the Lagrange time derivative for this case, (6.84a) and (6.84b) just represent the quantum force undergone by each subsystem,

$$m_x\frac{d^2 x}{dt^2} = -\frac{\partial V_{\text{eff}}}{\partial x}, \quad m_y\frac{d^2 y}{dt^2} = -\frac{\partial V_{\text{eff}}}{\partial y}. \tag{6.85}$$

Now, assume $m_y \gg m_x$. Under this hypothesis, it is expected that the subsystem y will behave almost classically. For example, if the full system is represented initially by a Gaussian wave packet, it will not display an important spreading along the y direction and, therefore, the second space-derivatives of S and R along this direction will be negligible. This allows us to reexpress (6.84a) and (6.84b) as

$$\frac{\partial}{\partial t}\left(\frac{\partial \tilde{S}}{\partial x}\right) + \frac{1}{m_x}\frac{\partial \tilde{S}}{\partial x}\frac{\partial^2 \tilde{S}}{\partial x^2} + \frac{1}{m_y}\frac{\partial \tilde{S}}{\partial y}\frac{\partial^2 \tilde{S}}{\partial y \partial x} = -\frac{\partial \tilde{V}_{\text{eff}}}{\partial x}, \tag{6.86a}$$

$$\frac{\partial}{\partial t}\left(\frac{\partial \tilde{S}}{\partial y}\right) + \frac{1}{m_x}\frac{\partial \tilde{S}}{\partial x}\frac{\partial^2 \tilde{S}}{\partial x \partial y} = -\frac{\partial \tilde{V}_{\text{eff}}}{\partial y}, \tag{6.86b}$$

where $\tilde{S}$ and $\tilde{R}$ represent the approximate values of S and R, respectively, under this assumption. In these equations, $\tilde{V}_{\text{eff}}$ is the corresponding approximate effective potential, with

$$\tilde{Q}(x, t|y) = -\frac{\hbar^2}{2m_x}\frac{1}{\tilde{R}}\frac{\partial^2 \tilde{R}}{\partial x^2}, \tag{6.87}$$

where $(x, t|y)$ means that $\tilde{Q}$ depends on y implicitly, through a sort of parametrization. On the other hand, (6.82a) becomes the approximate continuity equation,

$$\frac{\partial \tilde{R}^2}{\partial t} + \frac{\partial}{\partial x}\left(\frac{\tilde{R}^2}{m_x}\frac{\partial \tilde{S}}{\partial x}\right) + \frac{1}{m_y}\frac{\partial \tilde{S}}{\partial y}\frac{\partial \tilde{R}^2}{\partial y} = 0. \tag{6.88}$$

Evaluating (6.86a) and (6.88) along the quasi-classical trajectory $y(t)$ allows us to define the pseudo-Lagrangian time derivative operator

$$\frac{d}{dt} = \frac{\partial}{\partial t} + v_y\frac{\partial}{\partial y}. \tag{6.89}$$

and, therefore, to reexpress those equations as

$$\frac{d}{dt}\left(\frac{\partial \tilde{S}}{\partial x}\right) + \left(\frac{1}{m_x}\frac{\partial \tilde{S}}{\partial x}\right)\left(\frac{\partial^2 \tilde{S}}{\partial x^2}\right) = -\frac{\partial \tilde{V}_{\text{eff}}}{\partial x}, \tag{6.90a}$$

$$\frac{d\tilde{R}^2}{dt} + \frac{\partial}{\partial x}\left(\tilde{R}^2\frac{1}{m_x}\frac{\partial \tilde{S}}{\partial x}\right) = 0, \tag{6.90b}$$

which satisfy the pseudo-Schrödinger equation

$$i\hbar\frac{d\tilde{\Psi}(x, y(t), t)}{dt} = \left[-\frac{\hbar^2}{2m_x}\frac{\partial^2}{\partial x^2} + V(x, y(t))\right]\tilde{\Psi}(x, y(t), t), \tag{6.91}$$

where $\tilde{\Psi} = \tilde{R}e^{i\tilde{S}/\hbar}$. As can be noticed, the dimensionality of the full quantum problem has now reduced to the subspace dimensionality associated with the subsystem x, for the classical-like subsystem y acts as a time-dependent parameter (the external potential V has become time-dependent in virtue of this parametrization). On the other hand, subsystem y evolves according to the quasi-classical Newtonian equation

$$m_y \frac{d^2 y}{dt^2} = \frac{\partial}{\partial y} \left[V(x, y, t) + \tilde{Q}(x, t|y) \right],$$ (6.92)

which arises from (6.86b) after the applying the pseudo-Lagrangian operator (note that it can also be obtained from (6.85) after the corresponding approximation is considered) and is integrated after getting the solution $\tilde{\Psi}(x, y(t), t)$ from (6.91).

6.3.4 Reduced Quantum Trajectories

An interesting description of Bohmian mechanics arises from the field of decoherence and the theory of open quantum systems. Here, in order to extract useful information about the system of interest, one usually computes its associated reduced density matrix by tracing out the total density matrix, $\hat{\rho}_t$, over the environment degrees of freedom. In the configuration representation and for an environment constituted by N particles, the system reduced density matrix is obtained after integrating $\hat{\rho}_t \equiv |\Psi\rangle_{tt}\langle\Psi|$ over the $3N$ environment degrees of freedom, $\{\mathbf{r}_i\}_{i=1}^N$,

$$\tilde{\rho}(\mathbf{r}, \mathbf{r}', t) = \int \langle \mathbf{r}, \mathbf{r}_1, \mathbf{r}_2, \dots \mathbf{r}_n | \Psi(t) \rangle \langle \Psi(t) | \mathbf{r}', \mathbf{r}_1, \mathbf{r}_2, \dots \mathbf{r}_n \rangle d\mathbf{r}_1 d\mathbf{r}_2 \cdots d\mathbf{r}_n.$$ (6.93)

The system (reduced) quantum current density can be derived from this expression, being

$$\tilde{\mathbf{J}}(\mathbf{r}, t) = \frac{\hbar}{m} \text{Im}[\nabla_{\mathbf{r}} \tilde{\rho}(\mathbf{r}, \mathbf{r}', t)] \Big|_{\mathbf{r}'=\mathbf{r}},$$ (6.94)

which satisfies the continuity equation

$$\dot{\tilde{\rho}} + \nabla \tilde{\mathbf{J}} = 0.$$ (6.95)

In (6.95), $\tilde{\rho}$ is the diagonal element (i.e., $\tilde{\rho} \equiv \tilde{\rho}(\mathbf{r}, \mathbf{r}, t)$) of the reduced density matrix and gives the measured intensity [184].

Taking into account (6.94) and (6.95), it is possible to define the velocity field, $\dot{\mathbf{r}}$, associated with the (reduced) system dynamics as

$$\tilde{\mathbf{J}} = \tilde{\rho} \dot{\mathbf{r}},$$ (6.96)

which is analogous to the Bohmian velocity field. Now, from (6.96), a new class of quantum trajectories is defined, which are the solutions to the equation of motion

$$\dot{\mathbf{r}} \equiv \frac{\hbar}{m} \frac{\text{Im}[\nabla_{\mathbf{r}} \tilde{\rho}(\mathbf{r}, \mathbf{r}', t)]}{\text{Re}[\tilde{\rho}(\mathbf{r}, \mathbf{r}', t)]} \Big|_{\mathbf{r}'=\mathbf{r}}.$$ (6.97)

These new trajectories are the so-called *reduced quantum trajectories* [177, 178], which are only related to the system reduced density matrix. As shown in [177], the dynamics described by (6.95) leads to the correct intensity (whose time-evolution is described by (6.97)) when the statistics of a large number of particles is considered. Moreover, it is also straightforward to show that (6.97) reduces to the well-known expression for the velocity field in Bohmian mechanics when there is no interaction with the environment.

6.4 Extended Madelung Formulation for Dissipative Systems

A simple and straightforward generalization of Madelung's hydrodynamical formulation of wave mechanics to dissipative systems can be reached [185] by considering the following replacements

$$\frac{\partial S}{\partial t} \rightarrow \frac{\partial S}{\partial t} + f(x, t), \tag{6.98a}$$

$$\frac{\partial S}{\partial x} \rightarrow \frac{\partial S}{\partial x} + g(x, t), \tag{6.98b}$$

in the continuity and Hamilton–Jacobi equations, with f and g being two functions to be determined. In general, the force friction resulting from these replacements is

$$F_f = \frac{\partial g}{\partial t} - \frac{\partial f}{\partial x}, \tag{6.99}$$

and, from Ehrenfest's theorem, the expectation values will satisfy

$$m\frac{d\langle v \rangle}{dt} = -\left\langle \frac{\partial V}{\partial x} \right\rangle + \langle F_f \rangle. \tag{6.100}$$

Now, in particular, for the choice $g = 0$ and $f = \gamma(S - \langle S \rangle)$ the resulting friction force acting on the particle is

$$F_f(x, t) = -\gamma \frac{\partial S}{\partial x} = -\gamma m v, \tag{6.101}$$

thus recovering Kostin's nonlinear wave equation [186, 187].

This hydrodynamic description of quantum mechanics has been developed considering dissipation for distributions in configuration space and phase space, and their moments [47, 188, 189]. Master equations are then obtained for the reduced system (after tracing over the environmental coordinates) by an extension of the Liouville–von Neuman equation including the dissipation operator or dissipator $\mathcal{D}$ [190],

$$\frac{\partial \rho}{\partial t} = -i\hbar \left[H, \rho\right] + \mathcal{D}\rho. \tag{6.102}$$

Caldeira and Leggett introduced a master equation like (6.102), although it is not of Linblad type since positivity of the reduced density is not guaranteed [191]. An infinite hierarchy of kinetic equations in terms of moments of the distribution arises which can be truncated by suitable approximations. The dissipation operator does not act upon the diagonal elements of ρ. Decoherence effects are finally obtained following this scheme which has been developed as an alternative to wave packet propagation.

6.5 Quantum Stochastic Trajectories. The Beable Interpretation

After Bell's theorem [8, 9] and subsequent experiments [10–12, 167], certain interpretations of quantum mechanics different from the standard Copenhagen interpretation have received much attention. The basic idea behind one of these approaches consists of dealing with classical concepts—mainly positions and momenta—or *beables*, as termed by Bell [192], but following originally the arguments of de Broglie and Bohm. This approach is radically opposite to one of the fundamental pillars of the standard version of quantum mechanics, the *non-commensurability of non-commuting observables*. Two of the better known beable interpretations are [121, 193]: the causal and the stochastic interpretations. In this kind of analysis, one of the most important questions arises, the concept of nonlocality. Classical mechanics is a local theory and if classical concepts are used in quantum mechanics, one has to understand how it works in order to avoid misinterpretations. The first attempts in the stochastic version appeared in the 1950s and were suggested by Schrödinger leading Fürth to study the formal analogy between the Brownian motion and the Schrödinger equation considered like a diffusion equation. More recently, the works by Kershaw [194], Comisar [119], Nelson [195] and Olavo [196] tried to explain the origin of stochasticity in the quantum world, although Grabert et al. [197] showed that quantum mechanics is not equivalent to a Markovian diffusion process. Nelson's approach has predominantly been most cited in the literature. By introducing a new osmotic velocity and different accelerations, Nelson obtained the continuity and Hamilton–Jacobi equations of Bohmian mechanics. Right after the causal theory came out, Bohm and Vigier [198] also introduced a stochastic formalism, where the quantum system is immersed in a fluid displaying some random behavior leading to an additional component in the mean velocity; even the fluid could be avoided and only to consider such an extra random contribution. Like Einstein, they introduced the so-called osmotic velocity in the diffusion process in order to balance the diffusion current to obtain the velocity of the Bohm theory. The osmotic velocity is pushing the particle to regions of maximum probability and is derivable from the potential $D \ln \rho$, D being the diffusion coefficient. In this treatment the authors remarked the difference between the probability density of particles and the quantum distribution seen as the square modulus of the wave function. The first one is ruled by the Fokker–Planck equation and the second one by the conservative continuity equation. At equilibrium, both coincide.

It is quite natural to write an Itô stochastic differential equation in one dimension in the Bohmian framework for the position as [193]

$$dx(t) = \left(\frac{D}{R} \frac{\partial R}{\partial x} + \frac{1}{m} \frac{\partial S}{\partial x} \right) dt + \sqrt{D}\, dW(t), \qquad (6.103)$$

where $W(t)$ is a Wiener process, D is an arbitrary but sufficiently small diffusion constant proportional to $\hbar$, when the wave function is written in polar form; the osmotic velocity is given by $D(\partial R/\partial x)/R$. For $D = 0$, (6.103) reduces to the causal expression for the velocity. The probability density of particles always obeys the standard continuity equation.

Obviously, this is not the only way to obtain quantum stochastic trajectories. There are at least several routes for such a goal. The first one could consist of carrying out a time evolution of the wave function in terms of the Caldirola–Kanai Hamiltonian discussed in Chaps. 2 and 5 for a pure dissipative dynamics, extracting quantum dissipative trajectories from the phase of the wave function when expressed in polar form according to (6.1). In the second route, the quantum stochastic trajectories could be obtained from solving the Itô or Stratonovich differential equation, as shown in Chap. 5, for stochastic wave functions expressed in polar form. The terminology quantum trajectories is somewhat confusing here since in the context of stochastic wave functions (see Chap. 5) is also used. Several types of stochastic differential equations can be considered [199], even in the non-Markovian case [200]. Finally, the third route could be based on a generalization of the Langevin equation by substituting the interaction potential by the effective interaction potential which would include the quantum potential. Analogously, the same procedure could be followed by using the van der Pol and Duffing equations mentioned in Chap. 2, (2.4) and (2.5), respectively.

Vink [193] also proposed to give a beable status to all observables or physical quantities. These physical quantities which can be discrete or continuous are considered as discrete by assuming that the separation between two consecutive values is very small, of the order of Planck scale. For a set of commuting observables with discrete eigenvalues, the continuity equation in such a representation can be expressed as a master equation,

$$\frac{\partial P_n}{\partial t} = \sum_m (T_{nm} P_m - T_{mn} P_n), \qquad (6.104)$$

where the transition matrix T gives rise to a time-dependent probability distribution of the set of observables $P_n(t)$. The right-hand side of (6.104) is proportional to a source matrix, J_{nm}, which has to fulfill a series of general conditions [193]. In particular, its elements have to follow a Gaussian ansatz for the homogeneous master equation (i.e., the right-hand side of (6.104) has to be equal to zero). This theoretical scheme has been applied to several simple cases, such as a free particle on a circle, a spinning particle in a magnetic field, the harmonic oscillator or coherent states [199, 201].

In any case, a more detailed discussion on these very important issues is provided in Volume 2.

6.6 Are Bohmian Trajectories Real Particle Trajectories?

Bohmian mechanics constitutes a new paradigm to interpret and evaluate quantum processes and phenomena in terms of individual trajectories pursued by individual quantum systems. Now, the question that immediately arises is: what is a Bohmian particle? In other words, is it the "real" path followed by a quantum particle or displayed by a quantum degree of freedom? There is in the literature a controversy on this issue [202–207] and, actually, it is not far from the answer to this question. In order to provide here a certain answer—probably not final, but at least satisfactory—, let us first discuss two cases from classical mechanics. First, consider one wants to perform a measurement on a simple object. One measurement is irrelevant and, therefore, we need to consider a number of them (it is the same if there are many identical copies of the same system), being the value of such a measurement the outcome from the average of all the realizations considered. If the initial state is the same, in principle one should obtain the same outcome and, therefore, the measurement would be dispersionless. However, there are always certain deviations and therefore instead of a point on a sample space described by the system degrees of freedom involved in such a measurement, one would describe the outcome by a certain density distribution function, thus making use of statistical mechanics. Second, consider now a (classical) fluid. In principle, it is constituted by many different particles (e.g., atoms, ions, molecules, etc.), all the degrees of freedom being described by a set of differential coupled equations, with as many equations as degrees of freedom are involved. If one is not interested in a microscopic description of the fluid, but in a macroscopic one, instead of using such systems of equations, the well-known Euler or Navier–Stokes equations would be used, which describe phenomenologically the evolution of a continuous fluid without paying any attention to the particular (microscopic) dynamics of its constituents. This is essentially the basis of classical hydrodynamics. Now, in this latter case, any experimental study is carried out by putting some particles, namely *tracer particles*, in the fluid, so that they can help us to visualize its flow dynamics by moving along *streamlines*, i.e., lines along which the fluid current goes or energy is transported. For example, if the fluid is gaseous, one can use smoke; if it is a liquid, one can make use of tinny floating particles, e.g., pollen or charcoal dust, or another liquid, e.g., ink. In the case of the hydrodynamical approaches utilized to the universe dynamics in cosmology, the tracer particles can be stars, galaxies or clusters.

After this brief but worthy parenthesis, it is interesting to go back again to Bohmian mechanics and the meaning of a quantum particle there. In principle, by looking at one of those diffraction experiments carried out with electrons [67] or atoms [68], one notices that these quantum particles behave as in the first classical case, i.e., a single measurement or detection is meaningless, and many of them are necessary in order to visualize the diffraction pattern and then obtain information either about the diffracted particle or the diffracting object. In other words, individual particles behave like individual point-like particles, though their distributions displays a wave-like behavior, in accordance with Schrödinger's equation or its Bohmian equivalents (6.2). It is therefore clear that ensemble properties need of an ensemble description, i.e., a density distribution function, whose role is played in quantum mechanics by the probability density or, at a more elementary level, the wave function. This is in agreement with Born's statistical interpretation of quantum mechanics. However, if (individual) particles are regarded as moving along single trajectories, are these trajectories the ones obtained from (6.6)? Bohmian trajectories reproduce all the features of quantum mechanics and, therefore, one would be tempted to think that this is so. However, Bohmian equations are regarded as hydrodynamic equations, the corresponding trajectories obtained from (6.6) can (should) not be regarded as the trajectories pursued by real electrons, but rather as streamlines associated with the corresponding quantum fluid or paths along which quantum probability flows. Electrons may move or not like that, basically depending on the laminar or turbulent regimen displayed by the fluid, but surely not exactly as Bohmian trajectories. In this sense, it could be said that a Bohmian particle is a particle that obeys a Bohmian dynamics. Such a particle allows us to infer dynamical properties of the quantum fluid, which are usually "hidden when studied by means of the wave function formalism (see below). Therefore, Bohmian particles are the quantum equivalent of classical tracer particles.

To conclude this discussion, it is worth noticing that, although Bohmian mechanics applies to matter particles, trajectories for photons (see Chap. 7) have recently been inferred experimentally by means of weak measurements [208]. These experiments are in accordance with the Bohmian mechanics grounded on weak values and Bayesianism proposed by Wiseman [209], and could open new directions for Bohmian mechanics. In particular, with the information they provide it is possible to infer the paths of quantum trajectories and therefore to obtain new insights of the dynamics involved in quantum processes and phenomena.

References

1. Ballentine, L.E.: Ballentine: Quantum Mechanics: A Modern Development. World Scientific, Singapore (1998)
2. Zurek, W.H., Wheeler, J.A.: Quantum Theory of Measurement. Princeton University Press, Princeton, NJ (1983)
3. Belinfante, F.J.: A Survey of Hidden-Variables Theories. Pergamon, New York (1973)

4. Tarozzi, G., van der Merwe, A. (eds.): Open Questions in Quantum Physics. Reidel, Dordrecht (1985)
5. von Neumann, J.: Mathematische Grundlagen der Quantenmechanik. Springer, Berlin (1932)
6. Bohm, D.: A suggested interpretation of the quantum theory in terms of "hidden" variables. I. Phys. Rev. **85**, 166–179 (1952)
7. Bohm, D.: A suggested interpretation of the quantum theory in terms of "hidden" variables. II. Phys. Rev. **85**, 180–193 (1952)
8. Bell, J.S.: On the Einstein Podolsky Rosen paradox. Physics **1**, 195–200 (1964)
9. Bell, J.S.: On the problem of hidden variables in quantum mechanics. Rev. Mod. Phys. **38**, 447–452 (1966)
10. Aspect, A., Grangier, P., Roger, G.: Experimental tests of realistic local theories via Bell's theorem. Phys. Rev. Lett. **47**, 460–463 (1981)
11. Aspect, A., Grangier, P. Roger G.: Experimental realization of Einstein-Podolsky-Rosen-Bohm gedankenexperiment: a new violation of Bell's inequalities. Phys. Rev. Lett. **49**, 91–94 (1982)
12. Aspect, A., Dalibard, J., Roger, G.: Experimental test of Bell's inequalities using time-varying analyzers. Phys. Rev. Lett. **49**, 1804–1807 (1982)
13. Holland, P.R.: The Quantum Theory of Motion. Cambridge University Press, Cambridge (1993)
14. Goldstein, S.: Quantum theory without observers—Part I. Phys. Today **51**(3), 42–46 (1998)
15. Goldstein, S.: Quantum theory without observers—Part II. Phys. Today **51**(4), 38–42 (1998)
16. Wyatt, R.E.: Quantum Dynamics with Trajectories: Introduction to Quantum Hydrodynamics. Springer, New York (2005)
17. Chattaraj, P.K. (ed.): Quantum Trajectories. CRC Press, Taylor and Francis, NY (2010)
18. Oriols, X., Mompart, J. (eds.): Applied Bohmian Mechanics: From Nanoscale Systems to Cosmology. Pan Standford Publishing, Singapore (2011)
19. Weiner, J.H., Askar, A.: Particle method for the numerical solution of the time-dependent Schrödinger equation. J. Chem. Phys. **54**, 3534–3541 (1971)
20. Dey, B.K., Askar, A., Rabitz, H.: Multidimensional wave packet dynamics with the fluid dynamical formulation of the Schrödinger equation. J. Chem. Phys. **109**, 8770–8782 (1998)
21. Sales Mayor, F., Askar, A., Rabitz, H.A.: Quantum fluid dynamics in the Lagrangian representation and applications to photodissociation problems. J. Chem. Phys. **111**, 2423–2435 (1999)
22. Wyatt, R.E.: Quantum wave packet dynamics with trajectories: application to reactive scattering. J. Chem. Phys. **111**, 4406–4413 (1999)
23. Wyatt, R.E.: Quantum wavepacket dynamics with trajectories: wavefunction synthesis along quantum paths. Chem. Phys. Lett. **313**, 189–197 (1999)
24. Lopreore, C., Wyatt, R.E.: Quantum wave packet dynamics with trajectories. Phys. Rev. Lett. **82**, 5190–5193 (1999)
25. Wyatt, R.E., Bittner, E.R.: Quantum wave packet dynamics with trajectories: implementation with adaptive Lagrangian grids. J. Chem. Phys. **113**, 8898–8907 (2000)
26. Hughes, K.H., Wyatt, R.E.: Wavepacket dynamics on dynamically adapting grids: application of the equidistribution principle. Chem. Phys. Lett. **366**, 336–342 (2002)
27. Hughes, K.H., Wyatt, R.E.: Wavepacket dynamics on arbitrary Lagrangian–Eulerian grids: application to an Eckart barrier. Phys. Chem. Chem. Phys. **5**, 3905–3910 (2003)
28. Trahan, C.J., Wyatt, R.E.: An arbitrary Lagrangian–Eulerian approach to solving the quantum hydrodynamic equations of motion: equidistribution with "smart" springs. J. Chem. Phys. **118**, 4784–4790 (2003)
29. Kendrick, B.K.: A new method for solving the quantum hydrodynamic equations of motion. J. Chem. Phys. **119**, 5805–5817 (2003)
30. Kendrick, B.K.: Quantum hydrodynamics: Application to N-dimensional reactive scattering. J. Chem. Phys. **121**, 2471–2482 (2004)

31. Zhao, Y., Makri, N.: Bohmian versus semiclassical description of interference phenomena. J. Chem. Phys. **119**, 60–67 (2003)
32. Makri, N.: Forward–backward quantum dynamics for time correlation functions. J. Phys. Chem. A **108**, 806–812 (2004)
33. Liu, J., Makri, N.: Monte Carlo Bohmian dynamics from trajectory stability properties. J. Phys. Chem. A **108**, 5408–5416 (2004)
34. Bittner, E.R.: Quantum initial value representations using approximate Bohmian trajectories. J. Chem. Phys. **119**, 1358–1364 (2003)
35. Garashchuk, S., Rassolov, V.A.: Semiclassical dynamics with quantum trajectories: formulation and comparison with the semiclassical initial value representation propagator. J. Chem. Phys. **118**, 2482–2490 (2003)
36. Poirier, B.: Reconciling semiclassical and Bohmian mechanics. I. Stationary states. J. Chem. Phys. **121**, 4501–4515 (2004)
37. Trahan, C., Poirier, B.: Reconciling semiclassical and Bohmian mechanics. II. Scattering states for discontinuous potentials. J. Chem. Phys. **124**, 034115(1–18) (2006)
38. Trahan, C., Poirier, B.: Reconciling semiclassical and Bohmian mechanics. III. Scattering states for continuous potentials. J. Chem. Phys. **124**, 034116(1–14) (2006)
39. Poirier, B.: Reconciling semiclassical and Bohmian mechanics. V. Wavepacket dynamics. J. Chem. Phys. **128**, 164115(1–15) (2008)
40. Babyuk, D., Wyatt, R.E.: Coping with the node problem in quantum hydrodynamics: the covering function method. J. Chem. Phys. **121**, 9230–9238 (2004)
41. Garashchuk, S., Rassolov, V.A.: Modified quantum trajectory dynamics using a mixed wave function representation. J. Chem. Phys. **120**, 8711–8715 (2004)
42. Gindensperger, E., Meier, C., Beswick, J.A.: Mixing quantum and classical dynamics using Bohmian trajectories. J. Chem. Phys. **113**, 9369–9372 (2000)
43. Gindensperger, E., Meier, C., Beswick, J.A.: Quantum-classical including continuum states using quantum trajectories. J. Chem. Phys. **116**, 8–13 (2002)
44. Gindensperger, E., Meier, C., Beswick, J.A., Heitz, M.-C.: Quantum-classical description of rotational diffractive scattering using Bohmian trajectories: comparison with full quantum wave packet results. J. Chem. Phys. **116**, 10051–10059 (2002)
45. Meier, C., Beswick, J.A.: Femtosecond pump-probe spectroscopy of I_2 in a dense rare gas environment: a mixed quantum/classical study of vibrational decoherence. J. Chem. Phys. **121**, 4550–4558 (2004)
46. Meierk, C.: Mixed quantum-classical treatment of vibrational decoherence. Phys. Rev. Lett. **93**, 173003(1–4) (2004)
47. Burghardt, I., Cederbaum, L.S.: Hydrodynamic equations for mixed quantum states. I. General formulation. J. Chem. Phys. **115**, 10303–10311 (2001)
48. Burghardt, I., Cederbaum, L.S.: Hydrodynamic equations for mixed quantum states. II. Coupled electronic states. J. Chem. Phys. **115**, 10312–10322 (2001)
49. Burghardt, I., Parlant, G.: On the dynamics of coupled Bohmian and phase–space variables: a new hybrid quantum-classical approach. J. Chem. Phys. **120**, 3055–3058 (2004)
50. Burghardt, I., Møller, K.B., Parlant, G., Cederbaum, L.S., Bittner, E.: Quantum hydrodynamics: mixed states, dissipation, and a new hybrid quantum-classical approach. Int. J. Quantum Chem. **100**, 1153–1162 (2004)
51. Hughes, K.H., Parry, S.M., Parlant, G., Burghardt, I.: A hybrid hydrodynamic-liouvillian approach to mixed quantum-classical dynamics: application to tunneling in a double well. J. Phys. Chem. A **111**, 10269–10283 (2007)
52. Garashchuk, S., Rassolov, V.A.: Semiclassical dynamics based on quantum trajectories. Chem. Phys. Lett. **364**, 562–567 (2002)
53. Garashchuk, S., Rassolov, V.A.: Quantum dynamics with Bohmian trajectories: energy conserving approximation to the quantum potential. Chem. Phys. Lett. **376**, 358–363 (2003)
54. Garashchuk, S., Rassolov, V.A.: Energy conserving approximations to the quantum potential: dynamics with linearized quantum force. J. Chem. Phys. **120**, 1181–1190 (2004)

55. Garashchuk, S., Rassolov, V.A.: Bohmian dynamics on subspaces using linearized quantum force. J. Chem. Phys. **120**, 6815–6825 (2004)
56. Sanz, A.S., Borondo, F., Miret-Artés, S.: Causal trajectories description of atom diffraction by surfaces. Phys. Rev. B **61**, 7743–7751 (2000)
57. Sanz, A.S., Borondo, F., Miret-Artés, S.: On the classical limit in atom-surface diffraction. Europhys. Lett. **55**, 303–309 (2001)
58. Sanz, A.S.: Una visión causal de los procesos de dispersión cuánticos. Ph.D. Thesis, Universidad Autónoma de Madrid, Madrid (2003)
59. Sanz, A.S., Borondo, F., Miret-Artés, S.: Particle diffraction studied using quantum trajectories. J. Phys.: Condens. Matter **14**, 6109–6145 (2002)
60. Spurk, J.H.: Fluid Dynamics. Springer-Verlag, Berlin (1997)
61. Sanz, A.S., Miret-Artés, S.: Selective adsorption resonances: quantum and stochastic approaches. Phys. Rep. **451**, 37–154 (2007)
62. Sanz, A.S., Miret-Artés, S.: Aspects of nonlocality from a quantum trajectory perspective: a WKB approach to Bohmian mechanics. Chem. Phys. Lett. **445**, 350–354 (2007)
63. Goldstein, H.: Classical Mechanics. Addison-Wesley, Reading (1980)
64. Takabayasi, T.: On the formulation of quantum mechanics associated with classical pictures. Prog. Theor. Phys. **8**, 143–182 (1952)
65. Takabayasi, T.: Remarks on the formulation of quantum mechanics with classical pictures and on relations between linear scalar fields and hydrodynamical fluids. Prog. Theor. Phys. **9**, 187–222 (1953)
66. Sanz, A.S., Miret-Artés, S.: A trajectory-based understanding of quantum interference. J. Phys. A **41**, 435303(1–23) (2008)
67. Tonomura, A., Endo, J., Matsuda, T., Kawasaki, T., Ezawa, H.: Demonstration of single-electron buildup of an interference pattern. Am. J. Phys. **57**, 117–120 (1989)
68. Shimizu, F., Shimizu, K., Takuma, H.: Double-slit interference with ultracold metastable neon atoms. Phys. Rev. A **46**, R17–R20 (1992)
69. Dimitrova, T.L., Weis, A.: The wave-particle duality of light: a demonstration experiment. Am. J. Phys. **76**, 137–142 (2008)
70. Dimitrova, T.L., Weis, A.: Lecture demonstrations of interference and quantum erasing with single photons. Phys. Scr. **T135**, 014003(1–4) (2009)
71. Dimitrova, T.L., Weis, A.: Single photon quantum erasing: a demonstration experiment. Eur. J. Phys. **31**, 625–637 (2010)
72. Landauer, R., Martin, Th.: Barrier interaction time in tunneling. Rev. Mod. Phys. **66**, 217–228 (1994)
73. Muga, J.G., Sala Mayato, R., Egusquiza, I. (eds.): Time in Quantum Mechanics, vol. 1, 2nd edn. Springer Series Lecture Notes in Physics, vol. 734. Springer, Berlin (2008)
74. Muga, J.G., Ruschhaupt A., Campo, A. (eds.): Time in Quantum Mechanics, vol. 2. Springer Series Lecture Notes in Physics, vol. 789. Springer, Berlin (2009)
75. Guantes, R., Sanz, A.S., Margalef-Roig, J., Miret-Artés, S.: Atom-surface diffraction: a trajectory description. Surf. Sci. Rep. **53**, 199–330 (2004)
76. Madelung, E.: Quantentheorie in hydrodynamischer form. Z. Phys. **40**, 322–326 (1926)
77. London, F.: Planck's constant and low temperature transfer. Rev. Mod. Phys. **17**, 310–320 (1945)
78. Fetter, A.L., Svidzinsky, A.A.: Vortices in a trapped dilute Bose–Einstein condensate J. Phys.: Condens. Matter **13**, R135–R194 (2001)
79. McCullough, E.A., Wyatt, R.E.: Quantum dynamics of the collinear (H, H_2) reaction. J. Chem. Phys. **51**, 1253–1254 (1969)
80. McCullough, E.A., Wyatt, R.E.: Dynamics of the collinear $H + H_2$ reaction. I. Probability density and flux. J. Chem. Phys. **54**, 3578–3591 (1971)
81. McCullough, E.A., Wyatt, R.E.: Dynamics of the collinear $H + H_2$ reaction. II. Energy analysis. J. Chem. Phys. **54**, 3592–3600 (1971)

82. Hirschfelder, J.O., Tang, K.T.: Quantum mechanical streamlines. III. Idealized reactive atom–diatomic molecule collision. J. Chem. Phys. **64**, 760–785 (1976)
83. Bader, R.F.W.: Quantum topology of molecular charge distributions. III. The mechanics of an atom in a molecule. J. Chem. Phys. **73**, 2871–2883 (1980)
84. Gomes, J.A.N.F.: Delocalized magnetic currents in benzene. J. Chem. Phys. **78**, 3133–3139 (1983)
85. Gomes, J.A.N.F.: Topological elements of the magnetically induced orbital current densities. J. Chem. Phys. **78**, 4585–4591 (1983)
86. Lazzeretti, P.: Ring currents. Prog. Nuc. Mag. Res. Spect. **36**, 1–88 (2000)
87. Pelloni, S., Faglioni, F., Zanasi R., Lazzeretti, P.: Topology of magnetic-field-induced current-density field in diatropic monocyclic molecules. Phys. Rev. A **74**, 012506(1–8) (2006)
88. Pelloni, S., Lazzeretti, P., Zanasi, R.: Spatial ring current model of the [2.2]paracyclophane molecule. J. Phys. Chem. A **111**, 3110–3123 (2007)
89. Pelloni, S., Lazzeretti, P., Zanasi, R.: Topological models of magnetic field induced current density field in small molecules. Theor. Chem. Acc. **123**, 353–364 (2009)
90. Pelloni, S., Lazzeretti, P.: Spatial ring current model for the prismane molecule. J. Phys. Chem. A **112**, 5175–5186 (2008)
91. Pelloni, S., Lazzeretti, P.: Topology of magnetic-field induced electron current density in the cubane molecule. J. Chem. Phys. **128**, 194305(1–10) (2008)
92. Pelloni, S., Lazzeretti, P.: Ring current models for acetylene and ethylene molecules. Chem. Phys. **356**, 153–163 (2009)
93. García-Cuesta, I., Sánchez de Merás, A., Pelloni, S., Lazzeretti, P.: Understanding the ring current effects on magnetic shielding of hydrogen and carbon nuclei in naphthalene and anthracene. J. Comput. Chem. **30**, 551–564 (2009)
94. Landau, L.D., Lifschitz, E.M.: Fluid Mechanics. Pergamon Press, Oxford (1959)
95. Bialynicki-Birula, I., Cieplak, M., Kaminski, J.: Theory of Quanta, Chap. 9. Oxford University Press, Oxford (1992)
96. Dirac, P.A.M.: Quantised singularities in the electromagnetic field. Proc. Roy. Soc. Lond. A **133**, 60–72 (1931)
97. Riess, J.: Nodal structure, nodal flux fields, and flux quantization in stationary quantum states. Phys. Rev. D **2**, 647–653 (1970)
98. Hirschfelder, J.O., Goebel, C.J., Bruch, L.W.: Quantized vortices around wavefunction nodes. II. J. Chem. Phys. **61**, 5456–5459 (1974)
99. Wu, H., Sprung, D.W.L.: Inverse-square potential and the quantum vortex. Phys. Rev. A **49**, 4305–4311 (1994)
100. Bialynicki-Birula, I., Bialynicka-Birula, Z.: Magnetic monopoles in the hydrodynamic formulation of quantum mechanics. Phys. Rev. D **3**, 2410–2412 (1971)
101. Aharonov, Y., Bohm, D.: Significance of electromagnetic potentials in the quantum theory. Phys. Rev. **115**, 485–491 (1959)
102. Sanz, A.S., Borondo, F., Miret-Artés, S.: Quantum trajectories in atom-surface scattering with single adsorbates: the role of quantum vortices. J. Chem. Phys. **120**, 8794–8806 (2004)
103. Sanz, A.S., Borondo, F., Miret-Artés, S.: Role of quantum vortices in atomic scattering from single adsorbates. Phys. Rev. B **69**, 115413(1–5) (2004)
104. Rosen, N.: The relation between classical and quantum mechanics. Am. J. Phys. **32**, 597–600 (1964)
105. Rosen, N.: Quantum particles and classical particles. Found. Phys. **16**, 687–700 (1986)
106. Ghosh, S.K., Deb, B.M.: Densities, density-functionals and electron fluids. Phys. Rep. **92**, 1–44 (1982)
107. Sanz, A.S., Miret-Artés, S.: A causal look into the quantum Talbot effect. J. Chem. Phys. **126**, 234106(1–11) (2007)
108. Davidović, M., Arsenović, D., Bozić, M., Sanz, A.S., Miret-Artés, S.: Should particle trajectories comply with the transverse momentum distribution?. Eur. Phys. J. Special Topics **160**, 95–104 (2008)

109. Yang, C.-D.: Quantum dynamics of hydrogen atom in complex space. Ann. Phys. (N.Y.) **319**, 399–443 (2005)
110. Yang, C.-D.: Wave-particle duality in complex space. Ann. Phys. (N.Y.) **319**, 444–470 (2005)
111. Yang, C.-D.: Solving quantum trajectories in Coulomb potential by quantum Hamilton–Jacobi theory. Int. J. Quantum Chem. **106**, 1620–1639 (2006)
112. Yang, C. D.: On modeling and visualizing single-electron spin motion. Chaos, Solitons & Fractals **30**, 41–50 (2006)
113. Yang, C.-D.: Modeling quantum harmonic oscillator in complex domain. Chaos, Solitons & Fractals **30**, 342–362 (2006)
114. Yang, C.-D.: Quantum Hamilton mechanics: Hamilton equations of quantum motion, origin of quantum operators, and proof of quantization axiom. Ann. Phys. **321**, 2876–2926 (2006)
115. Yang, C.-D.: The origin and proof of quantization axiom $\mathbf{p} \rightarrow \hat{\mathbf{p}} = -i\hbar\nabla$ in complex spacetime. Chaos, Solitons & Fractals **32**, 274–283 (2007)
116. Yang, C.-D.: Complex tunneling dynamics. Chaos, Solitons & Fractals **32**, 312–345 (2007)
117. Yang, C.-D.: Quantum motion in complex space. Chaos, Solitons & Fractals **33**, 1073–1092 (2007)
118. Fürth, R.: Über Einige Beziehungen Zwischen Klassischer Statistik und Quantenmechanick. Z. Phys. **81**, 143–162 (1993)
119. Comisar, G.G.: Brownian motion of nonrelativistic quantum mechanics. Phys. Rev. **138**, B1332–B1337 (1965)
120. Bohm, D., Vigier, J.P.: Model of the causal interpretation of quantum theory in terms of a fluid with irregular fluctuations.. Phys. Rev. **96**, 208–216 (1954)
121. Bohm, D., Hiley, B.J.: Non-locality and locality in the stochastic interpretation of quantum mechanics. Phys. Rep. **172**, 93–122 (1989)
122. John, M.V.: Modified de Broglie–Bohm approach to quantum mechanics. Found. Phys. Lett. **15**, 329–343 (2002)
123. John, M.V.: Probability and complex quantum trajectories. Ann. Phys. (N.Y.) **324**, 220–231 (2010)
124. Sanz, A.S., Miret–Artés, S.: Interplay of causticity and vorticality within the complex quantum Hamilton–Jacobi formalism. Chem. Phys. Lett. **458**, 239–243 (2008)
125. Pauli, W.: Die allgemeine Prinzipien der Wellenmechanick. In: Geiger, H. , Scheel, K. (eds) Handbuch der Physik, vol. 24, part 1, 2nd edn. Springer-Verlag, Berlin (1933)
126. Gottfried, K.: Quantum Mechanics. W.A. Benjamin, New York (1966)
127. Floyd, E.R.: Bohr-Sommerfeld quantization with the effective action variable. Phys. Rev. D **25**, 1547–1551 (1982)
128. Floyd, E.R.: Modified potential and Bohm's quantum-mechanical potential. Phys. Rev. D **26**, 1339–1347 (1982)
129. Floyd, E.R.: Arbitrary initial conditions of nonlocal hidden-variables. Phys. Rev. D **29**, 1842–1844 (1984)
130. Floyd, E.R.: Closed-form solutions for the modified potential. Phys. Rev. D **34**, 3246–3249 (1986)
131. Floyd, E.R.: Where and why the generalized Hamilton–Jacobi representation describes microstates of the Schrödinger wave function. Found. Phys. Lett. **9**, 489–497 (1996)
132. Floyd, E.R.: Reflection time and the Goos–Hänchen effect for reflection by a semi-infinite rectangular barrier. Found. Phys. Lett. **13**, 235–251 (2000)
133. Floyd, E.R.: Interference, reduced action, and trajectories. Found. Phys. **37**, 1386–1402 (2007)
134. Floyd, E.R.: Welcher Weg? A trajectory representation of a quantum Young's diffraction experiment. Found. Phys. **37**, 1403–1420 (2007)
135. Faraggi, A.E., Matone, M.: Quantum transformations. Phys. Lett. A **249**, 180–190 (1998)
136. Faraggi, A.E., Matone, M.: The equivalence principle of quantum mechanics: uniqueness theorem. Phys. Lett. B **437**, 369–380 (1998)
137. Faraggi, A.E., Matone, M.: Equivalence principle, Planck length and quantum Hamilton–Jacobi equation. Phys. Lett. B **445**, 77–81 (1998)

138. Faraggi, A.E., Matone, M.: Equivalence principle: tunneling, quantized spectra and trajectories from the quantum HJ equation. Phys. Lett. B **445**, 357–365 (1999)
139. Faraggi, A.E., Matone, M.: Quantum mechanics from an equivalence principle. Phys. Lett. B **450**, 34–40 (1999)
140. Faraggi, A.E., Matone, M.: The equivalence postulate of quantum mechanics. Int. J. Mod. Phys. A **15**, 1869–2017 (2000)
141. Barker-Jarvis, J., Kabos, P.: Modified de Broglie approach applied to the Schrödinger and Klein–Gordon equation. Phys. Rev. A **68**, 042110(1–8) (2003)
142. Chou, C.-C., Sanz, A.S., Miret-Artés S., Wyatt, R.E.: Hydrodynamic view of wave-packet interference: quantum caves. Phys. Rev. Lett. **102**, 250401(1–4) (2009)
143. Chou, C.-C., Sanz, A.S., Miret-Artés, S., Wyatt, R.E.: Quantum interference within the complex quantum Hamilton–Jacobi formalism. Ann. Phys. (N.Y.) **325**, 2193–2211 (2010)
144. Leacock, R.A., Padgett, M.J.: Hamilton–Jacobi theory and the quantum action variable. Phys. Rev. Lett. **50**, 3–6 (1983)
145. Leacock, R.A., Padgett, M.J.: Hamilton–Jacobi/action-angle quantum mechanics. Phys. Rev. D **28**, 2491–2502 (1983)
146. Tannor, D.J.: Introduction to Quantum Mechanics: A Time Dependent Perspective. University Science Press, Sausalito (2006)
147. Boiron, M., Lombardi, M.: Complex trajectory method in semiclassical of wave packets. J. Chem. Phys. **108**, 3431–3444 (1998)
148. Goldfarb, Y., Degani, I., Tannor, D.J.: Bohmian mechanics with complex action: a new trajectory formulation of quantum mechanics. J. Chem. Phys. **125**, 231103(1–4) (2006)
149. Sanz, A.S., Miret-Artés, S.: Comment on "Bohmian mechanics with complex action: a new trajectory-based formulation of quantum mechanics" [J. Chem. Phys. **125**, 231103 (2006)]. J. Chem. Phys. **127**, 197101(1–3) (2007)
150. Goldfarb, Y., Degani, I., Tannor, D.J.: Response to "Comment on 'Bohmian mechanics with complex action: a new trajectory-based formulation of quantum mechanics' " [J. Chem. Phys. **127**, 197101 (2007)]. J. Chem. Phys. **127**, 197102(1–3) (2007)
151. Goldfarb, Y., Schiff, J., Tannor, D.J.: Unified derivation of Bohmian methods and the incorporation of interference effects. J. Phys. Chem. A **111**, 10416–10421 (2007)
152. Goldfarb, Y., Degani, I., Tannor, D.J.: Semiclassical approximation with zero velocity trajectories. Chem. Phys. **338**, 106–112 (2007)
153. Goldfarb, Y., Tannor, D.J.: Interference in Bohmian mechanics with complex action. J. Chem. Phys. **127**, 161101(1–4) (2007)
154. Chou, C.-C., Wyatt, R.E.: Computational method for the quantum Hamilton–Jacobi equation: bound states in one dimension. J. Chem. Phys. **125**, 174103(1–10) (2006)
155. Chou, C.-C., Wyatt, R.E.: Computational method for the quantum Hamilton–Jacobi equation: one-dimensional scattering problems. Phys. Rev. E **74**, 066702(1–9) (2006)
156. Chou, C.-C., Wyatt, R.E.: Quantum trajectories in complex space. Phys. Rev. A **76**, 012115(1–14) (2007)
157. Rowland, B.A., Wyatt, R.E.: Analysis of barrier scattering with real and complex quantum trajectories. J. Phys. Chem. A **111**, 10234–10250 (2007)
158. Wyatt, R.E., Rowland, B.A.: Quantum trajectories in complex space: multidimensional barrier transmission. J. Chem. Phys. **127**, 044103(1–12) (2007)
159. David, J.K., Wyatt, R.E.: Barrier scattering with complex-valued quantum trajectories: taxonomy and analysis of isochrones. J. Chem. Phys. **128**, 094102(1–9) (2008)
160. Bohr, N.: Über die Anwendung der Quantentheorie auf den Atombau I. Die Grundpostulate der Quantentheorie. Z. Physik **13**, 117–165 (1923)
161. Liboff, R.L.: The correspondence principle revisited. Phys. Today **37**, 50–55 (1984)
162. Berry, M.V.: Quantum chaology, not quantum chaos. Phys. Scr. **40**, 335–336 (1989)
163. Ehrenfest, P.: Bemerkung über die angenäherte Gültigkeit der klassischen Mechanik innerhalb der Quantenmechanik. Z. Phys. **45**, 455–457 (1927)
164. Elmore, W.C., Heald, M.A.: Physics of Waves. Dover Publications, New York (1985)

165. Schrödinger, E.: Discussion of probability relation between separated systems. Proc. Cambridge Phil. Soc. **31**, 555–563 (1935)
166. Schrödinger, E.: Probability relations between separated systems. Proc. Camb. Phil. Soc. **32**, 446–452 (1936)
167. Kwiat, P.G., Mattle, K., Weinfurter, H., Zeilinger, A., Sergiemko, V., Shih, Y.: New high-intensity source of polarization-entangled photon pairs. Phys. Rev. Lett. **75**, 4337–4341 (1995)
168. Nielsen, M., Chuang, I.: Quantum Computation and Quantum Information. Cambridge University Press, Cambridge (2000)
169. Zurek, W.H.: Decoherence and the transition from quantum to classical. Phys. Today **44**, 36–44 (1991)
170. Giulini, D., Kiefer, C., Kupsch, J., Stamatescu, I.O., Zeh, H.D.: Decoherence and the Appearence of a Classical World in Quantum Theory. Springer, Berlin (1996)
171. Joos, E., Zeh, H.D.: The emergence of classical properties through interaction with the environment. Z. Phys. B **59**, 223–243 (1985)
172. Dewdney, C.: Nonlocally corelated trajectories in 2-particle quantum mechanics. Found. Phys. **18**, 867–886 (1988)
173. Lam, M.M., Dewdney, C.: Locality and nonlocality in correlated two-particle interferometry. Phys. Lett. A **150**, 127–135 (1990)
174. Guay, E., Marchildon, L.: Two-particle interference in standard and Bohmian quantum mechanics. J. Phys. A **36**, 5617–5624 (2003)
175. Na, K., Wyatt, R.E.: Quantum hydrodynamic analysis of decoherence: quantum trajectories and stress tensor. Phys. Lett. A **306**, 97–103 (2002)
176. Na, K., Wyatt, R.E.: Quantum hydrodynamic analysis of decoherence. Phys. Scr. **67**, 169–180 (2003)
177. Sanz, A.S., Borondo, F.: A quantum trajectory description of decoherence. Eur. Phys. J. D **44**, 319–326 (2007)
178. Sanz, A.S., Borondo, F.: Contextuality, decoherence and quantum trajectories. Chem. Phys. Lett. **478**, 301–306 (2009)
179. Oriols, X.: Quantum-trajectory approach to time-dependent transport in mesoscopic system with electron–electron interactions. Phys. Rev. Lett. **98**, 066803(1–4) (2007)
180. Billing, G.D.: Classical path method in inelastic and reactive scattering. Int. Rev. Phys. Chem. **13**, 309–335 (1994)
181. Tully, J.C.: Molecular dynamics with electronic transitions. J. Chem. Phys. **93**, 1061–1071 (1990)
182. Tully, J.C.: Nonadiabatic molecular dynamics. Int. J. Quantum Chem. **40**(S25), 299–309 (1991)
183. Prezhdo, O.V., Brooksby, C.: Quantum backreaction through the Bohmian particles. Phys. Rev. Lett. **86**, 3215–3219 (2001)
184. Sanz, A.S., Borondo, F., Bastiaans, M.: Loss of coherence in double-slit diffraction experiments. Phys. Rev. A. **71**, 042103(1–7) (2005)
185. Razavy, M.: Classical and Quantum Dissipative Systems. Imperial College Press, London (2005)
186. Kostin, M.D.: On the Schrödinger–Langevin equation. J. Chem. Phys. **57**, 3589–3591 (1972)
187. Kostin, M.D.: Friction and dissipative phenomena in quantum mechanics. J. Stat. Phys. **12**, 145–151 (1975)
188. Burghardt, I., Möller, K.B.: Quantum dynamics for dissipative systems: a hydrodynamic perspective. J. Chem. Phys. **117**, 7409–7425 (2002)
189. Trahan, C.J., Wyatt, R.E.: Evolution of classical and quantum phase-space distributions: a new trajectory approach for phase–space hydrodynamics. J. Chem. Phys. **119**, 7017–7029 (2003)
190. Breuer, H.P., Petruccione, F.: The Theory of Open Quantum Systems. Oxford University Press, Oxford (2002)

191. Caldeira, A.O., Leggett, A.J.: Path integral approach to quantum Brownian motion. Physica A **121**, 587–616 (1983)
192. Bell, J.S.: Speakable and Unspeakable in Quantum Mechanics: Collected Papers on Quantum Philosophy. Cambridge University Press, Cambridge (1987)
193. Vink, J.C.: Quantum mechanics in terms of discrete beables. Phys. Rev. A **48**, 1808–1818 (1993)
194. Kershaw, D.: Theory of hidden variables. Phys. Rev. **136**, B1850–B1856 (1964)
195. Nelson, E.: Derivation of the Shrödinger equation from Newtonian mechanics. Phys. Rev. **150**, 1079–1085 (1966)
196. Olavo, L.S.F.: Foundations of quantum mechanics: connection with stochastic processes. Phys. Rev. A **61**, 052109(1–14) (2000)
197. Grabert, H., Hänggi, P., Talkner, P.: Is quantum mechanics equivalent to a classical stochastic process?. Phys. Rev. A **19**, 2440–2445 (1979)
198. Bohm, D., Vigier, J.P.: Model of the causal interpretation of quantum theory in terms of a fluid with irregular fluctuations. Phys. Rev. **96**, 208–216 (1954)
199. Santos, L.F., Escobar, C.O.: Enhanced diffusion and the continuous spontaneous localization model. Phys. Rev. A **60**, 2712–2715 (1999)
200. Bassi, A., Ferialdi, L.: Non-Markovian quantum trajectories: an exact result. Phys. Rev. Lett. **103**, 050403(1–4) (2009)
201. Lorenzen, F., de Ponte, M.A., Moussa, M.H.Y.: Extending Bell's beables to encompass dissipation, decoherence, and the quantum-to-classical transition through quantum trajectories. Phys. Rev. A **80**, 032101(1–8) (2009)
202. Englert, B.-G., Scully, M.O., Süssmann, G., Walther, H.: Surrealistic Bohm trajectories. Z. Naturforsch. A **47**, 1175–1186 (1992)
203. Dürr, D., Fusseder, F., Goldstein, S., Zhangí, N.: Comment on "Surrealistic Bohm trajectories". Z. Naturforsch. A **48**, 1261–1262 (1993)
204. Englert, B.-G., Scully, M.O., Süssmann, G., Walther, H.: Reply to Comment on "Surrealistic Bohm trajectories". Z. Naturforsch. A **48**, 1263–1264 (1993)
205. Becker, L.: On the supposed surrealism of Bohmian mechanics. Z. Naturforsch. A **52**, 533–538 (1997)
206. Scully, M.O.: Do Bohm trajectories always provide a trustworthy physical picture of particle motion? Phys. Scr. **T76**, 41–46 (1998)
207. Hiley, B.J., Callaghan, R.E.: Delayed-choice experiments and the Bohm approach. Phys. Scr. **74**, 336–348 (2006)
208. Kocsis, S., Braverman, B., Ravets, S., Stevens, M.J., Mirin, R.P., Shalm, L.K., Steinberg, A.M.: Observing the average trajectories of single photons in a two-slit interferometer. Science **332**, 1170–1173 (2011)
209. Wiseman, H.M.: Grounding Bohmian mechanics in weak values and bayesianism. New. J. Phys. **9**, 165(1–12) (2007)

Chapter 7
Trajectories in Optics

7.1 Introduction

As seen in Sect. 4.1, the back and forth debate about the nature of light, wave or corpuscle, lasted for quite a long time, ending with the complementarity principle. The discussion about the nature of interactions also developed alongside, from which the idea of field or force field came out. During his investigations on magnetism, Faraday realized about the importance of a field as a physical object. He noticed that electric and magnetic forces can be described as an effect produced by the presence of the corresponding fields, which are the entities that eventually govern the motion of particles (through lines of force). Actually, not only as responsible for particle motion, but Faraday also conferred them an independent physical reality based on the empirical fact that they carry energy. Later on, these ideas would influence Maxwell in his development of the first unified field theory in physics, namely the electromagnetic theory, based on a set of equations describing the dynamics of the electric and magnetic fields or, in brief, the electromagnetic field. By the end of the XIXth century, this field was understood as a collection of two vector fields in space—nowadays, it is assumed to be a single antisymmetric 2nd-rank tensor field in spacetime. Now, the introduction of the concept of field in optics (and, in general, in electromagnetism) results very interesting, for it brings this theory closer to quantum mechanics.

When studying interference and diffraction phenomena (see Sect. 4.3), the concept of ray can be used with the meaning of an auxiliary line connecting two end points regardless of its physical "reality". However, it can also be considered as the "actual" *path* pursued by light to go from one point to another, as happens when talking about waveguides (see Sect. 4.4). This distinction—as a tool or as a physical path—is nonetheless rather arbitrary; in the end the ray is not other thing than a model to describe the propagation of light according to a Newtonian conception. This description results very practical to connect the wave propagation around obstacles or through apertures with the intuitive idea of the rectilinear propagation, reflection or refraction of light. Indeed, this predates Newton and can be traced back to

A. S. Sanz and S. Miret-Artés, *A Trajectory Description of Quantum Processes.*
I. Fundamentals, Lecture Notes in Physics 850, DOI: 10.1007/978-3-642-18092-7_7,
© Springer-Verlag Berlin Heidelberg 2012

Euclid [1] and Hero of Alexandria [2, 3]. The former posed the basis of the geometrical treatment of light, while the latter first proposed a *principle of minimum distance* (when a light ray is reflected by a mirror, the path taken from the object to the observer's eye is the shortest path from all possible ones), thus anticipating Fermat [4]. Nevertheless, the ray model is an approximate description, valid only whenever the wavelength λ of the incident light is negligible compared with the dimensions of the objects it meets during its propagation (actually, the limit $\lambda \to 0$). This is the field of *geometric optics*, the short wavelength limit of physical optics or wave optics.

A brief account on geometric optics results interesting because, in spite of its approximate description of optical phenomena, it allows us to understand them in a very intuitive manner (with respect to our everyday experience of these phenomena, based on objects following well-defined paths). In order to go from the source to the illuminated object, light pursues a straight line unless the medium it traverses has a nonhomogeneous refractive index. In such a case, there is a curvature of the light ray, as happens with common mirages or also the mirages of astronomical objects, which make the size of these objects to look slightly higher above the horizon [5]. This bending or deflection of light rays can also be found when they pass near heavy celestial bodies due to the gravity of the latter [6], this process being known as *gravitational lensing* [7–9]. This strongly reminds us the behavior of particle trajectories under the action of potentials. Indeed, not only the mathematical description of both the curvature of light rays and the paths followed by classical particles are similar, but conceptually both rest on the same principle: *Fermat's principle* [4], which rules the path of light rays, transforms into *Hamilton's least action principle* when it is used to describe the dynamics of matter particles [10].

Despite the interpretative and conceptual advantages of the ray-based concept of light, this description fails when one tries to interpret the phenomena of physical optics—this concept can be taken into account to obtain diffraction and interference patterns, but such rays do not provide us with any clue on the way how the electromagnetic energy *flows*. The reason for this is that these rays do not follow electromagnetic energy *streamlines*. In order to understand this concept, let us consider for a while the quantum mechanical behavior of matter particles. Matter wave diffraction was observed long ago, starting with Davisson and Germer [11] and Stern et al. [12–15] around the 1930s, who studied electron and rare–gas–atom diffraction from crystals and surfaces. Nowadays, experiments range from medium-sized molecules [16, 17] to objects that can already be considered mesoscopic-sized [18–23]. In this kind of experiments, the electronic, atomic or molecular beams used are very intense. The particle flux per area unit and time unit is so large that no individual (electron, atom or molecule) arrivals can then be detected, but a relatively high intensity. On the contrary, there is another type of recent experiments, where the intensity is so small that particles can be collected basically one by one [24, 25]. This kind of experiments constitutes a nice manifestation of the statistical nature of quantum mechanics [26–29], where particles distribute according to the quantum probability density in the limit of a large number of detections. In the case of light, some recent experiments [30, 31] have shown that the typical interference patterns described by wave optics can be reconstructed from low intensity beams, i.e., single photon counts. In the case of matter waves, the individual evolution of particles can be described by means of Bohmian mechanics; in the case of light, it is clear that the usual concept of ray is not enough to describe the appearance of interference and diffraction patterns, but the model has to incorporate the wavy features of light. This is precisely the purpose behind the *hydrodynamical formulation of electromagnetism* [32–36]. Although the photon is a massless "quantum" particle and this entails conceptual troubles regarding the definition of an appropriate associated wave function [37], this formulation shows that the evolution of electromagnetic fields can be understood as the propagation of a fluid [33–48]. Within this scenario, similar in spirit to Madelung's approach to Schrödinger's wave mechanics [49], a trajectory-based picture of standard wave optics is possible. This representation arises (statistically) when the number of photon paths (detections) becomes relatively large [50]. Actually, this approach may increase its relevance after two recent experiments dealing with photon detection by weak measurements (see Chap. 5 and Appendix B). In one of them, a technique is reported to directly measure the photon wave function [51], while in the other, average photon trajectories are inferred experimentally from Young's two-slit experiment [52].

7.2 Geometric Optics: The Optics of Rays

There are circumstances where the propagation of light (or, in general, any form of electromagnetic radiation) can be described in a good approximation by neglecting its wavelength. For example, when sunlight illuminates an object, a sharp-edged shadow is observed, i.e., a shadow with very well-defined boundaries. This is the so-called "geometric shadow" and the model that allows us to describe the formation of these shadows as well as any other optical phenomenon with the "language" of geometry is *geometric optics*. Within this model, based on assuming the limit $\lambda \to 0$ (λ being the light wavelength), the basic element is the concept of *ray*, the path pursued by light when it travels from one space point to another one. This approximate model is strongly connected to our perception of well-defined shadows. This perception is based on the fact that usually the details of the boundaries between light and shadow cannot be appreciated, although such boundaries actually consist of a succession of alternating darker and lighter diffraction bands covering an extension of a few wavelengths. In the case of sunlight, in the visible range of the electromagnetic spectrum, the wavelength is negligible compared to the size of the objects illuminated. Therefore, such a diffraction effect cannot be perceived and it leads us to observe sharp-edged, geometric shadows.

The basic equation of geometric optics is the so-called *eikonal equation*, as first shown by Sommerfeld and Runge in 1911 [53], who reached this equation from the scalar wave equation in the limit $\lambda \to 0$. Here this procedure [54] will also be followed due to its similarity to the way how other important results are derived in optics and quantum mechanics, as seen in previous chapters (a more complete derivation can be done starting from Maxwell's equations, which take into account the vectorial nature of the electromagnetic field [55, 56]). Thus, let us assume an electromagnetic field in an inhomogeneous field (i.e., with refractive index depending on position). This field can be obtained from the scalar wave equation (4.7) expressed as

$$\nabla^2 \Psi(\mathbf{r}, t) - \frac{n^2(\mathbf{r})}{c^2} \frac{\partial^2 \Psi(\mathbf{r}, t)}{\partial t^2} = 0, \tag{7.1}$$

where $v(\mathbf{r}) = c/n(\mathbf{r})$ is the local value of the speed of light within the medium traversed. Assuming the refractive index is continuous and smoothly varying with position, a JWKB-like solution for (7.1) can be considered (see Sect. 7.3),

$$\Psi(\mathbf{r}, t) = A(\mathbf{r})e^{i[kS(\mathbf{r}) - \omega t]}, \tag{7.2}$$

where $k = \omega/c = 2\pi/\lambda$. In (7.2), the (real-valued) functions A and S have to be determined. Substituting (7.2) into (7.1), and then rearranging terms, yields

$$\nabla^2 A + \left[n^2 - (\nabla S)^2 \right] k^2 A = 0, \tag{7.3a}$$

$$\nabla \cdot (A^2 \nabla S) = 0, \tag{7.3b}$$

which are the optical analog of (3.131b) and (3.131a), respectively.

   In the limit of geometric optics, the wavelength of light is assumed to be negligible compared with the dimensions of interest, but also with respect to the local changes of the refractive index. In other words, the envelope A has to vary slowly compared with λ, which is the (spatial) scale measuring the local changes (oscillations) of the phase kS. This means that, for the distances of interest (in general, much greater than λ), $\nabla^2 A \approx 0$ and, therefore,

$$(\nabla S)^2 = n^2. \tag{7.4}$$

This is the *eikonal equation*, the basic equation of geometric optics. In this equation, S is the *eikonal*, a direct extension of the concept of optical path (remember from Chap. 4 that the phase appeared as a function of the coordinates), but not a path or ray itself. Accordingly, a *wavefront* is an eikonal such that $S(\mathbf{r}) = $ constant. Equation (7.4) can also be expressed as

$$\nabla S = n\hat{s}, \tag{7.5}$$

where $\hat{s}(\mathbf{r})$ is a unit vector pointing (at a position $\mathbf{r}$) along the direction perpendicular to the surface or wavefront S or, equivalently, along the direction along which S

propagates (at $\mathbf{r}$). The wavefront can be then obtained once the local refractive index $n(\mathbf{r})$ and a certain (initial) constant wavefront S_0 are given.

Once the wavefronts are known, one can determine the path followed by a *ray*. Rays are continuous curves, everywhere parallel to the local direction indicated by $\hat{\mathbf{s}}$ and perpendicular to wavefronts. Indeed, if $\mathbf{r}$ denotes the position of a point on a ray, unitary displacements along the ray with respect to its arc-length s will be equal to $\hat{\mathbf{s}}$ (i.e., $d\mathbf{r}/ds = \hat{\mathbf{s}}$). Therefore, from (7.5),

$$n\frac{d\mathbf{r}}{ds} = \nabla S. \tag{7.6}$$

This is the *equation of rays* of geometric optics. In order to understand the physical meaning of ray, consider the time-averaged[1] *Poynting vector* $\mathbf{P}(\mathbf{r})$ [56, 57] (see Sect. 7.5). In the limit of geometric optics, this vector can be expressed as

$$\mathbf{P} = \frac{c}{n^2}U\nabla S = U\mathbf{v}, \tag{7.7}$$

where $U(\mathbf{r})$ is the time-averaged electromagnetic energy density and $\mathbf{v} = v\hat{\mathbf{s}} = (c/n)\hat{\mathbf{s}}$. According to (7.7), the time-averaged Poynting vector (i.e., the flow of electromagnetic energy carried by light or, in general, an electromagnetic field) propagates along the direction indicated by the velocity field $\mathbf{v}$, parallel to $\hat{\mathbf{s}}$ and perpendicular to the wavefront S. From the second equality in (7.7), an alternative form for the equation of rays can be defined, as

$$\frac{d\mathbf{r}}{d\tau} = \mathbf{v}, \tag{7.8}$$

which is in terms of a proper time τ $(ds = (c/n)d\tau)$. From (7.8), which is totally equivalent to (7.6), but considering displacements along the ray path instead of path arc-lengths, it is apparent that light rays are just streamlines along which electromagnetic energy flows.

Equation (7.6) involves the eikonal. The path pursued by a ray can also be determined without taking the eikonal into account as follows. The directional derivative along the ray can be expressed as

$$\frac{d}{ds} = \hat{\mathbf{s}} \cdot \nabla. \tag{7.9}$$

Using (7.6) and (7.5), (7.9) becomes

---

[1] For the sake of simplicity, here time-averaged quantities will be assumed in order to avoid time-dependence. More specifically, this allows us to describe stationary electromagnetic fluxes associated with light (electromagnetic radiation in the visible range of the spectrum), neglecting the fast oscillations of the corresponding electric and magnetic fields. However, in general, the formulations presented in this chapter can also be applied to describe the evolution of time-dependent electromagnetic fields.

$$\frac{d}{ds}\left(n\frac{d\mathbf{r}}{ds}\right) = \nabla n. \tag{7.10}$$

Therefore, if the refractive index is known the ray path can be obtained by solving either (7.10) or, equivalently, the set of coupled equations

$$\frac{d\mathbf{r}}{ds} = \hat{\mathbf{s}}, \tag{7.11a}$$

$$\frac{d}{ds}\left(n\hat{\mathbf{s}}\right) = \nabla n, \tag{7.11b}$$

with initial conditions $\mathbf{r}_0$ and $\hat{\mathbf{s}}_0$. As it is apparent after a quick look back to Chap. 1, the following strong connections or analogies are found:

• Equation (7.6) and Jacobi's law of motion (1.12a).
• Equation (7.10) and Newton's equation of motion (1.5).
• Equation (7.11) and Hamilton's equations of motion (1.8).

Indeed, classical mechanics is often considered the same limit approximation to quantum mechanics as geometric optics is a limit for wave optics.

   As an example to illustrate how rays arise, consider a homogeneous medium (constant n) and a propagating plane wavefront whose normal with respect to some prefixed origin is defined by the direction cosines α, β and γ. According to (7.5), the eikonal will be

$$S(\mathbf{r}) = S_0 + n(\alpha x + \beta y + \gamma z), \tag{7.12}$$

while the rays are any straight line characterized by a slope along the direction $\hat{\mathbf{s}} = \alpha\hat{\mathbf{i}} + \beta\hat{\mathbf{j}} + \gamma\hat{\mathbf{k}}$. In the case of the ray path, (7.6) becomes

$$\frac{d\hat{\mathbf{s}}}{ds} = 0, \tag{7.13}$$

which means that $\hat{\mathbf{s}} = \hat{\mathbf{s}}_0$, i.e., the ray follows a straight line.

   Once S is known along the $\hat{\mathbf{s}}$ direction, the component of ∇A along the same direction can also be found from (7.3a), since this equation can be again expressed as

$$\frac{1}{A}\frac{dA}{ds} = -\frac{1}{2}\frac{\nabla \cdot (n\hat{\mathbf{s}})}{n}. \tag{7.14}$$

Thus, the variation of the amplitude A along a particular ray can be determined, but not the transversal change when moving from one ray to another, where discontinuities (singularities) in A may actually appear.

7.2.1 Fermat's Principle

The formal grounds of geometric optics lay on *Fermat's principle* [4]. According to this principle, the path followed by a ray when it goes from one space point to another is such that it makes stationary (usually, a minimum) the time of transit of the associated wave. From the calculus of variations [57] (see Appendix A), a necessary and sufficient condition for the definite integral

$$I(g) \equiv \int_{\sigma_1}^{\sigma_2} F d\sigma \qquad (7.15)$$

to be an extremum or stationary value (maximum, minimum or inflection point with zero slope) is that the *Euler–Lagrange differential equation*

$$\frac{\partial F}{\partial g} - \frac{d}{d\sigma}\frac{\partial F}{\partial g'} = 0 \qquad (7.16)$$

is satisfied for the function $F[g(\sigma), g'(\sigma); \sigma]$ of the independent variable σ and the unknown functions $g(\sigma)$ and $g'(\sigma) \equiv dg/d\sigma$, also subject to the constraint that g has some prescribed values at σ_1 and σ_2. In the case of Fermat's principle, I represents the elapsed time for a wave to go from one prescribed point A to another one B. This transit time for the wave to propagate along the path connecting the two points, A and B, is

$$\Delta t = \int_A^B dt = \int_A^B \frac{n(\mathbf{r})}{c} ds, \qquad (7.17)$$

with $ds^2 = dx^2 + dy^2 + dz^2$. Since the transit time depends explicitly on the path considered, (7.17) can be put in terms of a certain parameter σ which measures the distance along any arbitrary path and is such that $\sigma(A) = 0$ and $\sigma(B) = 1$, for convenience. With this, (7.17) becomes

$$\Delta t = \int_0^1 F(x, y, z, x', y', z') d\sigma = \int_0^1 \frac{n(x, y, z)}{c}\sqrt{x'^2 + y'^2 + z'^2} d\sigma, \quad (7.18)$$

with $x' = dx/d\sigma$ and the same for y' and z'. The Euler–Lagrange equation for the x-coordinate is then

$$\sqrt{x'^2 + y'^2 + z'^2}\,\frac{\partial n}{\partial x} - \frac{d}{d\sigma}\left(\frac{nx'}{\sqrt{x'^2 + y'^2 + z'^2}}\right) = 0, \qquad (7.19)$$

which can be conveniently rearranged to yield

$$\frac{d(n\alpha)}{ds} = \frac{\partial n}{\partial x}, \qquad (7.20)$$

where $\alpha = x'/\sqrt{x'^2 + y'^2 + z'^2}$ is the direction cosine of the vector $\mathbf{r}'$ along the x-axis (i.e., the angle associated with the path element $ds = \sqrt{x'^2 + y'^2 + z'^2}d\sigma$). As can be readily noticed, by proceeding in the same way with y and z, one reaches (7.6). This shows the equivalence between the ray equation obtained from geometric optics and its formal derivation appealing to Fermat's principle.

7.2.2 Hamiltonian Analogy and Optical Paths

The success reached by Fermat in optics led Hamilton in 1834 to rationalize classical mechanics in terms of the language of optics, i.e., introducing ideas such as Fermat's principle and the surfaces of constant phase (namely, wavefronts S). Hamilton realized that there is a certain analogy between the Newtonian trajectories, described by particles under the action of an external potential, and the optical paths pursued by light rays in continuum media with variable refractive index. In this sense, classical mechanics can be seen as the limit of a wavy motion in a configuration space where the trajectories are perpendicular to surfaces with constant action, as well as the geometric optics is the limit of the wave optics—based on it, Schrödinger would later on derive his wave version of quantum mechanics. This correspondence between classical mechanics and optics is known as the *Hamiltonian analogy* [56, 58]. When this analogy is used in optics (the other way around), it is known as the *Lagrangian Formulation of optics*, which constitutes a useful tool to compute, for example, the path pursued by rays in nonhomogeneous media [10, 59–62].

The Hamiltonian analogy remained in a sort of impasse until 1927, when Busch [63–65] used it to explain the focalizing effect of electromagnetic fields on rays in terms of optics. Almost by the same time, Schrödinger used the Hamiltonian analogy to derive his wave version of quantum mechanics (see Sect. 3.2.1), passing from geometric optics to the wave optics of matter particles and establishing a more direct meaning to de Broglie's concept of wavelength for matter particles. Furthermore, Schrödinger also used this tool to argue that the concept of classical trajectory is not compatible with wave mechanics.

The application of the Hamiltonian analogy to the derivation of Schrödinger's equation was seen in Sect. 3.2.1. Now its use will be illustrated within a different context, showing how the trajectory of a charged particle can be obtained from a typical optical problem by introducing a point-to-point variable refractive index. Thus, consider an electron moving under the action of a time-independent electric field, $\mathbf{E}$. The electron motion is ruled by Newton's laws,

$$\frac{d\mathbf{p}}{dt} = e\mathbf{E} = -\nabla V, \tag{7.21}$$

where V is the electric potential generated by the field $\mathbf{E}$ and $\mathbf{p}$ is the electron momentum. This equation can be generalized to any velocity $\mathbf{v}$ by means of the expression for the relativistic momentum,

$$\mathbf{p} = \frac{m\mathbf{v}}{\sqrt{1-\beta^2}}, \tag{7.22}$$

where $\beta = v/c$, c is the speed of light and m is the electron mass at rest. Equation (7.21) can be decomposed in two scalar equations according to the tangent and perpendicular directions of the electron motion. Thus, if $\mathbf{p} = p\mathbf{u}_\parallel$ and $\mathbf{v} = v\mathbf{u}_\parallel$, where $\mathbf{u}_\parallel$ is a unit vector in the direction of the electron motion, then

$$\frac{d\mathbf{p}}{dt} = \frac{dp}{dt}\mathbf{u}_\parallel + pv\frac{d\mathbf{u}_\parallel}{du}. \tag{7.23}$$

Here, $d\mathbf{u}_\parallel/du$ is a vector pointing along the normal direction to the electron motion, with modulus equal to the curvature of the trajectory, $1/\rho$. Hence it can be expressed as $\mathbf{u}_\perp/\rho$, with $\mathbf{u}_\perp$ being the unit vector in this direction. Substituting this into (7.23) yields

$$\frac{d\mathbf{p}}{dt} = \frac{dp}{dt}\mathbf{u}_\parallel + \frac{pv}{\rho}\mathbf{u}_\perp, \tag{7.24}$$

which, when compared with (7.21), shows that the center of curvature of the trajectory will be contained (at each time) by the plane defined by tangent to the trajectory, $\mathbf{u}_\parallel$, and the direction of the electric field. Finally, projecting ∇V onto both the tangent and normal directions of motion, $\mathbf{u}_\parallel$ and $\mathbf{u}_\perp$, respectively, the electron equations of motion are

$$\frac{dp}{dt} = -\mathbf{u}_\parallel \cdot \nabla V, \tag{7.25a}$$

$$\frac{pv}{\rho} = -\mathbf{u}_\perp \cdot \nabla V. \tag{7.25b}$$

Note that (7.25a) gives the electron position as a function of time along its trajectory, while (7.25b) is the trajectory equation.

After multiplying (7.25a) by v and then integrating the resulting equation under the condition that, initially, the electron has an energy E and is at rest,

$$E = \frac{mc^2}{\sqrt{1-\beta^2}} + V, \tag{7.26}$$

or, equivalently,

$$E = mc^2\sqrt{1 + \left(\frac{p}{mc}\right)^2} + V. \tag{7.27}$$

Thus, the electron momentum is expressed as a function of the coordinates,

$$p = mc\sqrt{\left[1 + \frac{(E_0 - V)}{mc^2}\right]^2 - 1}, \tag{7.28}$$

where mc^2 is the electron energy at rest and E_0 its non-relativistic energy, $E = mc^2 + E_0$. From (7.28), note how in the non-relativistic case $\beta \ll 1$ the momentum approaches the well-known kinematic relation $p = \sqrt{2m(E_0 - V)}$.

Now, consider the component (7.25b), where v and V can be expressed as functions of p through (7.22) and (7.27). A very simple law is obtained,

$$\frac{1}{\rho} = \frac{\mathbf{u}_\perp \cdot \nabla p}{p} = \mathbf{u}_\perp \cdot \nabla (\ln p). \tag{7.29}$$

This relation is identical to the equation obtained in geometric optics to describe the propagation of rays in nonhomogeneous media when the momentum p is substituted by the medium refractive index n,

$$\frac{1}{\rho} = \mathbf{u}_\perp \cdot \nabla (\log n). \tag{7.30}$$

An immediate consequence of this result is the appearance of a formal relationship or analogy between electron trajectories and light rays, with the momentum being analogous to a variable refractive index. Since ρ is always positive, when a light ray moves within a variable refractive index medium, it bends over towards those regions with increasing n. Similarly, electrons will move towards regions with larger momentum.

7.3 The JWKB Approximation

The *Jeffreys–Wentzel–Kramers–Brillouin approximation* [66–70] or, in brief, JWKB approximation, also constitutes an intermediate step between the optics of rays and the wave optics. In brief, this approximation essentially consists of "dressing" the rays of geometric optics with a wave and, therefore, see how (plane) waves developing along a set of rays give rise to the phenomena of wave optics. From a more quantitative viewpoint, this approximation allows us to find approximate solutions of a wave equation (regardless of what the wave is describing) in nonhomogeneous media with a gradual variation of the inhomogeneity (i.e., no abrupt discontinuities). This approximation and its application in quantum mechanics already appeared in Chap. 3. In particular, it was formerly used to describe tunneling in disintegration processes through α-particle decay [71, 72] and then further developed to find solutions of both the time-independent and the time-dependent Schrödinger equations.

In order to understand how this approximation works in optics, let the one-dimensional scalar wave equation be

$$\frac{\partial^2 \Psi(x,t)}{\partial x^2} - \frac{1}{v^2(x)} \frac{\partial^2 \Psi(x,t)}{\partial t^2} = 0, \tag{7.31}$$

where the propagation velocity v depends on the position, as it happens when dealing with nonhomogeneous media (space-dependent refractive indexes). The purpose of the JWKB approximation is to provide an approximate solution to (7.31) for an arbitrary function proceeding as follows. If v was independent of the position, a solution of this equation would be a traveling wave,

$$\Psi(x,t) = Ae^{i(kx-\omega t)}, \tag{7.32}$$

with $k = \omega/c$. Now, if the medium properties slowly vary with position, the solution to (7.31) should look pretty much like (7.32), but with the quantities A and k being dependent on the position. Thus, consider a solution of the form

$$\Psi(x,t) = A(x)e^{i[S(x)-\omega t]}. \tag{7.33}$$

After substituting this ansatz into (7.31), rearranging terms and then separating the real and imaginary parts of the resulting equation, one finds

$$\frac{d^2A}{dx^2} + \left[\frac{\omega^2}{v^2(x)} - \left(\frac{dS}{dx}\right)^2\right]A = 0, \tag{7.34a}$$

$$2\frac{dS}{dx}\frac{dA}{dx} + A\frac{d^2S}{dx^2} = 0. \tag{7.34b}$$

Solving (7.31) is therefore equivalent to find the functions $A(x)$ and $S(x)$ which satisfy this system of real, coupled differential equations. In a homogeneous medium, $d^2A/dx^2 = 0$. Thus, if the medium properties vary slowly with x, it can be assumed that $d^2A/dx^2 \approx 0$ and, therefore, at a first approximation, from (7.34a),

$$\frac{dS}{dx} \approx \frac{\omega}{v(x)} \equiv k(x) = \frac{2\pi}{\lambda(x)} \quad \longrightarrow \quad S(x) \approx 2\pi \int_{x_0}^{x} \frac{dx}{\lambda(x)}, \tag{7.35}$$

where $\lambda(x)$ is a local wavelength depending on position and x_0 is some reference point at which the amplitude of the wave is known. Substituting now (7.35) into (7.34b),

$$A(x) = A(x_0)\sqrt{\frac{v(x)}{v(x_0)}}, \tag{7.36}$$

and replacing this result and (7.35) in (7.33) yields

$$\Psi(x,t) = A(x_0)\sqrt{\frac{v(x)}{v(x_0)}}\, e^{i\left[\int_{x_0}^{x} k(x)dx - \omega t\right]}. \tag{7.37}$$

As it can be noticed, the amplitude varies in proportion to the local wave velocity and the phase depends on a local wavelength.

In order to find the range of validity of this approximation, let us consider that the exact solution to (7.31) is such that

$$\frac{dS}{dx} = \frac{2\pi}{\lambda(x)}\left[1 + \epsilon(x)\right],\tag{7.38}$$

where $\epsilon(x)$ is some perturbation dependent on position. Substituting (7.36) and (7.38) into (7.34a) gives

$$\left(1 + \frac{\epsilon}{2}\right)\epsilon = \frac{\lambda\lambda''}{16\pi^2} - \frac{(\lambda')^2}{32\pi^2},\tag{7.39}$$

where $\lambda' \equiv d\lambda/dx$. Under the assumption of a slowly varying medium with x, the second term on the right-hand side of (7.39) is going to be negligible in comparison to the former. Thus, considering ϵ is a small perturbation (i.e., $|\epsilon| \ll 1$),

$$|\epsilon| \approx \frac{(\lambda')^2}{32\pi^2} \ll 1.\tag{7.40}$$

This condition can also be rewritten as $\delta\lambda \ll 4\sqrt{2}\pi\delta x$. Accordingly, the JWKB approximation is going to work well whenever the variations of the local wavelength ($\delta\lambda$) are relatively small when compared with the spacial scales considered (δx). Or, in other words, the variations of the wave velocity (which is a measure of the properties of the medium) are very small over many wavelengths.

7.4 Optical Singularities: Caustics and Rainbows

Short wavelength fields are approximated by families of trajectories or rays which are normal to wavefronts. Under certain conditions, such trajectories can be focused in a given region of the space along envelopes where the number of rays is very high (infinite intensity). These envelopes formed by the coalescence of rays are called *caustics* and the deflection angles for which caustics occur are the *rainbow angles*. Quoting Berry [73, 74]:

> The caustic is one of the few things in geometrical optics that has any physical reality. Wavefronts and rays are not realizable; they are just convenient symbols on which we can hang our ideas. The caustic on the other hand is real and becomes visible by blowing a cloud of smoke in the region of the focus of a lens.

On the other hand, it is precisely on caustics where the trajectory/ray picture breaks down, since the wave equation only provides finite solutions for the intensity.

In many optical problems, the intensity is given by the square modulus of a general Fraunhofer diffraction integral of the type

$$\int g(x)e^{ikf(x)}\tag{7.41}$$

and asymptotic approximations are obtained for very large values of k. These rapidly oscillating integrals are then evaluated by the well known *stationary phase method*

[75, 76]. The basic idea is to replace such an integral by a Gaussian integral around the critical points or stationary phase points of $f(x)$. If the phase is not pure imaginary, the *steepest descent approximation* [56] to the corresponding integrals is applied. When two or more stationary phase points coalesce on a caustic such methods for the evaluation of integrals are no longer valid and the so called *catastrophe theory* developed mainly by Thom and Zeeman has to be applied [75, 77, 78]. So, in a mathematical language, caustics correspond to singularities of gradient maps.

Very interesting and important work has been developed in connection with the formation of diffraction patterns in particle-surface scattering problems. These patterns show a close relation with rainbows and caustics [79, 80]. Generally speaking, it could be said that caustics constitute the "skeleton" of diffraction patterns. A more detailed account on these issues can be found in [81].

7.5 Hydrodynamical Trajectories in Optics: Photon Paths

The extensive use of such trajectory-based descriptions in problems involving matter particles strikingly contrasts with its lack in analogous experiments with light or, for a better analogy, with photons. In particular, this is quite remarkable by inspecting the use of Bohmian mechanics to provide intuitive pictures of quantum processes in the last years—as well as to develop alternative numerical tools aimed at describing quantum properties without having to deal with the wave function first. In optics, however, not much attention has been paid to the development of similar trajectory-based pictures, except for the use of the rays from geometric optics (dressed by waves) as a crude way to understand diffraction phenomena under different circumstances. Obviously, this is not the way one should follow, since such rays make sense within geometric optics, but not within wave optics. To some extent, this is analogous to the mechanisms in terms of classical trajectories provided to explain quantum processes. However, unlike quantum mechanics, finding an analogous to Bohmian mechanics in optics is a more subtle issue, because it is connected with the need for finding the analogous to the wave function. From a quick look at the literature [33, 34, 38, 39, 43–47], one readily notices that defining a wave function for a photon is not a trivial issue, but something that gets directly into a thoughtful conceptual debate. Nevertheless, Lundeen et al. [52] have recently reported a direct technique to detect the photon wave function based on a combination of weak and strong measurements on complementary observables, which might render some light on this debate.

Nevertheless, putting aside fundamental issues and adopting a more pragmatic viewpoint, one can tackle the problem as follows. In optics it is well known that, like quantum matter particles, the observed patterns also arise as a consequence of a statistics of photons [51]. Indeed, some recent experiments [30–32] where the two-slit interference pattern is reconstructed photon by photon show very nicely this behavior. This is exactly the same observed in quantum mechanics if, instead of photons, electrons, for example, are considered [24]. Now, in quantum mechanics, the pattern of spots is described by the probability density[2] $\rho = |\Psi|^2$, so the basic elements in Bohmian mechanics are a "guiding" wave function Ψ and the hydrodynamical flow carrying ρ throughout space, namely the probability current density $\mathbf{J}$. In optics (and, in general, in electromagnetism), the patterns displayed (i.e., the photon distributions) are proportional to the electromagnetic energy (or energy density). Therefore, it is reasonable to associate here the electromagnetic field with a sort of guiding field, where the Poynting vector plays the role of the carrier of the electromagnetic energy through space. This is one of the former ideas considered in the literature, which can be traced back to Braunbek and Laukien [82] and Prosser [83, 84], and now is again in fashion [85–90]. This approach, which will be described in more detail throughout this Section, is based on a "classical" treatment of electromagnetism, where one is in the limit of a large number of photons—i.e., the number of photon is a non well-defined quantity, contrary to what happens in the limit of quantum optics [51]. Similarly, Ghose et al. [91, 92] proposed a way to attack the problem of dealing with photon trajectories, but within the relativistic limit.

To start with, consider an electromagnetic field in vacuum. The electric and magnetic fields (components) can then be expressed as harmonic waves,

$$\tilde{\mathbf{E}}(\mathbf{r}, t) = \mathbf{E}(\mathbf{r})e^{-i\omega t}, \quad \tilde{\mathbf{H}}(\mathbf{r}, t) = \mathbf{H}(\mathbf{r})e^{-i\omega t}. \tag{7.42}$$

Without loss of generality, a stationary electromagnetic field will be assumed (see footnote 1 in p. 235), although the formulation can also be applied to describe time-dependent electromagnetic fields. Thus, from now on consider the time-averaged electric and magnetic fields, thus avoiding the oscillating contribution (which does not change the flow dynamics in this case). This is the same as only paying attention to the spatial part of these fields, which satisfy the time-independent form of Maxwell's equations (4.1),

$$\nabla \cdot \mathbf{E}(\mathbf{r}) = 0, \tag{7.43a}$$

$$\nabla \cdot \mathbf{H}(\mathbf{r}) = 0, \tag{7.43b}$$

$$\nabla \times \mathbf{E}(\mathbf{r}) = i\omega\mu_0\mathbf{H}(\mathbf{r}), \tag{7.43c}$$

$$\nabla \times \mathbf{H}(\mathbf{r}) = -i\omega\epsilon_0\mathbf{E}(\mathbf{r}), \tag{7.43d}$$

as well as the boundary conditions associated with the particular problem under study. As in Chap. 4, in this Section the convention of complex electric and magnetic fields will also be considered in order to highlight the similarities with matter waves, although the same conclusions also hold for real fields. From (7.43) it can be readily shown that both $\mathbf{E}(\mathbf{r})$ and $\mathbf{H}(\mathbf{r})$ satisfy the vector Helmholtz equations,

---

[2] It is worth stressing that this is the case when dealing with single-count experiments. If experiments are carried out with high intensity beams (constituted by either matter particles or photons), the diffraction pattern is described by the transverse flux (cross-section), which is obtained from the probability current density rather than from the probability density.

$$\nabla^2 \mathbf{E}(\mathbf{r}) + k^2 \mathbf{E}(\mathbf{r}) = 0, \tag{7.44a}$$

$$\nabla^2 \mathbf{H}(\mathbf{r}) + k^2 \mathbf{H}(\mathbf{r}) = 0, \tag{7.44b}$$

with $k = \omega/c$. These equations are the time-independent analogues of (4.2). The electromagnetic energy flow lines are now obtained from the real part of the time-averaged complex Poynting vector,

$$\mathbf{P}(\mathbf{r}) = \frac{1}{2} \mathrm{Re} \left\{ \mathbf{E}(\mathbf{r}) \times \mathbf{H}^*(\mathbf{r}) \right\}, \tag{7.45}$$

which describes the flow of the time-averaged electromagnetic energy density through space,

$$U(\mathbf{r}) = \frac{1}{4} \left[\epsilon_0 \mathbf{E}(\mathbf{r}) \cdot \mathbf{E}^*(\mathbf{r}) + \mu_0 \mathbf{H}(\mathbf{r}) \cdot \mathbf{H}^*(\mathbf{r}) \right], \tag{7.46}$$

as the solutions or characteristics of the equation

$$\frac{d\mathbf{r}}{ds} = \frac{1}{c} \frac{\mathbf{P}(\mathbf{r})}{U(\mathbf{r})}, \tag{7.47}$$

where ds is the infinitesimal element of arc-length or metric distance. The solutions obtained from (7.47) can also be seen as rays in analogy to those derived from the eikonal equation, but taking into account a very important difference: the former are more general solutions which include wavy (diffraction) features. These general rays which spread out are lines along which the electromagnetic energy flows (just as in Bohmian mechanics quantum trajectories are lines along which probability flows, as said in Sect. 6.5)—in this regard, although the rays from geometric optics may also carry electromagnetic energy (when derived from the eikonal equation), they are more localized paths. Also, within optical electromagnetism, since the electromagnetic energy flows along these rays and, on the other hand, such an energy is carried by photons, the concept of *photon path* [88] will be used when referring to them. Of course, it is important to clarify that they should not be identified with the actual path that a photon would follow, just as a Bohmian trajectory cannot not be identified with the real trajectory that a quantum particle would pursue (see discussion in Sect. 6.5). Furthermore, notice that the rays or streamlines obtained from (7.47) are stationary, i.e., they somehow remain "frozen" in space, because the electromagnetic field is permanent (time-independent). If the time-dependence of the field was considered, these rays would describe trajectories evolving in time, thus accounting for the temporary dependence of the electromagnetic field, which would be therefore represented by a kind of pulse of (electromagnetic) energy.

In order to find a functional form for (7.47), consider for the sake of simplicity a problem characterized by symmetry such that the electric and magnetic fields are independent of the z-coordinate. From elementary electromagnetism [56], it is known that a problem completely independent of one Cartesian coordinate is essentially scalar, since it can be formulated in terms of a single dependent variable. Thus, taking into account

$$\frac{\partial \mathbf{H}}{\partial z} = \mathbf{0} = \frac{\partial \mathbf{E}}{\partial z} \tag{7.48}$$

in (7.43c) and (7.43d), two independent sets of equations are obtained,

$$\frac{\partial E_z}{\partial y} = \frac{i\omega}{\epsilon_0 c^2} H_x, \tag{7.49a}$$

$$\frac{\partial E_z}{\partial x} = -\frac{i\omega}{\epsilon_0 c^2} H_y, \tag{7.49b}$$

$$\frac{\partial H_y}{\partial x} - \frac{\partial H_x}{\partial y} = -i\omega\epsilon_0 E_z, \tag{7.49c}$$

and

$$\frac{\partial H_z}{\partial y} = -i\omega\epsilon_0 E_x, \tag{7.50a}$$

$$\frac{\partial H_z}{\partial x} = i\omega\epsilon_0 E_y, \tag{7.50b}$$

$$\frac{\partial E_y}{\partial x} - \frac{\partial E_x}{\partial y} = \frac{i\omega}{\epsilon_0 c^2} H_z. \tag{7.50c}$$

The set (7.49) only involves H_x, H_y and E_z, and therefore is commonly referred to as a case of E-*polarization*. On the other hand, the set (7.50), which only involves E_x, E_y and H_z, is referred to as H-*polarization*. More specifically, as it is inferred from the two first lines of the set of equations (7.49), in the case of E-polarization the electric field is polarized along the z-direction, while the magnetic field is confined to the XY-plane, i.e., $E_x = E_y = H_z = 0$, with the components of the magnetic field satisfying

$$H_x = -\frac{i\epsilon_0 c^2}{\omega}\frac{\partial E_z}{\partial y}, \quad H_y = \frac{i\epsilon_0 c^2}{\omega}\frac{\partial E_z}{\partial x}. \tag{7.51}$$

Substituting these expressions into (7.49c) yields

$$\frac{\partial^2 E_z}{\partial x^2} + \frac{\partial^2 E_z}{\partial y^2} + k^2 E_z = 0. \tag{7.52}$$

Thus,

$$\mathbf{E}_E(\mathbf{r}) = \mathbf{E}_\parallel(\mathbf{r}) = E_z\hat{\mathbf{z}}, \tag{7.53a}$$

$$\mathbf{H}_E(\mathbf{r}) = \mathbf{H}_\perp(\mathbf{r}) = H_x\hat{\mathbf{x}} + H_y\hat{\mathbf{y}}, \tag{7.53b}$$

with E_z satisfying Helmholtz's equation, according to (7.52), and where the symbols $\parallel$ and $\perp$ are used to specify which field is polarized parallel to the z-direction and which one perpendicularly, respectively. Similarly, in the case of H-polarization the magnetic field is polarized along the z direction and the electric one confined to the plane XY, i.e., $H_x = H_y = E_z = 0$, with the components of the electric field being

$$E_x = \frac{i}{\omega\epsilon_0}\frac{\partial H_z}{\partial y}, \quad E_y = -\frac{i}{\omega\epsilon_0}\frac{\partial H_z}{\partial x}. \tag{7.54}$$

Substituting now these into (7.50c) yields

$$\frac{\partial^2 H_z}{\partial x^2} + \frac{\partial^2 H_z}{\partial y^2} + k^2 H_z = 0. \tag{7.55}$$

This allows to characterize H-polarization as

$$\mathbf{E}_H(\mathbf{r}) = \mathbf{E}_\perp(\mathbf{r}) = E_x\hat{\mathbf{x}} + E_y\hat{\mathbf{y}}, \tag{7.56a}$$

$$\mathbf{H}_H(\mathbf{r}) = \mathbf{H}_\parallel(\mathbf{r}) = H_z\hat{\mathbf{z}}, \tag{7.56b}$$

with H_z satisfying the Helmholtz equation (7.55). Therefore, any general (time-independent) solution will be expressible as

$$\mathbf{E}(\mathbf{r}) = \alpha\mathbf{E}_\perp(\mathbf{r}) + \beta e^{i\phi}\mathbf{E}_\parallel(\mathbf{r}) = \frac{i\alpha}{\omega\epsilon_0}\left[\nabla \times \mathbf{H}_\parallel(\mathbf{r})\right] + \beta e^{i\phi}\mathbf{E}_\parallel(\mathbf{r}), \tag{7.57a}$$

$$\mathbf{H}(\mathbf{r}) = \beta e^{i\phi}\mathbf{H}_\perp(\mathbf{r}) + \alpha\mathbf{H}_\parallel(\mathbf{r}) = -\frac{i\beta e^{i\phi}}{\omega\mu_0}\left[\nabla \times \mathbf{E}(\mathbf{r})_\parallel\right] + \alpha\mathbf{H}_\parallel(\mathbf{r}), \tag{7.57b}$$

where α and β are real quantities and the phase shift between both components is given by ϕ.

Since E_z and H_z satisfy Helmholtz's equation, consider that both are proportional to a scalar field $\Psi(\mathbf{r})$, which also satisfies Helmholtz's equation, such that

$$\mathbf{H}_\parallel = \Psi(\mathbf{r})\hat{\mathbf{z}}, \tag{7.58a}$$

$$\mathbf{E}_\parallel = \frac{\omega\mu_0}{k}\Psi(\mathbf{r})\hat{\mathbf{z}}, \tag{7.58b}$$

where (7.43d) has been used to obtain the correct dimensionality in the right-hand side of (7.58b). If (7.58) are substituted into (7.57), the latter relations become

$$\mathbf{E}(\mathbf{r}) = \frac{i\alpha}{\omega\epsilon_0} \frac{\partial \Psi}{\partial y} \hat{\mathbf{x}} - \frac{i\alpha}{\omega\epsilon_0} \frac{\partial \Psi}{\partial x} \hat{\mathbf{y}} + \frac{k\beta e^{i\phi}}{\omega\epsilon_0} \Psi \hat{\mathbf{z}}, \tag{7.59a}$$

$$\mathbf{H}(\mathbf{r}) = -\frac{i\beta e^{i\phi}}{k} \frac{\partial \Psi}{\partial y} \hat{\mathbf{x}} + \frac{i\beta e^{i\phi}}{k} \frac{\partial \Psi}{\partial x} \hat{\mathbf{y}} + \alpha \Psi \hat{\mathbf{z}}, \tag{7.59b}$$

with the time-dependent counterparts being

$$\tilde{\mathbf{E}}(\mathbf{r}, t) = \left[\frac{i\alpha}{\omega\epsilon_0} \frac{\partial \Psi}{\partial y} \hat{\mathbf{x}} - \frac{i\alpha}{\omega\epsilon_0} \frac{\partial \Psi}{\partial x} \hat{\mathbf{y}} + \frac{k\beta e^{i\phi}}{\omega\epsilon_0} \Psi \hat{\mathbf{z}} \right] e^{-i\omega t}, \tag{7.60a}$$

$$\tilde{\mathbf{H}}(\mathbf{r}, t) = \left[-\frac{i\beta e^{i\phi}}{k} \frac{\partial \Psi}{\partial y} \hat{\mathbf{x}} + \frac{i\beta e^{i\phi}}{k} \frac{\partial \Psi}{\partial x} \hat{\mathbf{y}} + \alpha \Psi \hat{\mathbf{z}} \right] e^{-i\omega t}. \tag{7.60b}$$

Equations (7.60) are general time-dependent solutions for a problem which can be described in terms of superpositions, such as (7.57a) and (7.57b). Once this set of equations is established, the problem reduces to finding Ψ and its propagation along x and y (remember that this set of equations is based upon the hypothesis that the electric and magnetic fields do not depend on the z-coordinate), which is just a boundary condition problem.

Next, different cases will be analyzed depending on the choice of the particular form assigned to the electric and magnetic fields.

7.5.1 Polarized Plane Waves

Consider the electric and magnetic fields propagate along the y-direction. In this case, the simplest form for Ψ is a plane wave,

$$\Psi(\mathbf{r}) = e^{iky}. \tag{7.61}$$

Introducing it into (7.60) yields

$$\tilde{\mathbf{E}}(\mathbf{r}, t) = \mathbf{E}(\mathbf{r}) e^{-i\omega t} = \left(\frac{k}{\omega\epsilon_0} \right) \left[-\alpha e^{iky} \hat{\mathbf{x}} + \beta e^{i(ky+\phi)} \hat{\mathbf{z}} \right] e^{-i\omega t}, \tag{7.62a}$$

$$\tilde{\mathbf{H}}(\mathbf{r}, t) = \mathbf{H}(\mathbf{r}) e^{-i\omega t} = \left[\beta e^{i(ky+\phi)} \hat{\mathbf{x}} + \alpha e^{iky} \hat{\mathbf{z}} \right] e^{-i\omega t}. \tag{7.62b}$$

In these expressions, the effect of polarization on the fields can be explicitly observed. This effect eventually leads to the particular topologies exhibited by the photon paths.

In (7.62), it was already inferred how polarization is going to play an important role in the interference patterns observed and, therefore, in the topology displayed by the photon paths. Let us consider the electric field (7.62a) (the magnetic field shows the same polarization properties because of their relationship), whose real components are

$$\tilde{E}_x^r = a_x \cos(ky - \omega t), \tag{7.63a}$$

$$\tilde{E}_z^r = a_z \cos(ky - \omega t + \phi), \tag{7.63b}$$

where $a_x = -(k/\omega\epsilon_0)\alpha$ and $a_z = (k/\omega\epsilon_0)\beta$. These field components can be recast as

$$\frac{\tilde{E}_x^r}{a_x} \cos\phi - \frac{\tilde{E}_z^r}{a_z} = \sin(ky - \omega t) \sin\phi, \tag{7.64a}$$

$$\frac{\tilde{E}_x^r}{a_x} \sin\phi = \cos(ky - \omega t) \sin\phi. \tag{7.64b}$$

After squaring and rearranging terms in these expressions, they can be recast in

$$\left(\frac{\tilde{E}_x^r}{a_x}\right)^2 + \left(\frac{\tilde{E}_z^r}{a_z}\right)^2 - 2\left(\frac{\tilde{E}_x^r}{a_x}\right)\left(\frac{\tilde{E}_z^r}{a_z}\right)\cos\phi = \sin^2\phi. \tag{7.65}$$

According to (7.65) several cases are possible depending on the value of the phase shift ϕ:

1. When $\phi = 0$ or π,

$$\left(\frac{\tilde{E}_x^r}{a_x} \mp \frac{\tilde{E}_z^r}{a_z}\right)^2 = 0 \quad \Rightarrow \quad \frac{\tilde{E}_x^r}{a_x} = \pm\frac{\tilde{E}_z^r}{a_z}. \tag{7.66}$$

   This is the case of *linear polarization*, which is independent of the particular value of both α and β.
2. When $\phi = \pm\pi/2$,

$$\left(\frac{\tilde{E}_x^r}{a_x}\right)^2 + \left(\frac{\tilde{E}_z^r}{a_z}\right)^2 = 1. \tag{7.67}$$

   This is the case of *elliptic polarization*. Here, in particular, if $\alpha = \beta$, there is *circular polarization*.
3. For any other value of ϕ, there is always elliptic polarization.

The particular values of ϕ, α and β defining the polarization state of the incident beam are very important regarding the observation of interference patterns in two-slit interference experiments with polarized light [93–96], for they are ruled by the so-called *Arago–Fresnel laws* [97–99]. But they are also going to be very important with respect to the topology of the corresponding photon paths, as shown below.

Regarding the photon paths, they are now obtained from the Poynting vector,

$$\mathbf{P}(\mathbf{r}) = \frac{1}{2}\text{Re}\left\{\mathbf{E}(\mathbf{r}) \times \mathbf{H}^*(\mathbf{r})\right\} = \frac{1}{2}\left(\frac{k}{\omega\epsilon_0}\right)(\alpha^2 + \beta^2)\hat{\mathbf{y}}, \tag{7.68}$$

and the electromagnetic energy density,

$$U(\mathbf{r}) = \frac{1}{4} \left[\epsilon_0 \mathbf{E}(\mathbf{r}) \cdot \mathbf{E}^*(\mathbf{r}) + \mu_0 \mathbf{H}(\mathbf{r}) \cdot \mathbf{H}^*(\mathbf{r}) \right] = \frac{\mu_0}{2} (\alpha^2 + \beta^2), \qquad (7.69)$$

where the electric and magnetic fields here refer to the time-independent parts of (7.62). Substituting (7.68) and (7.69) into (7.47),

$$\frac{d\mathbf{r}}{ds} = \hat{\mathbf{y}}, \qquad (7.70)$$

which, after integration, yields

$$x(s) = x_0, \quad z(s) = z_0, \qquad (7.71\text{a})$$

$$y(s) = y_0 + s. \qquad (7.71\text{b})$$

That is, provided there are no variations of the position in x or z, the photon paths are straight lines parallel to the y-axis. That is, the electromagnetic energy evolves as predicted by geometric optics, as a flow that evolves parallel to the y-axis regardless of the polarization state.

7.5.2 Young's Two-Slit Experiment

Young's two-slit experiment is viewed as a paradigm of interference phenomena. It is therefore worth being treated here from the viewpoint of photon paths. In this case, the electromagnetic scalar field Ψ behind the two slits can be described, in general, as

$$\Psi(\mathbf{r}) = c_1 \psi_1(\mathbf{r}) + c_2 \psi_2(\mathbf{r}), \qquad (7.72)$$

where ψ_1 and ψ_2 represent the electromagnetic beams leaving each slit. Substituting this field into (7.59), and then into (7.45), one thus obtains the photon paths corresponding to two-slit diffraction process. Now, depending on the polarization state of each outgoing beam, different cases can be discussed. Here, in particular, the case where both partial beams leave the slits with the same polarization state will be analyzed. This allows to consider Ψ as a whole instead of dealing with ψ_1 and ψ_2 explicitly all the way.

Regarding the slits, assume that both are on a screen parallel to the XZ-plane and placed at $y = 0$. The slits are parallel to the z-axis and their width along this direction is much larger than along the x-direction (i.e., $w_z \gg w_x$). Because of this assumption, as in Sect. 7.5.1, the electromagnetic energy should be independent of the z-coordinate. That is, except near the upper or lower borders of the slits, one should observe exactly the same interference pattern regardless of this coordinate.

Taking such assumptions into account, after substitution of (7.59) into (7.45), the components of the Poynting vector along the different directions can be expressed as

$$P_x = \frac{i(\alpha^2 + \beta^2)}{4\omega\epsilon_0}\left(\Psi\frac{\partial\Psi'^*}{\partial x} - \Psi^*\frac{\partial\Psi'}{\partial x}\right), \tag{7.73a}$$

$$P_y = \frac{i(\alpha^2 + \beta^2)}{4\omega\epsilon_0}\left(\Psi\frac{\partial\Psi^*}{\partial y} - \Psi^*\frac{\partial\Psi}{\partial y}\right), \tag{7.73b}$$

$$P_z = \frac{i\alpha\beta\sin\phi}{2k\omega\epsilon_0}\left(\frac{\partial\Psi}{\partial x}\frac{\partial\Psi^*}{\partial y} - \frac{\partial\Psi}{\partial y}\frac{\partial\Psi^*}{\partial x}\right). \tag{7.73c}$$

Proceeding similarly with (7.46) leads us to the electromagnetic energy density,

$$U = \frac{(\alpha^2 + \beta^2)}{4\omega^2\epsilon_0}\left(\frac{\partial\Psi}{\partial x}\frac{\partial\Psi^*}{\partial x} + \frac{\partial\Psi}{\partial y}\frac{\partial\Psi^*}{\partial y} + k^2\Psi\Psi^*\right), \tag{7.74}$$

which determines the interference pattern at the observation screen. Since the polarization part (prefactor in terms of α and β) and the space part (depending on Ψ) are factorized in (7.74), the interference pattern observed will not depend on the polarization state, in agreement with Arago–Fresnel laws [98–100]. Substituting now these equations into (7.47) the corresponding path equations along each direction are

$$\frac{dx}{ds} = ik\left(\frac{\Psi\frac{\partial\Psi^*}{\partial x} - \Psi^*\frac{\partial\Psi}{\partial x}}{\frac{\partial\Psi}{\partial x}\frac{\partial\Psi^*}{\partial x} + \frac{\partial\Psi}{\partial y}\frac{\partial\Psi^*}{\partial y} + k^2\Psi\Psi^*}\right), \tag{7.75a}$$

$$\frac{dy}{ds} = ik\left(\frac{\Psi\frac{\partial\Psi^*}{\partial y} - \Psi^*\frac{\partial\Psi}{\partial y}}{\frac{\partial\Psi}{\partial x}\frac{\partial\Psi^*}{\partial x} + \frac{\partial\Psi}{\partial y}\frac{\partial\Psi^*}{\partial y} + k^2\Psi\Psi^*}\right), \tag{7.75b}$$

$$\frac{dz}{ds} = \frac{2i\alpha\beta\sin\phi}{(\alpha^2 + \beta^2)}\left(\frac{\frac{\partial\Psi}{\partial x}\frac{\partial\Psi^*}{\partial y} - \frac{\partial\Psi}{\partial y}\frac{\partial\Psi^*}{\partial x}}{\frac{\partial\Psi}{\partial x}\frac{\partial\Psi^*}{\partial x} + \frac{\partial\Psi}{\partial y}\frac{\partial\Psi^*}{\partial y} + k^2\Psi\Psi^*}\right). \tag{7.75c}$$

As can be noticed in these expressions, all the information about the polarization state of both interfering beams is contained in the prefactor of (7.75c). This is the reason why in the case here analyzed the particular form of Ψ (a coherent superposition) is not relevant. In other words, this simply means that, if both diffracted beams have the same polarization state, they will give rise to interference, in accordance with Arago–Fresnel laws.

For linear polarization [87], (7.75c) vanishes and the photon path equations can be readily solved by simply parameterizing, for example, y as a function of x, i.e.,

$$\frac{dy}{dx} = \frac{\left(\Psi \frac{\partial \Psi^*}{\partial y} - \Psi^* \frac{\partial \Psi}{\partial y}\right)}{\left(\Psi \frac{\partial \Psi^*}{\partial x} - \Psi^* \frac{\partial \Psi}{\partial x}\right)}, \qquad (7.76)$$

while the solution of (7.75c) is simply $z = z_0$ (i.e., no motion or evolution along the z-direction). On the contrary, in the case of elliptic (or circular) polarization, the z-component does play an important role, as can be noticed when the photon path equations are computed,

$$\frac{dy}{dx} = \frac{\left(\Psi \frac{\partial \Psi^*}{\partial y} - \Psi^* \frac{\partial \Psi}{\partial y}\right)}{\left(\Psi \frac{\partial \Psi^*}{\partial x} - \Psi^* \frac{\partial \Psi}{\partial x}\right)}, \qquad (7.77a)$$

$$\frac{dz}{dx} = \frac{2\alpha\beta \sin\phi}{(\alpha^2 + \beta^2)k} \left[\frac{\frac{\partial \Psi}{\partial x}\frac{\partial \Psi^*}{\partial y} - \frac{\partial \Psi}{\partial y}\frac{\partial \Psi^*}{\partial x}}{\Psi \frac{\partial \Psi^*}{\partial x} - \Psi^* \frac{\partial \Psi}{\partial x}} \right]. \qquad (7.77b)$$

As can be seen, (7.77a) remains as in the case of linear polarization, described by (7.76). However, although neither the electric field nor the magnetic one depend on the z-coordinate, the photon paths are going to display features in this direction as an effect of having elliptical polarization, which only vanish for linear polarization, when $\phi = 0$ or π. This means that the projection of the photon paths onto the XY-plane is exactly as in the case of linear polarization (the typical trajectories for the two-slit experiment [88]). To understand which kind of motion is expected in this direction (or, in other words, how it is going to manifest in the corresponding photon paths), let us go back to (7.73c). Rearranging terms and using (7.73a) and (7.73b), this equation can be rewritten as

$$P_z = \frac{\alpha\beta \sin\phi}{(\alpha^2 + \beta^2)k}\left(\frac{\partial S_y}{\partial x} - \frac{\partial S_x}{\partial y}\right) = \left[\frac{\alpha\beta}{(\alpha^2 + \beta^2)k}\right]\boldsymbol{\zeta} \cdot \hat{\mathbf{z}}, \qquad (7.78)$$

where

$$\boldsymbol{\zeta} \equiv \begin{pmatrix} \hat{\mathbf{x}} & \hat{\mathbf{y}} & \hat{\mathbf{z}} \\ \dfrac{\partial}{\partial x} & \dfrac{\partial}{\partial y} & 0 \\ S_x & S_y & 0 \end{pmatrix}. \qquad (7.79)$$

According to (7.78), the presence of a polarization state gives rise to a flow along the z-direction in terms of the vorticity manifested by the fields P_x and P_y, which may lead the photon paths to display loops out of the XY-plane. Nodal structures and other singularities and topological structures may then appear, as was already shown by Nye [100, 101] within the context of wave dislocations [102] as well as within the context of the Riemann–Silberstein complex formulation of Maxwell's equations [11, 103, 104]. Experimentally, what one would observe on the XZ plane is simply the typical fringe-like interference pattern constituted by dark and light parallel stripes, which results from the accumulation of trajectories arriving to this plane. Note that (7.74) describes the interference pattern and is the result of transporting the electromagnetic energy density from the slits to some observation screen in accordance with the guidance or continuity equation [56]

$$\mathbf{P}(\mathbf{r}) = U(\mathbf{r})\mathbf{v}, \tag{7.80}$$

which is an alternative way to express (7.47), where $\mathbf{v}$ is an effective vector velocity field that transports the electromagnetic energy density through space in the form of the Poynting vector (i.e., the electromagnetic energy current density). However, from (7.79), all the arrivals at a certain height z_f will not arise from positions at the slits at the same height $z_0 = z_f$, but there is a flux upwards and downwards which breaks the longitudinal (along the z-direction) symmetry of the experiment when it is studied from the viewpoint of photon paths.

As an illustration, a set of photon paths for linear polarization are displayed in Fig. 7.1, where the wavelength is $\lambda = 500$ nm and the inter-slit distance is $d = 20\lambda = 10$ μm (the slit width is equal to their separation, d, as in a Ronchi grating). The axes are normalized as follows: the x-axis is normalized to the inter-slit distance, d, while the y-axis is normalized to the so-called *Talbot distance* [105], $L_T = d^2/\lambda = 200$ μm (this distance is the one associated with the repetition of the diffraction pattern produced by a periodic grating; see Volume 2). Photon paths present the same kind of features which can also be observed with Bohmian trajectories for matter particles diffracted under similar conditions [106, 107]:

1. The transition between the near-field or Fresnel region (see bottom panel in Fig. 7.1) and the far-field or Fraunhofer region is smooth.
2. They do not cross the symmetry axis of the slit system.

In the Fresnel region (see bottom panel in Fig. 7.1), the paths display the typical single-diffraction effects until the two diffracted beams meet at about $y = 0.4L_T$ and interference becomes active. Far beyond the Fresnel region, the topology of the paths becomes *stationary*, i.e., the angular distribution of the path density (intensity) does not change regardless of the distance from the two slits (for example, the angular densities at $y = 2$ and $y = 3$ are exactly the same). Within this region, the paths align along the well-known Fraunhofer diffraction directions,

$$\theta_n \approx n\frac{\lambda}{d}, \quad n = 0, \pm 1, \pm 2, \ldots. \tag{7.81}$$

Fig. 7.1 Photon paths
behind two identical slits.
Top: Paths covering the near
and far field space regions;
the accumulation of paths
along certain directions is a
manifestation of the
diffraction maxima in the
Fraunhofer or far field
region. *Bottom*: Enlargement
of the paths shown in the top
panel in the Fresnel or near
field region. The numerical
values of the parameters
used here are $\lambda = 500$ nm,
$d = 20\lambda = 10\,\mu$m
and $L_T = d^2/\lambda = 200\,\mu$m

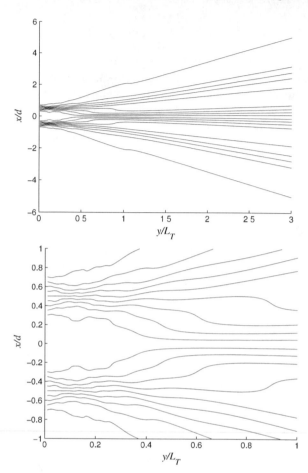

Equivalently, if the observation plane is at a distance $y = L$, the x positions around which the intensity will be maximal are $x_n \approx L\theta_n$.

The theory presented here not only provides the paths for photons, but it also allows to reproduce the experimental results if a counting on arrival positions is carried out, as happens with matter particles [107, 108]. This is shown in Fig. 7.2, where the intensity pattern associated with (7.74) is reproduced when a histogram is built of the arrivals of paths distributed initially according to a uniform distribution inside the slits. The agreement between the continuous energy density and the histogram is very apparent, as it is also when looking at the trajectories in the top panel of Fig. 7.1: they distribute in accordance with the corresponding intensities. For example, the central peak consists of twice the number of trajectories that those contributing to the two surrounding peaks.

In a recent experimental work, Kocsis et al. [53] have reported a reconstruction of photon paths inferred from the two-slit experiment carrying out weak measurements.

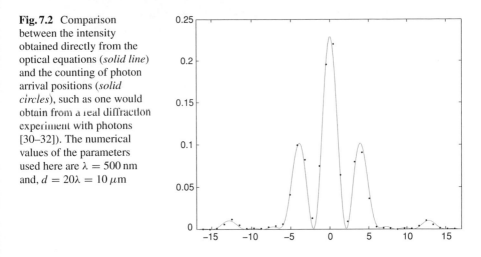

Fig. 7.2 Comparison between the intensity obtained directly from the optical equations (*solid line*) and the counting of photon arrival positions (*solid circles*), such as one would obtain from a real diffraction experiment with photons [30–32]). The numerical values of the parameters used here are $\lambda = 500$ nm and, $d = 20\lambda = 10\,\mu$m

These paths are in agreement with the behavior displayed by the photon paths of Fig. 7.1.

7.5.3 Two-Slit Diffraction and Optical Erasure

In Sect. 7.5.2, a typical example of Young two-slit experiment (with the polarization state of the two diffracted beams being the same) has been discussed. By the beginning of the XIXth century Arago and Fresnel generalized Young's interference experiment by considering different cases of polarization, which led to a series of laws on polarization and interference, namely the aforementioned Arago–Fresnel laws [98–100]. According to these laws, if two diffracted beams with different polarization states interfere, the visibility of the interference pattern will decrease. Actually, if the polarization states are orthogonal the pattern will disappear totally— for example, no interference pattern will be observed if the two interfering beams are linearly polarized in orthogonal directions [108–110] or both are elliptically polarized, but one is left-handed and the other right-handed [110, 111]. In the case of low intensity beams (i.e., within the domain of quantum optics), polarized light is used in experiments such as the so-called *interference quantum eraser* [112, 113], which are very important nowadays because of their implications at a fundamental level [114] and also from the viewpoint of quantum information [115]. According to the conventional or standard viewpoint, this kind of experiments are commonly interpreted following "which-way"-like arguments. However, as seen above, if the same experiment is revisited with the photon path formalism, one can observe that the outcomes are a consequence on how the electromagnetic energy flow is influenced when modifying the polarization properties of each slit, thus giving rise to a partial or a total suppression of the interference pattern.

In order to study the effects described by the Arago–Fresnel laws, consider now the explicit form for the superposition (7.72). Thus, when this superposition is substituted into (7.59), then

$$\mathbf{E(r)} = \frac{i\alpha c_1}{\omega\epsilon_0}\frac{\partial\psi_1}{\partial y}\hat{\mathbf{x}} - \frac{i\alpha c_1}{\omega\epsilon_0}\frac{\partial\psi_1}{\partial x}\hat{\mathbf{y}} + \frac{k\beta e^{i\phi}c_1}{\omega\epsilon_0}\psi_1\hat{\mathbf{z}}$$

$$+ \frac{i\alpha c_2}{\omega\epsilon_0}\frac{\partial\psi_2}{\partial y}\hat{\mathbf{x}} - \frac{i\alpha c_2}{\omega\epsilon_0}\frac{\partial\psi_2}{\partial x}\hat{\mathbf{y}} + \frac{k\beta e^{i\phi}c_2}{\omega\epsilon_0}\psi_2\hat{\mathbf{z}}, \tag{7.82a}$$

$$\mathbf{H(r)} = -\frac{i\beta e^{i\phi}c_1}{k}\frac{\partial\psi_1}{\partial y}\hat{\mathbf{x}} + \frac{i\beta e^{i\phi}c_1}{k}\frac{\partial\psi_1}{\partial x}\hat{\mathbf{y}} + \alpha c_1\psi_1\hat{\mathbf{z}}$$

$$- \frac{i\beta e^{i\phi}c_2}{k}\frac{\partial\psi_2}{\partial y}\hat{\mathbf{x}} + \frac{i\beta e^{i\phi}c_2}{k}\frac{\partial\psi_2}{\partial x}\hat{\mathbf{y}} + \alpha c_2\psi_2\hat{\mathbf{z}}, \tag{7.82b}$$

where the vector components have been separated purposely in terms of each diffracted beams. Now, let us suppose that, after passing through slit 1, the electromagnetic field becomes E-polarized, which is the same as selecting the β components of the electromagnetic field in slit 1 (i.e., only these components pass through this slit). Similarly, after passing through slit 2, the electromagnetic field is H-polarized, thus selecting the α components. With this, (7.82) become

$$\mathbf{E(r)} = \frac{i\alpha c_2}{\omega\epsilon_0}\frac{\partial\psi_2}{\partial y}\hat{\mathbf{x}} - \frac{i\alpha c_2}{\omega\epsilon_0}\frac{\partial\psi_2}{\partial x}\hat{\mathbf{y}} + \frac{k\beta e^{i\phi}c_1}{\omega\epsilon_0}\psi_1\hat{\mathbf{z}}, \tag{7.83a}$$

$$\mathbf{H(r)} = -\frac{i\beta e^{i\phi}c_1}{k}\frac{\partial\psi_1}{\partial y}\hat{\mathbf{x}} + \frac{i\beta e^{i\phi}c_1}{k}\frac{\partial\psi_1}{\partial x}\hat{\mathbf{y}} + \alpha c_2\psi_2\hat{\mathbf{z}}. \tag{7.83b}$$

Before computing the photon paths, it is interesting to note the following feature about the interference pattern. Let us express (7.74) in terms of the two diffracted beams, i.e.,

$$U = \frac{(\alpha^2+\beta^2)c_1^2}{4\omega^2\epsilon_0}\left(\frac{\partial\psi_1}{\partial x}\frac{\partial\psi_1^*}{\partial x} + \frac{\partial\psi_1}{\partial y}\frac{\partial\psi_1^*}{\partial y} + k^2\psi_1\psi_1^*\right)$$

$$+ \frac{(\alpha^2+\beta^2)c_2^2}{4\omega^2\epsilon_0}\left(\frac{\partial\psi_2}{\partial x}\frac{\partial\psi_2^*}{\partial x} + \frac{\partial\psi_2}{\partial y}\frac{\partial\psi_2^*}{\partial y} + k^2\psi_2\psi_2^*\right)$$

$$+ \frac{(\alpha^2+\beta^2)c_1c_2}{4\omega^2\epsilon_0}\left(\frac{\partial\psi_1}{\partial x}\frac{\partial\psi_2^*}{\partial x} + \frac{\partial\psi_1}{\partial y}\frac{\partial\psi_2^*}{\partial y} + k^2\psi_1\psi_2^*\right)$$

$$+ \frac{(\alpha^2+\beta^2)c_1c_2}{4\omega^2\epsilon_0}\left(\frac{\partial\psi_2}{\partial x}\frac{\partial\psi_1^*}{\partial x} + \frac{\partial\psi_2}{\partial y}\frac{\partial\psi_1^*}{\partial y} + k^2\psi_2\psi_1^*\right). \tag{7.84}$$

In shorthand notation, (7.84) can also be recast as

$$U = U_1 + U_2 + U_{12}, \tag{7.85}$$

where U_1 and U_2 are the electromagnetic energy densities associated with the partial waves ψ_1 or ψ_2, i.e., the first and second terms in (7.84), respectively. On the other hand, U_{12}, the last two terms in (7.84), is the electromagnetic energy density arising form the *interference* of these waves. Thus, assuming orthogonality between the interfering beams (E-polarized components passing only through slit 1, while the H-polarized ones pass through slit 2), the electromagnetic energy density becomes

$$U = \frac{\beta^2 c_1^2}{4\omega^2 \epsilon_0} \left(\frac{\partial \psi_1}{\partial x} \frac{\partial \psi_1^*}{\partial x} + \frac{\partial \psi_1}{\partial y} \frac{\partial \psi_1^*}{\partial y} + k^2 \psi_1 \psi_1^* \right)$$
$$+ \frac{\alpha^2 c_2^2}{4\omega^2 \epsilon_0} \left(\frac{\partial \psi_2}{\partial x} \frac{\partial \psi_2^*}{\partial x} + \frac{\partial \psi_2}{\partial y} \frac{\partial \psi_2^*}{\partial y} + k^2 \psi_2 \psi_2^* \right)$$
$$= U_1 + U_2. \tag{7.86}$$

That is, the electromagnetic energy density detected behind the two slits satisfies the law of addition of intensities which rules incoherent light, in accordance with Arago–Fresnel laws. On the contrary, if the diffracted beams have the same polarization state, they will interfere constructively and full interference will be observed. For beams with different polarizations (but without reaching the orthogonal case), interference can also be observed, but with a lower visibility.

In order to study now the photon paths, (7.73) is expressed explicitly as functions of ψ_1 and ψ_2, which yields

$$P_x = \frac{i(\alpha^2 + \beta^2)}{4\omega\epsilon_0} \left[c_1^2 \left(\psi_1 \frac{\partial \psi_1^*}{\partial x} - \psi_1^* \frac{\partial \psi_1}{\partial x} \right) + c_2^2 \left(\psi_2 \frac{\partial \psi_2^*}{\partial x} - \psi_2^* \frac{\partial \psi_2}{\partial x} \right) \right]$$
$$+ \frac{i(\alpha^2 + \beta^2) c_1 c_2}{4\omega\epsilon_0} \left[\left(\psi_1 \frac{\partial \psi_2^*}{\partial x} - \psi_2^* \frac{\partial \psi_1}{\partial x} \right) + \left(\psi_2 \frac{\partial \psi_1^*}{\partial x} - \psi_1^* \frac{\partial \psi_2}{\partial x} \right) \right], \tag{7.87a}$$

$$P_y = \frac{i(\alpha^2 + \beta^2)}{4\omega\epsilon_0} \left[c_1^2 \left(\psi_1 \frac{\partial \psi_1^*}{\partial y} - \psi_1^* \frac{\partial \psi_1}{\partial y} \right) + c_2^2 \left(\psi_2 \frac{\partial \psi_2^*}{\partial y} - \psi_2^* \frac{\partial \psi_2}{\partial y} \right) \right]$$
$$+ \frac{i(\alpha^2 + \beta^2) c_1 c_2}{4\omega\epsilon_0} \left[\left(\psi_1 \frac{\partial \psi_2^*}{\partial y} - \psi_2^* \frac{\partial \psi_1}{\partial y} \right) + \left(\psi_2 \frac{\partial \psi_1^*}{\partial y} - \psi_1^* \frac{\partial \psi_2}{\partial y} \right) \right], \tag{7.87b}$$

$$P_z = \frac{i\alpha\beta \sin\phi}{2k\omega\epsilon_0} \left[c_1^2 \left(\frac{\partial \psi_1}{\partial x} \frac{\partial \psi_1^*}{\partial y} - \frac{\partial \psi_1}{\partial y} \frac{\partial \psi_1^*}{\partial x} \right) + c_2^2 \left(\frac{\partial \psi_2}{\partial x} \frac{\partial \psi_2^*}{\partial y} - \frac{\partial \psi_2}{\partial y} \frac{\partial \psi_2^*}{\partial x} \right) \right]$$
$$+ \frac{i\alpha\beta \sin\phi c_1 c_2}{2k\omega\epsilon_0} \left[\left(\frac{\partial \psi_1}{\partial x} \frac{\partial \psi_2^*}{\partial y} - \frac{\partial \psi_1}{\partial y} \frac{\partial \psi_2^*}{\partial x} \right) + \left(\frac{\partial \psi_2}{\partial x} \frac{\partial \psi_1^*}{\partial y} - \frac{\partial \psi_2}{\partial y} \frac{\partial \psi_1^*}{\partial x} \right) \right],$$
$$\tag{7.87c}$$

or, in shorthand notation,

$$P_x = P_{x,1} + P_{x,2} + P_{x,12}, \tag{7.88a}$$

$$P_y = P_{y,1} + P_{y,2} + P_{y,12}, \tag{7.88b}$$

$$P_z = P_{z,1} + P_{z,2} + P_{z,12}. \tag{7.88c}$$

As it is apparent from these expressions, the electromagnetic energy flux also shows the same structure as the electromagnetic energy density—there is a separate contribution from each slit as well as an interference term. As in Sect. 7.5.2, the photon paths are computed after substitution of (7.87) and (7.86) into (7.47).

By varying the state of the polarizers one may obtain a similar effect to *decoherence*, i.e., the interference features will start to disappear, this giving rise to a kind of erasure of the information that makes the trajectories to display their characteristic wiggly behavior [116, 117]. Hence, a particular case of interest is that of interference beams with orthogonal polarization after crossing the slits. In this case, (7.87) reduce to

$$P_x = \frac{i\beta^2 c_1^2}{4\omega\epsilon_0}\left(\psi_1\frac{\partial\psi_1^*}{\partial x} - \psi_1^*\frac{\partial\psi_1}{\partial x}\right) + \frac{i\alpha^2 c_2^2}{4\omega\epsilon_0}\left(\psi_2\frac{\partial\psi_2^*}{\partial x} - \psi_2^*\frac{\partial\psi_2}{\partial x}\right), \tag{7.89a}$$

$$P_y = \frac{i\beta^2 c_1^2}{4\omega\epsilon_0}\left(\psi_1\frac{\partial\psi_1^*}{\partial y} - \psi_1^*\frac{\partial\psi_1}{\partial y}\right) + \frac{i\alpha^2 c_2^2}{4\omega\epsilon_0}\left(\psi_2\frac{\partial\psi_2^*}{\partial y} - \psi_2^*\frac{\partial\psi_2}{\partial y}\right), \tag{7.89b}$$

$$P_z = 0, \tag{7.89c}$$

which give rise to the photon path equation

$$\frac{dy}{dx} = \frac{\beta^2 c_1^2\left(\psi_1\frac{\partial\psi_1^*}{\partial y} - \psi_1^*\frac{\partial\psi_1}{\partial y}\right) + \alpha^2 c_2^2\left(\psi_2\frac{\partial\psi_2^*}{\partial y} - \psi_2^*\frac{\partial\psi_2}{\partial y}\right)}{\beta^2 c_1^2\left(\psi_1\frac{\partial\psi_1^*}{\partial x} - \psi_1^*\frac{\partial\psi_1}{\partial x}\right) + \alpha^2 c_2^2\left(\psi_2\frac{\partial\psi_2^*}{\partial x} - \psi_2^*\frac{\partial\psi_2}{\partial x}\right)}. \tag{7.90}$$

When this equation is compared to (7.76), the stationarity along the z-direction (i.e., no motion along this direction) is not caused by having linearly polarized diffracted beams, but because their polarization states are orthogonal, regardless whether they are linear or elliptical. However, unlike the case of linear polarization, here there is not a final interference pattern. This is because, as mentioned above, the orthogonality of the polarization states gives rise to two independent (non interfering) electromagnetic energy fluxes across the slits, as indicated by (7.86). Nevertheless, the fact of having simple addition of electromagnetic energy densities does not mean that the photon paths will not be influenced by the presence of both slits (electromagnetic energy fluxes). On the contrary, both will strongly influence the topology of the photon paths, as can be inferred from (7.90). In particular, it can be shown [118] that, for equations like (7.90), the trajectories exiting through one slit never cross the trajectories coming from the other one, and, if both slits are identical, there is perfect symmetry between both groups of trajectories with respect to the axis $y = 0$ (assuming the center of each slit is at the same distance from $y = 0$).

7.6 Photon Paths and Optical Schrödinger-like Equations

As seen in Sect. 4.6.1, when classical electromagnetism is reformulated in terms of the Riemann–Silberstein [119–121] (see Sect. 4.6.1), the electromagnetic energy density (7.46) can be expressed as

$$U = \Xi \cdot \Xi^*. \tag{7.91}$$

Moreover, under the absence of free electric charges and charge current densities, the flow of this energy density across the space, described by the Poynting vector (7.45), reads as

$$\mathbf{P} = ic\mathbf{F} \times \mathbf{F}^*, \tag{7.92}$$

where the relation

$$\nabla(\mathbf{A} \times \mathbf{B}) = \mathbf{B} \cdot \nabla \times \mathbf{A} - \mathbf{A} \cdot \nabla \times \mathbf{B} \tag{7.93}$$

for any two vectors, $\mathbf{A}$ and $\mathbf{B}$, has been used. Taking into account these two results, the photon paths are obtained now in terms of Ξ by substituting (7.91) and (7.92) into (7.47), which yields

$$\frac{d\mathbf{r}}{ds} = i \left(\frac{\Xi \times \Xi^*}{\Xi \cdot \Xi^*} \right). \tag{7.94}$$

Since conditions of stationarity were assumed above (i.e., the electromagnetic field covers the whole space and is only affected by the boundary conditions of the problem considered), (7.94) transports the time-averaged electromagnetic energy density, U, as described by the (also time-averaged) Poynting vector, $\mathbf{P}$. That is, in agreement with (7.80).

References

1. Burton, H.E.: The optics of Euclid. J. Opt. Soc. Am. **35**, 357–372 (1945)
2. Schmidt, W., Nix, L.: Heronis Alexandrini: Opera Quae Supersunt Omnia. Vol. 2: Mechanica et Catoptrica. Teubner, Leipzig (1900)
3. Mihas, P.: The problem of focusing and real images. Eur. J. Phys. **29**, 539–553 (2008)
4. Tannery, P., Henry, C. (eds.): Œuvres de Fermat, pp. 354–457. Gauthier-Villars, Paris (1894)
5. Minnaert, M.: The Nature of Light and Colour in the Open Air. Dover, New York (1954)
6. Misner, C.W., Thorne, K.S., Wheeler, J.A.: Gravitation. W. H. Freeman, San Francisco (1973)
7. Wambsganss, J.: Gravitational lenses: the uses of strong and weak lenses. Nature **386**, 27–28 (1997)
8. Wambsganss, J.: Gravitational lensing in astronomy. Living Rev. Relativ. **1**, 1–80 (1998). http://www.livingreviews.org/lrr-1998-12

9. Wambsganss, J.: Gravitational lensing. Astrophys. Space Sci. **278**, 123–128 (2001)
10. Evans, J., Rosenquist, M.: "$F = ma$" optics. Am. J. Phys. **54**, 876–883 (1986)
11. Davisson, C., Germer, L.H.: Diffraction of electrons by a crystal of nickel. Phys. Rev. **30**, 705–740 (1927)
12. Knauer, F., Stern, O.: Intensitätsmessungen an Molekularstrahlen von Gasen. Z. Phys. **53**, 766–778 (1929)
13. Estermann, I., Stern, O.: Beugung von Molekularstrahlen. Z. Phys. **61**, 95–125 (1930)
14. Estermann, I., Frisch, R., Stern, O.: Monochromasierung der de Broglie–Wellen von Moleku- larstrahlen. Z. Phys. **73**, 348–365 (1930)
15. Estermann, I., Frisch, R., Stern, O.: Versuche mit monochromatischen de Broglie–Wellen von Molekularstrahlen. Phys. Z. **32**, 670–675 (1931)
16. Cáceres, J.O., Morato, M., González-Ureña, A.: Interaction of polar molecules with resonant radio frequency electric fields: imaging of the NO molecular beam splitting. J. Phys. Chem. A **110**, 13643–13645 (2006)
17. González-Ureña, A., Requena, A., Bastida, A., Zúñiga, J.: On the interaction of a beam of polar molecules with a static and a resonant RF field as a source of molecular interferences. Eur. Phys. J. D **49**, 297–303 (2008)
18. Arndt, M., Nairz, O., Voss-Andreae, J., Keller, C., Van der Zouw, G., Zeilinger, A.: Wave- particle duality of C60 molecules. Nature **401**, 680–682 (1999)
19. Nairz, O., Arndt, M., Zeilinger, A.: Experimental challenges in fullerene interferometry. J. Mod. Opt. **47**, 2811–2821 (2000)
20. Nairz, O., Brezger, B., Arndt, M., Zeilinger, A.: Diffraction of complex molecules by structures made of light. Phys. Rev. Lett. **87**, 160401(1–4) (2001)
21. Brezger, B., Hackermller, L., Uttenthaler, S., Petschinka, J., Arndt, M., Zeilinger, A.: Matter- wave interferometer for large molecules. Phys. Rev. Lett. **88**, 100404(1–4) (2002)
22. Nairz, O., Arndt, M., Zeilinger, A.: Quantum interference experiments with large molecules. Am. J. Phys. **71**, 319–325 (2003)
23. Hackermüller, L., Uttenthaler, S., Hornberger, K., Reiger, E., Brezger, B., Zeilinger, A., Arndt, M.: Wave nature of biomolecules and fluorofullerenes. Phys. Rev. Lett. **91**, 090408(1– 4) (2003)
24. Tonomura, A., Endo, J., Matsuda, T., Kawasaki, T., Ezawa, H.: Demonstration of single- electron buildup of an interference pattern. Am. J. Phys. **57**, 117–120 (1989)
25. Shimizu, F., Shimizu, K., Takuma, H.: Double-slit interference with ultracold metastable neon atoms. Phys. Rev. A **46**, R17–R20 (1992)
26. Born, M.: Zur Quantenmechanik der Stoßvorgänge. Z. Phys. **37**, 863–867 (1926)
27. Born, M.: Quantenmechanik der Stoßvorgänge. Z. Phys. **38**, 803–840 (1926)
28. Born, M.: Physical aspects of quantum mechanics. Nature **119**, 354–357 (1927)
29. Born, M.: Quantenmechanik und Statistik. Naturwissenschaften **15**, 238–242 (1927)
30. Dimitrova, T.L., Weis, A.: The wave-particle duality of light: a demonstration experiment. Am. J. Phys. **76**, 137–142 (2008)
31. Dimitrova, T.L., Weis, A.: Single photon quantum erasing: a demonstration experiment. Eur. J. Phys. **31**, 625–637 (2010)
32. Bialynicki-Birula, I.: On the wave function of the phonon. Acta Phys. Pol. A **86**, 97–107 (1994)
33. Bialynicki-Birula, I.: The photon wave function. In: Eberly, J.H., Mandel, L., Wolf, E. (eds.) Coherence and Quantum Optics, vol. 7, pp. 313–322. Plenum Press, New York (1996)
34. Bialynicki-Birula, I.: Photon wave function. Prog. Opt. **36**, 245–294 (1996)
35. Bialynicki-Birula, I.: Hydrodynamics of relativistic probability flows. In: Infeld, E., Zelazny, R., Galkowski, A. (eds.) Nonlinear Dynamics, Chaotic and Complex Systems, pp. 64–71. Cambridge University Press, Cambridge (1997)
36. Sipe, J.E.: Photon wave functions. Phys. Rev. A **52**, 1875–1883 (1995)
37. Scully, M.O., Zubairy, M.S.: Quantum Optics. Cambridge University Press, Cambridge (1997)

38. Landau, L., Peierls, R.: Quantenelektrodynamik im Konfigurationsraum. Z. Phys. **62**, 188–200 (1930)
39. Dirac, P.A.M.: The Principles of Quantum Mechanics. Clarendon Press, Oxford (1958)
40. Cook, R.J.: Photon dynamics. Phys. Rev. A **25**, 2164–2167 (1982)
41. Cook, R.J.: Lorents covariance of photon dynamics. Phys. Rev. A **26**, 2754–2760 (1982)
42. Inagaki, T.: Quantum-mechanical approach to a free photon. Phys. Rev. A **49**, 2839–2843 (1994)
43. Kobe, D.H.: A relativistic Schrödinger-like equation for a photon and its second quantization. Found. Phys. **29**, 1203–1231 (1999)
44. Berry, M.V.: Riemann–Silberstein vortices for paraxial waves. J. Opt. A **6**, S175–S177 (2004)
45. Holland, P.R.: Hydrodynamic construction of the electromagnetic field. Proc. R. Soc. A **461**, 3659–3679 (2005)
46. Raymer, M.G., Smith, B.J.: The Maxwell wave function of the photon. Proc. SPIE **5866**, 293–297 (2005)
47. Smith, B.J., Raymer, M.G.: Photon wave functions: wave-packet quantization of light and coherence theory. New J. Phys. **9**, 414(1–37) (2007)
48. Zhi-Yong, W., Cai-Dong, X., Ole, K.: The first-quantized theory of photons. Chin. Phys. Lett. **24**, 418–420 (2007)
49. Madelung, E.: Quantentheorie in hydrodynamischer Form. Z. Phys. **40**, 322–326 (1926)
50. Loudon, R.: The Quantum Theory of Light. Oxford University Press, Oxford (1983)
51. Lundeen, J.S., Sutherland, B., Patel, A., Stewart, C., Bamber, C.: Direct measurement of the quantum wavefunction. Nature **474**, 188–191 (2011)
52. Kocsis, S., Braverman, B., Ravets, S., Stevens, M.J., Mirin, R.P., Shalm, L.K., Steinberg, A.M.: Observing the average trajectories of single photons in a two-slit interferometer. Science **332**, 1170–1173 (2011)
53. Sommerfeld, A., Runge, I.: Anwendung der Vektorrechnung auf die Grundlagen der geometrischen Optik. Ann. Phys. Leipzig **35**, 290–298 (1911)
54. Elmore, W.C., Heald, M.A.: Physics of Waves. Dover, New York (1969)
55. Born, M., Wolf, E.: Principles of Optics, 7th edn. Pergamon Press, Oxford (2002)
56. Jackson, J.D.: Classical Electrodynamics, 3rd edn. Wiley, New York (1998)
57. Courant, R., Hilbert, D.: Methods of Mathematical Physics, vol. 1. Interscience Publishers, New York (1953)
58. Goldstein, H.: Classical Mechanics. Addison-Wesley, Reading (1980)
59. Evans, J., Nandi, K.K., Islam, A.: The optical-mechanical analogy in general relativity: exact Newtonian forms for the equations of motion of particles and photons. Gen. Relat. Gravit. **28**, 413–439 (1996)
60. Nandi, K.K., Migranov, N.G., Evans, J.C., Amedeker, M.K.: Planetary and light motions from Newtonian theory: an amusing exercise. Eur. J. Phys. **27**, 429–435 (2006)
61. Ambrosiniy, D., Ponticiello, A., Schirripa Spagnolo, G., Borghi, R., Gori, F.: Bouncing light beams and the Hamiltonian analogy. Eur. J. Phys. **18**, 284–289 (1997)
62. Calvo, M.L., Pérez-Ríos, J.: Dynamic programming revisited: a generalized formalism for arbitrary ray trajectories in inhomogeneous optical media with radial dependence. J. Opt. A **11**, 125403(1–9) (2009)
63. Busch, H.: Über die Wirkungsweise der Konzentrierungsspule bei der Braunschen Röhre. Arch. Elektrotech. **18**, 583–594 (1927)
64. Busch, H., Brüche, E. (eds.): Beiträge zur Elektronenoptik. Barth, Leipzig (1937)
65. Hawkes, P.: Recent advances in electron optics and electron microscopy, vol. 29, pp. 837–855. Annales de la Foundation, Louis de Broglie (2004)
66. Jeffreys, H.: On certain approximate solutions of linear differential equations of the second order. Proc. London Math. Soc. **23**(2), 428–436 (1925)
67. Wentzel, G.: Verallgemeinerung der Quantenbedingungen für die Zwecke der Wellenmechanik. Z. Phys. **38**, 518–529 (1926)
68. Kramers, H.A.: Wellenmechanik und halbzahlige Quantisierung. Z. Phys. **39**, 828–840 (1926)

69. Brillouin, L.: La mécanique ondulatoire de Schrödinger; une méthode générale de résolution par approximations successives. Comptes Rendus **183**, 24–26 (1926)
70. Brillouin, L.: Sur un type gènéral de problèmes, permettant la séparation des variables dans la mécanique ondulatoire de Schrödinger. Comptes Rendus **183**, 270–271 (1926)
71. Gamow, G.: Zur Quantentheorie des Atomkernes. Z. Phys. **51**, 204–212 (1928)
72. Gurney, R.W., Condon, E.U.: Wave mechanics and radioactive disintegration. Nature **122**, 439–439 (1928)
73. Stavroudis, O.N.: The Optics of Rays, Wavefronts and Caustics. Academic Press, New York (1972)
74. Berry, M.V.: Waves and Thom's theorem. Adv. Phys. **25**, 1–26 (1976)
75. Bakhoom, N.G.: Expansions of a certain integral function. Proc. Lond. Math. Soc. **35**(2), 83–100 (1933)
76. Erdérlyi, A.: Asymptotic Expansions. Dover, New York (1956)
77. Berry, M.V., Upstill, C.: Catastrophe optics: morphologies of caustics and their diffraction patterns. Prog. Opt. **18**, 259–317 (1980)
78. Gilmore, R.: Catastrophe Theory for Scientists and Engineers. Dover, New York (1981)
79. Berry, M.V.: Cusped rainbows and incoherence effects in the rippling-mirror model for particles scattering from surfaces. J. Phys. A. Math. Gen. **8**, 566–584 (1975)
80. Garibaldi, U., Levi, A.C., Spadacini, R., Tommei, G.E.: Quantum theory of atom-surface scattering:diffraction and rainbow. Surf. Sci. **48**, 649–675 (1975)
81. Guantes, R., Sanz, A.S., Margalef-Roig, J., Miret-Artés, S.: Atom-surface diffraction: a trajectory description. Surf. Sci. Rep. **53**, 199–330 (2004)
82. Braunbek, W., Laukien, G.: Einzelheiten zur Halbebenen-Beugung. Optik. **9**, 174–179 (1952)
83. Prosser, R.D.: The interpretation of diffraction and interference in terms of energy flow. Int. J. Theor. Phys. **15**, 169–180 (1976)
84. Prosser, R.D.: Quantum theory and the nature of interference. Int. J. Theor. Phys. **15**, 181–193 (1976)
85. Wünscher, T., Hauptmann, H., Herrmann, F.: Which way does the light go? Am. J. Phys. **70**, 599–606 (2002)
86. Hesse, E.: Modelling diffraction during ray tracing using the concept of energy flow lines. J. Quant. Spect. Rad. Trans. **109**, 1374–1838 (2008)
87. Davidović, M., Sanz, A.S., Arsenović, D., Božić, M., Miret-Artés, S.: Electromagnetic energy flow lines as possible paths of photons. Phys. Scr. T **135**, 014009(1–5) (2009)
88. Sanz, A.S., Davidović, M., Božić, M., Miret-Artés, S.: Understanding interference experiments with polarized light through photon trajectories. Ann. Phys. (N.Y.) **325**, 763–784 (2010)
89. Božić, M., Davidović, M., Dimitrova, T.L., Miret-Artés, S., Sanz, A.S., Weis, A.: Generalized Arago–Fresnel laws: the EME-flow-line description. J. Russ. Laser Res. **31**, 117–128 (2010)
90. Gondran, M., Gondran, A.: Energy flow lines and the spot of Poisson–Arago. Am. J. Phys. **78**, 598–602 (2010)
91. Ghose, P., Home, D.: On boson trajectories in the Bohm model. Phys. Lett. A **191**, 362–364 (1994)
92. Ghose, P., Majumdar, A.S., Guha, S., Sau, J.: Bohmian trajectories for photons. Phys. Lett. A **290**, 205–213 (2001)
93. Henry, M.: Fresnel–Arago laws for interference in polarized light: a demonstration experiment. Am. J. Phys. **49**, 690–691 (1981)
94. Kanseri, B., Bisht, N.S., Kandpal, H.C., Rath, S.: Observation of the Fresnel and Arago laws using the Mach–Zehnder interferometer. Am. J. Phys. **76**, 39–42 (2008)
95. Kanseri, B., Bisht, N.S., Rath, S., Kandpal, H.C.: A modified version of Young's interferometer to study the Fresnel and Arago interference laws. Eur. J. Phys. **30**, 835–844 (2009)
96. Rodríguez-Lara, B.M., Ricardez-Vargas, I.: Interference with polarized light beams: generation of spatially varying polarization. Am. J. Phys. **77**, 1135–1143 (2009)

97. Arago, D.F.J., Fresnel, L.: Sur L'Action que les rayons de lumière polarisés exercent les uns sur les autres. Ann. Chimie Physique **X**, 288–305 (1819)
98. Collett, E.: Mathematical formulation of the interference laws of Fresnel and Arago. Am. J. Phys. **39**, 1483–1495 (1971)
99. Barakat, R.: Analytic proofs of the Arago–Fresnel laws for the interference of polarized light. J. Opt. Soc. Am. A **10**, 180–185 (1993)
100. Nye, J.F.: Polarization effects in the diffraction of electromagnetic waves: the role of dislocations. Proc. R. Soc. Lond. A **387**, 105–132 (1983)
101. Nye, J.F.: Lines of circular polarization in electromagnetic wave fields. Proc. R. Soc. Lond. A **389**, 279–290 (1983)
102. Nye, J.F., Berry, M.V.: Dislocations in wave trains. Proc. R. Soc. Lond. A **336**, 165–190 (1974)
103. Bialynicki-Birula, I., Bialynicka-Birula, Z.: Vortex lines of the electromagnetic field. Phys. Rev. A **67**, 062114(1–8) (2003)
104. Kaiser, G.: Helicity, polarization and Riemann–Silberstein vortices. J. Opt. A **6**, S243–S245 (2004)
105. Sanz, A.S., Miret-Artés, S.: A causal look into the quantum Talbot effect. J. Chem. Phys. **126**, 234106(1–11) (2007)
106. Sanz, A.S., Borondo, F., Miret-Artés, S.: Causal trajectories description of atom diffraction by surfaces. Phys. Rev. B **61**, 7743–7751 (2000)
107. Sanz, A.S., Borondo, F., Miret-Artés, S.: Particle diffraction studied using quantum trajectories. J. Phys.: Condens. Matter **14**, 6109–6145 (2002)
108. Hunt, J.L., Karl, G.: Interference with polarized light beams. Am. J. Phys. **38**, 1249–1250 (1970)
109. Henneberger, W.C., Zitter, R.N.: Polarized double-slit diffraction. Am. J. Phys. **51**, 464–465 (1983)
110. Ferguson, J.L.: A simple, bright demonstration of the interference of polarized light. Am. J. Phys. **52**, 1141–1142 (1984)
111. Pescetti, D.: Interference between elliptically polarized light beams. Am. J. Phys. **40**, 735–740 (1972)
112. Scully, M.O., Englert, B.G., Walther, H.: Quantum optical tests of complementarity. Nature **351**, 111–116 (1991)
113. Walborn, S.P., Terra-Cunha, M.O., Pádua, S., Monken, C.H.: Double-slit quantum eraser. Phys. Rev. A **65**(1–6), 033818 (2002)
114. Scarani, V., Suarez, A.: Introducing quantum mechanics: one-particle interferences. Am. J. Phys. **66**, 718–721 (1998)
115. Nielsen, M.A., Chuang, I.L.: Quantum Computation and Quantum Information. Cambridge University Press, Cambridge (2000)
116. Sanz, A.S., Borondo, F.: A quantum trajectory description of decoherence. Eur. Phys. J. D **44**, 319–326 (2007)
117. Sanz, A.S., Borondo, F.: Contextuality, decoherence and quantum trajectories. Chem. Phys. Lett. **478**, 301–305 (2009)
118. Sanz, A.S., Miret-Artés, S.: A trajectory-based understanding of quantum interference. J. Phys. A **41**, 435303(1–23) (2008)
119. Silberstein, L.: Elektromagnetische Grundgleichungen in bivectorieller Behandlung. Ann. Phys. (Leipzig) **22**, 579–586 (1907)
120. Silberstein, L.: Nachtrag zur Abhandlung über "Elektromagnetische Grundgleichungen in bivectorieller Behandlung". Ann. Phys. (Leipzig) **24**, 783–784 (1907)
121. Bateman, H.: The Mathematical Analysis of Electrical and Optical Wave Motion on the Basis of Maxwell's Equations. Dover, New York (1955)

Appendix A
Calculus of Variations

The calculus of variations appears in several chapters of this volume as a means to formally derive the fundamental equations of motion in classical mechanics as well as in quantum mechanics. Here, the essential elements involved in the calculus of variations are briefly summarized.[1]

Consider a functional $\mathcal{F}$ depending on a function y of a single variable x (i.e., $y = y(x)$) and its first derivative $y' = dy/dx$. Moreover, this functional is defined in terms of the line or path integral

$$\mathcal{F}[I] = \int_{x_a}^{x_b} I(y, y', x) dx. \tag{A.1}$$

Accordingly, the value of $\mathcal{F}$ will depend on the path chosen to go from x_a to x_b. The central problem of the *calculus of variations* [1–3] consists of determining the path $y(x)$ that makes $\mathcal{F}$ an extremum (a maximum, a minimum or a saddle point). In other words, this is equivalent to determining the conditions for which (A.1) acquires a stationary value or, equivalently, is invariant under first-order variations (or perturbations) of the path $y(x)$, i.e.,

$$\delta \mathcal{F} = \delta \int_{x_a}^{x_b} I \, dx = \int_{x_a}^{x_b} \delta I \, dx = 0. \tag{A.2}$$

Let us thus define the quantities $\delta y = Y(x) - y(x)$ and $\delta I = I(Y, Y', x) - I(y, y', x)$, where $Y(x)$ denotes the perturbed path. Variations are taken with respect to the same x value, so $\delta x = 0$. It is straightforward to show that $\delta y' = d(\delta y)/dx$ and therefore

---

[1] The brief description of the essential elements involved in the calculus of variations presented here can be complemented with more detailed treatments, which can be found in well-known textbooks on mathematical physics, e.g., [1–3].

A. S. Sanz and S. Miret-Artés, *A Trajectory Description of Quantum Processes.*
I. Fundamentals, Lecture Notes in Physics 850, DOI: 10.1007/978-3-642-18092-7,
© Springer-Verlag Berlin Heidelberg 2012

$$\delta I = \left(\frac{\partial I}{\partial y} + \frac{\partial I}{\partial y'} \frac{d}{dx} \right) \delta y. \tag{A.3}$$

Substituting this expression into (A.2) and then integrating by parts yields

$$\int_{x_a}^{x_b} \left(\frac{\partial I}{\partial y} - \frac{d}{dx} \frac{\partial I}{\partial y'} \right) \delta y \, dx = 0, \tag{A.4}$$

since, at the boundaries, $\delta y(x_a) = \delta y(x_b) = 0$. Because δy is an arbitrary, infinitesimal increment, it can be chosen so that the integrand in (A.4) vanishes. This leads to the well-known *Euler–Lagrange equation*,

$$\frac{\partial I}{\partial y} - \frac{d}{dx} \frac{\partial I}{\partial y'} = 0. \tag{A.5}$$

The function y satisfying this equation, if it exists, is said to be an *extremal curve* or *extremal*.

Equation (A.5) can also be recast as

$$\frac{\partial I}{\partial x} - \frac{d}{dx} \left(I - y' \frac{\partial I}{\partial y'} \right) = 0, \tag{A.6}$$

which arises after taking into account the dependence of I on x, y and y' as well as the fact that

$$\frac{d}{dx} = \frac{\partial}{\partial x} + y' \frac{\partial}{\partial y} + y'' \frac{\partial}{\partial y'}. \tag{A.7}$$

Equation (A.6) is useful whenever I does not depend explicitly on x, for it becomes

$$I - y' \frac{\partial I}{\partial y'} = \text{constant}, \tag{A.8}$$

which is also an extremal.

Consider now that I depends on several functions $y_1, y_2, \ldots, y_N$ of x and their respective derivatives, $y'_1, y'_2, \ldots, y'_N$. Then, proceeding in a similar way, a functional

$$\mathcal{F}[I] = \int_{x_a}^{x_b} I(y_1, y_2, \ldots, y_N, y'_1, y'_2, \ldots, y'_N, x) dx \tag{A.9}$$

can be defined, which becomes an extremum or stationary when the set of Euler–Lagrange equations

$$\frac{\partial I}{\partial y_i} - \frac{d}{dx} \frac{\partial I}{\partial y'_i} = 0, \qquad i = 1, 2, \ldots, N, \tag{A.10}$$

is satisfied. However, it could happen that the search for an extremum condition is subject to a constraint, as in the so-called *isoperimetric problems* (e.g., determining the closed plane curve of maximum area and fixed perimeter). In such cases, given a set $J_1, J_2, \ldots, J_M$ of constraining conditions that depend on x and the $y_i (i = 1, 2, \ldots, N)$, the set of N equations (A.10) is replaced by the set of $N + M$ equations

$$
\begin{cases}
\dfrac{\partial I}{\partial y_i} - \dfrac{d}{dx}\dfrac{\partial I}{\partial y_i'} + \displaystyle\sum_{j=1}^{M} \lambda_j(x)\dfrac{\partial J_j}{\partial y_i} = 0 \\
J_j(x, y_1, y_2, \ldots, y_N) = 0
\end{cases}
. \qquad (A.11)
$$

The λ_j functions are the so-called *Lagrange undetermined multipliers*, M unknown functions of x (or constants) which have to be determined in order to obtain a full (complete) solution to the problem.

If the constraints in (A.11) are specified by a set of M functional integral constraints,

$$
\mathcal{F}_j = \int_{x_a}^{x_b} J_j(y_1, y_2, \ldots, y_N, y_1', y_2', \ldots, y_N', x)dx = c_j, \qquad (A.12)
$$

where all c_j are constant and the $\mathcal{F}_j$ are extrema for the y_i, a function

$$
K = I + \sum_{j=1}^{M} \lambda_j J_j \qquad (A.13)
$$

can be defined. Proceeding as before, one finds that these functions have to satisfy the Euler–Lagrange equation

$$
\frac{\partial K}{\partial y_i} - \frac{d}{dx}\frac{\partial K}{\partial y_i'} = 0, \qquad i = 1, 2, \ldots, N, \qquad (A.14)
$$

as well as the integral constraints (A.12).

In the particular case of mechanical systems, when the variational principle is applied, power series expansions up to the third order in the displacement are often considered. In these series expansions, the zeroth-order term gives us the action integral along the reference trajectory; the second-order is called the first variation, which vanishes for any path due to the stationarity condition; the third-order or second variation provides us with information about the nature of the stationary value (maximum, minimum or saddle point) by analyzing the eigenvalues of the matrix associated with the corresponding quadratic form in the displacements.

The formalism described above is rather general. As seen in Chap. 3, for example, it is closely related to the formal derivation of Schrödinger's wave equation. In this case, instead of several functions y_i of a single variable x, one considers a function ψ of several variables x_i. These functions are usually called *field functions* or *fields*. Furthermore, a subtle conceptual difference can be found in the application of the calculus of variations in classical and in quantum mechanics. In classical mechanics, it is tightly connected to the concept of energy (Hamiltonian); different solutions are then obtained from its application, namely the classical trajectories. In quantum mechanics, though, this idea is extended to functionals of a *single* dependent variable (the *wave function* field) and several independent variables, thus generalizing the classical case. Thus, rather than keeping constant the energy along a given path, energy conservation appears in the calculation of the average or expectation value of such an *observable*.

References

1. Morse, P.M., Feshbach, H.: Methods of Theoretical Physics. McGraw-Hill, New York (1953)
2. Margenau, H., Murphy, G.M.: The Mathematics of Physics and Chemistry, 2nd edn. Van Nostrand, New York (1956)
3. Arnold, V.I.: Mathematical Methods of Classical Mechanics. Springer, New York (1989)

Appendix B
Stochastic Processes

The theory of stochastic processes is of fundamental interest to understand the theory of open classical and quantum systems, as seen in Chaps. 2, 5 and 6. In this regard, in order to make this volume self-contained as far as possible and to better understand the dynamics of open systems, some elementary concepts on probability theory and stochastic processes are introduced here.[1]

B.1 Random or Stochastic Variables

A probability or measure space is defined by three mathematical objects $(\Omega, \mathcal{A}, P)$. In this triad, Ω represents the set of elementary outcomes or *sample space*, with such elements usually denoted by ω, with $\omega \in \Omega$; $\mathcal{A}$ is a *collection* or *field of events*; and P is a *probability measure*.

The field of events $\mathcal{A}$ is also called a σ-algebra of events. It consists of a family or set of subsets of Ω, such that:

1. $\Omega \in \mathcal{A}$ and $\emptyset \in \mathcal{A}$.
2. If $A \in \mathcal{A}$, then $\bar{A} = \Omega - A \in \mathcal{A}$.
3. If $A, B \in \mathcal{A}$, then $A \cup B \in \mathcal{A}, A \cap B \in \mathcal{A}$ and $A - B \in \mathcal{A}$.

In brief, the σ-algebra forms a closed system under the union ($\cup$), intersection ($\cap$) and complementary set operations.

A probability measure is simply a map $P : \mathcal{A} \to \mathbb{R}$, which assigns a real number to each event A of the σ-algebra. The probability of an event A, denoted by $P(A)$, satisfies the Kolmogorov axioms of probability, from which a series of consequences arise. For example, the probability $P(A)$ has a numeric bound for all A, namely $0 \le P(A) \le 1$. Thus, $P(\Omega) = 1$ characterizes a *certain event* and $P(\emptyset) = 0$ an *impossible event*; $P(\bar{A}) = 1 - P(A)$ represents the probability of the *complementary event* of A. If A_i is a countable collection of non-overlapping (or *mutually exclusive*)

---

[1] For further reading on these issues, the interested reader may consult more specialized works, e.g., [1–10].

A. S. Sanz and S. Miret-Artés, *A Trajectory Description of Quantum Processes. I. Fundamentals*, Lecture Notes in Physics 850, DOI: 10.1007/978-3-642-18092-7, © Springer-Verlag Berlin Heidelberg 2012

sets, i.e., $A_i \cap A_j = \emptyset$ for $i \neq j$, then $P(A_1 \cup A_2 \cup \cdots) = \sum_i P(A_i)$. On the contrary, given A and B, if $A \cap B \neq \emptyset$, the non-vanishing probability $P(A \cap B)$ represents the *joint probability* that any element of the intersection occurs; two events are called independent when $P(A \cap B) = P(A)P(B)$. Similarly, the *conditional probability* that an event A occurs, given another event B also occurs (i.e., $P(B) \neq 0$), is defined as $P(A|B) = P(A \cap B)/P(B)$.

A *random* or *stochastic variable* X is a function from the sample space Ω into the set of real (or complex) numbers with the property $A = \{\omega/X(\omega) \leq x\} \in \mathcal{A}$, with $x \in \mathbb{R}$ being a *realization* of X. This random variable or function is said to be measurable with respect to the σ-algebra of $\mathcal{A}$, this definition often being expressed in terms of the inverse image, as $A = X^{-1}((-\infty, x]) \in \mathcal{A}$, with $x \in \mathbb{R}$. Accordingly, the probability of an observable event of $\mathcal{A}$ is defined as $P_X(A) = P(X^{-1}(A)) = P(\{\omega/X(\omega) \in \mathcal{A}\})$. In order to formalize this statement, the σ-algebra of *Borel sets* of $\mathbb{R}$, denoted by $\mathcal{B}$, is introduced, which is defined as the smallest σ-algebra containing all subsets of the form $(-\infty, x)$, with $x \in \mathbb{R}$—in particular, it contains all open and closed intervals of the real axis. A function is then measurable when, for any Borel set $B \in \mathcal{B}$, the pre-image $A = X^{-1}(B)$ belongs to the σ-algebra $\mathcal{A}$ of events. The *distribution function* or *probability distribution* of the random variable is defined as $F_X(x) = P(\{\omega/X(\omega) \leq x\}) = P(X \leq x)$, with $x \in \mathbb{R}$, which satisfies

1. $F_X(-\infty) = 0$.
2. $F_X(+\infty) = 1$.
3. $F_X(x)$ is an increasing monotonically right continuous function.

In many applications, continuous random variables are often found, with their *probability density* being defined as $p_X(x) = dF_X(x)/dx$ or, equivalently, as

$$F_X(x) = \int_{-\infty}^{x} p_X(x')dx', \tag{B.1}$$

so that

$$\begin{aligned} dF_X(x) = F_X(x + dx) - F_X(x) &= p_X(x)dx \\ &= P(x \leq X < x + dx) = P(d\omega). \end{aligned} \tag{B.2}$$

For random vectors or an arbitrary collection of d random variables, similar functions and densities can be defined in $\mathbb{R}^d$—if $d = 2$, they are called bivariate distributions, while for any general d, they are multivariate ones. This information provides a complete characterization of the random variable X. Thus, if the density of a random variable exists, the probability that x will be contained in the interval $(x, x + dx)$ goes to zero with dx. Therefore, the probability that X has exactly an x value is zero. Sets containing one single point as well as any set only containing a countable number of points have zero probability. In probability theory, all equalities are at best only *almost certainly true*, *almost surely* or *with probability one*. Very often X drops in F_X and p_X.

Functions of random variables, $Y = g(X)$, can also be defined. Thus, if F_X is the probability distribution of X, then $P_Y(B) = P_X(g^{-1}(B))$ and

$$p_Y(y) = \int \delta(y - g(x))p_X(x)dx, \tag{B.3}$$

where δ denotes the Dirac δ-function; if $Y = X_1 + X_2$, then $g(x_1, x_2) = x_1 + x_2$; and, if X_1 and X_2 are independent, then

$$p_Y(y) = \int p_{X_1}(x_1)p_{X_2}(y - x_1)dx_1, \tag{B.4}$$

which is the convolution of the densities associated with X_1 and X_2.

Formally, the average, expectation or mean value of a discrete random variable X is defined as

$$E\{X\} = \sum_i x_i P(A_i), \tag{B.5}$$

which is also often denoted as $\bar{X}$ or $\langle X \rangle$. On the other hand, if X is a continuous random variable,

$$E\{X\} = \int_\Omega X(\omega)dP(\omega) = \int_\Omega X(\omega)P(d\omega), \tag{B.6}$$

which can also be expressed in a more familiar form as

$$E\{X\} = \int_{\mathbb{R}} x dF_X(x) = \int_{\mathbb{R}} x p_X(x)dx. \tag{B.7}$$

The mean value (B.7) is also known as the *first moment* of the distribution function; higher nth-order moments $E\{X^n\}$ can be defined in a similar fashion provided $p_X(x)$ exists. In this regard, it is worth mentioning that the knowledge of all moments is not a sufficient condition to uniquely determine $p_X(x)$. From (B.7), the expectation value of functions of a random variable can also be defined. For example, if the mean value is known, the central moments of a random variable are defined as

$$E\{(\delta X)^r\} = E\{(X - E\{X\})^r\} = \int_{\mathbb{R}} (x - E\{X\})^r p_X(x)dx. \tag{B.8}$$

The second central moment, σ^2, defined as

$$\text{Var}(X) = \Delta X = E\{(X - E\{X\})^2\} = E(X^2) - E(X)^2, \tag{B.9}$$

also denoted as ΔX, is known as the *variance, mean-square deviation* or *fluctuation*. This moment is a measure of the width of the probability density or, in other words, of the fluctuations of X with respect to its mean value, δX. The number and location of the extrema of the probability distribution are important, because their maxima are the so-called most probable states (which is a local property).

When there are two random variables, it is very often interesting to know if they are correlated or not. A measure of their degree of correlation is given by the so-called *covariance*, defined as

$$
\begin{aligned}
\sigma_{XY} &= E\{(X - E\{X\})(Y - E\{Y\})\} \\
&= \int_{\mathbb{R}} (x - E\{X\})(y - E\{Y\})p_{XY}(x, y)dxdy,
\end{aligned}
\tag{B.10}
$$

where $p_{XY}(x, y)$ is the *joint probability density*. If X and Y are independent (or uncorrelated), then $\sigma_{XY} = 0$ (the opposite is not true in general). For several random variables, a *covariance matrix* can be defined, which is symmetric and positive semi-definite. The off-diagonal elements are called covariances and are a measure of the linear dependence of two random variables. Hence, if the variables are pairwise independent, the covariance matrix will be diagonal. Correlation coefficients between two random variables are then defined by their covariance divided by the square root of the product of their respective variances, i.e., $C(X, Y) = \sigma_{XY}/\sqrt{\Delta X \Delta Y}$. It is clear that $0 \leq |C(X, Y)| \leq 1$.

The *characteristic function* of a random variable X is defined as the Fourier transform of its probability density,

$$
G(\xi) = E\{e^{i\xi X}\} = \int p_X(x)e^{i\xi x}dx,
\tag{B.11}
$$

for a real number ξ—note the close similarity between this form and the transformations that allow to pass in quantum mechanics from the configuration to the momentum space and vice versa. This function is also called the generating function of all moments of the random variable, since the nth derivative of $G(\xi)$ evaluated at $\xi = 0$ gives the nth moment, $E\{X^n\}$—of course, this relies on the implicit assumption that $p_X(x)$ is sufficiently regular and therefore the exponential admits a Taylor series expansion. From the Fourier inversion formula, $p_X(x)$ can be determined with probability one. As it can be shown, a sequence of probability densities converges to a limiting probability density if the corresponding characteristic functions converge to the characteristic function associated with the limiting probability density. A straightforward generalization to more variables can easily be carried out. In such a case, if all of them are independent, the corresponding characteristic function will be the product of the individual characteristic functions. Similarly, the logarithm of the characteristic function generates all the *cumulants* of X. The first cumulant is the mean value, while the second cumulant is the covariance of two random variables. Finally, X can also be given by the integral of a stochastic process.

When two random variables are not statistically independent, there is some information about one of them with respect to the other. In such a case, one can define the so-called *marginal distributions*,

$$p_X(x) = \int_{\mathbb{R}} p_{XY}(x, y)dy, \tag{B.12a}$$

$$p_Y(y) - \int_{\mathbb{R}} p_{XY}(x, y)dx. \tag{B.12b}$$

The conditional expectation value of the random variable X is defined as

$$E\{X|B\} = \int_{\Omega} X(\omega)P(d\omega|B), \tag{B.13}$$

although it can also be written as

$$E\{X|B\}P(B) = \int_B X(\omega)P(d\omega). \tag{B.14}$$

Many times it is necessary to establish a set of conditions with respect to a collection of events or to the history of events. This is done, for example, to make the best possible prediction of the actual random variable X knowing the available or previous information. The sub-σ-algebra $\mathcal{C}$ generated by $\{A_i\}$, with $\mathcal{C} \subset \mathcal{A}$, is precisely the available information. The best possible prediction or estimate of X is another random variable Y which has to be $\mathcal{C}$-measurable and fulfills the condition

$$\int_{A_i} Y(\omega)P(d\omega) = \int_{A_i} X(\omega)P(d\omega), \tag{B.15}$$

with $A_i \subset \mathcal{C}$. According to the *Radon–Nikodym theorem* [11, 12], there is a random variable Y with the above properties, such that it is almost surely unique and can be expressed as

$$Y = E\{X|\mathcal{C}\} = \int_{\Omega} X(\omega)P(d\omega|\mathcal{C}). \tag{B.16}$$

The most important properties of conditional expectations are:

- $E\{E\{X|\mathcal{C}\}\} = E\{X\}$.
- If $X \geq 0$, then $E\{X|\mathcal{C}\} \geq 0$.
- If X is measurable with respect to $\mathcal{C}$, then $E\{X|\mathcal{C}\} = X$.
- If $E\{X\} < \infty$ and $E\{Y\} < \infty$, then $E\{aX + bY|\mathcal{C}\} = aE\{X|\mathcal{C}\} + bE\{Y|\mathcal{C}\}$, with a and b being constant.
- If X and $\mathcal{C}$ are independent, then $E\{X|\mathcal{C}\} = E\{X\}$.
- If $\mathcal{C}_1$ and $\mathcal{C}_2$ are sub-σ-algebras of $\mathcal{A}$, such that $\mathcal{C}_1 \subset \mathcal{C}_2 \subset \mathcal{A}$, then $E\{E\{X|\mathcal{C}_2\}|\mathcal{C}_1\} = E\{E\{X|\mathcal{C}_1\}|\mathcal{C}_2\} = E\{X|\mathcal{C}_1\}$.

A particular class of conditional expectation values is obtained when the conditioning is considered with respect to another random variable. In this case, $E\{X|\mathcal{C}\} =$

$E\{X|Z\}$, when the σ-algebra C is generated by the random variable Z. Since $E\{X|Z\}$ is a random variable, it can be shown by means of a theorem that $E\{X|Z\} = h(Z)$, i.e., the corresponding random variable can be written as a measurable function h of Z. This function is real-valued and almost surely uniquely defined. One can therefore write $E\{X|Z = z\} = h(z)$ and, if the marginal probability density

$$p_Z(z) = \int_{\mathbb{R}} p_{XZ}(x, z)dx \tag{B.17}$$

is positive, then

$$p(x|z) = \frac{p_{XZ}(x, z)}{p_Z(z)}. \tag{B.18}$$

From these expressions, one finds

$$E\{x|z\} = \int_{\mathbb{R}} xp(x|z)dx \tag{B.19a}$$

$$E\{E\{x|z\}\} = E\{x\} = \int_{\mathbb{R}} xp_{XZ}(x, z)dxdz. \tag{B.19b}$$

One of the most widely used probability distributions is the *Gaussian* or *normal distribution*. According to the *central limit theorem*, any random variable X given by the sum of N statistically independent and identically distributed random variables becomes Gaussian or normally distributed. That is, in the limit $N \to \infty$ and provided the first and second moments do not diverge (in $p_X(x)$ very often X drops in most of textbooks),

$$p(x) = \frac{1}{\sqrt{2\pi\sigma^2}}e^{-x^2/2\sigma^2}. \tag{B.20}$$

Within this context, the concept of limiting distribution readily appears, not related to the regular behavior of the moments of the distribution itself, but to stability. It can be shown that a probability distribution can only be a limiting distribution if it is a *stable* or *Lévy distribution*. In this class of distributions, the logarithm of their characteristic functions must satisfy a certain mathematical expression [7] and display long-range, inverse power-law tails which may lead to a divergence of even the lowest order moments. For example, the second moment of the *Cauchy* or *Lorentzian distribution* is infinite, going as $|x^{-(1+\alpha)}|$ when $0 < \alpha < 2$ and $|x| \to \infty$.

Another important distribution is the *binomial distribution*, which is a discrete probability distribution accounting for the number of successful events with probability p in a sequence of n independent experiments with two possible outcomes (e.g., yes/no, 0/1). This type of distribution describes, for example, a random walker on a line, stepping forward and backwards randomly. The binomial distribution converges to a Gaussian one in the limit of a large number of jumps, as can be easily shown using

Stirling's formula. An "intermediate" case between both distributions is the *Poisson distribution*, which is continuous and appears in the joint limit $p \to 0$ and $n \to \infty$, but keeping constant the quantity $\lambda = pN$.

The concept of limit of a sequence of random variables arises naturally in many different physical situations, though there is not a unique way to define it. Some of these definitions are:

1. The *almost certainly* or *surely limit*,

$$X = \lim_{n \to \infty} X_n, \qquad (B.21)$$

i.e., X_n converges almost surely to X, is the simplest definition. In brief, this limit is more explicitly expressed as

$$X = \text{as--}\lim_{n \to \infty} X_n. \qquad (B.22)$$

2. The *mean square limit* or limit in the mean,

$$X = \text{ms--}\lim_{n \to \infty} X_n, \qquad (B.23)$$

which implies that

$$\text{ms--}\lim_{n \to \infty} E\{(X_n - X)^2\} = 0. \qquad (B.24)$$

3. The *stochastic limit* or *limit in probability*,

$$X = \text{st--}\lim_{n \to \infty} X_n. \qquad (B.25)$$

According to this definition

$$\lim_{n \to \infty} P(|X_n - X| < \epsilon) = 0 \qquad (B.26)$$

for positive ϵ.

4. The *limit in distribution*,

$$E\{f(X)\} = \lim_{n \to \infty} E\{f(X_n)\}, \qquad (B.27)$$

for any continuous bound function $f(x)$.

B.2 Stochastic Processes

A family of random variables indexed by the parameter time is known as a *random* or *stochastic process*, denoted by a collection of real (or complex) numbers

$\{X_t(\omega)\}$ or $\{X(\omega, t)\}$. If time is fixed and ω varies over the sample space, the random variable is measurable in the sense that the pre-images of any Borel set in $\mathbb{R}$ must belong to the σ-algebra of events of the probability space considered. Conversely, if ω is fixed and t varies on a given interval T, the function is real-valued on the time axis. This is called a *realization*, *stochastic trajectory* or *sample path of the stochastic process* X_t. A stochastic process can then be regarded as a map $X : \Omega \times T \to \mathbb{R}$. Multivariate stochastic processes X_t are defined as vector stochastic processes with a given number of components, each one being a real-valued stochastic process.

A stochastic process is characterized by a hierarchy of joint distribution functions $F(x, t) = P(X_t \leq x)$, $F(x_1, t_1; x_2, t_2) = P(X_{t_1} \leq x_1; X_{t_2} \leq x_2)$, etc., which satisfies two properties:

1. *Symmetry*: two distribution functions differing by a permutation of n time values are equal.
2. *Compatibility*: the lower members of the hierarchy can be obtained from the higher ones.

In terms of the probability density, if

$$F(x_1, t_1; \cdots, x_n, t_n) = \int_{-\infty}^{x_1} \cdots \int_{-\infty}^{x_n} dx'_1 \cdots dx'_n p(x'_1, t_1; \cdots; x'_n, t_n), \qquad (B.28)$$

the compatibility property then reads as

$$p(x_1, t_1; \cdots; x_m, t_m) = \int_{\mathbb{R}} \cdots \int_{\mathbb{R}} dx'_{m+1} \cdots dx'_n p(x_1, t_1; \cdots; x_n, t_n), \qquad (B.29)$$

for $m < n$. According to Kolmogorov's fundamental theorem, for every hierarchy of joint distribution functions satisfying both properties, there exist a probability space $(\Omega, \mathcal{A}, P)$ and a stochastic process X_t defined on it, which possess the given distribution functions.

Two stochastic processes are said to be equivalent if they have an identical hierarchy of joint distribution functions—which does not mean that the realization of the two processes are identical. A stochastic process is said to have almost surely continuous sample paths if

$$P(\{\omega/X_t(\omega) \text{ is a continuous function on time}\}) = 1. \qquad (B.30)$$

Taking into account the several definitions given above for the limit of a random variable sequence, a series of definitions of continuity for a stochastic process can also be established:

1. X_t is *continuous in probability* if for every t and positive ϵ

$$\lim_{s \to t} P(|X_s - X_t| > \epsilon) = 0. \qquad (B.31)$$

2. X_t is *continuous in mean square* if for every t

$$\lim_{s \to t} E\{(X_s - X_t)^2\} = 0. \qquad (B.32)$$

3. X_t is *continuous almost surely* if for every t

$$P(\{\omega / \lim_{s \to t} X_s(\omega) = X_t(\omega)\}) = 1. \tag{B.33}$$

In any of the three definitions, the continuity of the sample path is guaranteed only if the probability that a discontinuity in such paths occurs at a given time is zero

A stochastic process is called *stationary* if all its finite-dimensional probability densities are invariant with respect to time shifts. Thus, the expectation value of a stationary stochastic process will be constant with time and the two-event (or two-dimensional) probability density will only depend on the time difference,

$$p(x_1, t_1; x_2, t_2) = p(x_1, x_2; t_2 - t_1). \tag{B.34}$$

Taking this into account, a (two-event) correlation function can be defined as

$$C(|t_2 - t_1|) = E\{(X_{t_1} - E\{X_{t_1}\})(X_{t_2} - E\{X_{t_2}\})\}, \tag{B.35}$$

from which the memory or correlation time of a stationary stochastic process X_t is given by

$$\tau_{\text{corr}} = \frac{1}{C(0)} \int_0^\infty C(\tau) d\tau, \tag{B.36}$$

with $C(\tau) = E\{(X_t - E\{X_t\})(X_{t+\tau} - E\{X_{t+\tau}\})\}$. This timescale can be considered as a measure for the rapidity of the stochastic process fluctuations. For example, short-memory or short-correlation times imply faster decreasing correlation functions. When dealing with realizations or stochastic trajectories, time averages can also be computed. For random variables, their mean value is defined as

$$\langle x \rangle = \lim_{T \to \infty} \frac{1}{2T} \int_{-T}^T x(t) dt \tag{B.37}$$

and the corresponding autocorrelation function as

$$C(\tau) = \langle x(t)x(t + \tau) \rangle = \lim_{T \to \infty} \frac{1}{2T} \int_{-T}^T x(t)x(t + \tau) dt. \tag{B.38}$$

If time averages and ensemble averages are equal, the stochastic process is said to be ergodic.

According to the spectral decomposition theorem, X_t can be written as a Fourier integral with random coefficients

$$X_t = \int_{\mathbb{R}} e^{ivt} dZ_v, \tag{B.39}$$

where Z_v is a stochastic process in the complex space. It has uncorrelated increments of zero mean value. As it can be shown,

$$C(\tau) = \int_{\mathbb{R}} e^{i\nu\tau} S(\nu)d\nu, \tag{B.40}$$

where $S(\nu)$ is known as the *power* or *frequency spectrum*, which satisfies the property $E\{dZ_\nu dZ_\nu^*\} = \int S(\nu)d\nu$. This quantity is a measure of the mean square power with which an oscillation of frequency ν contributes to the process X_t. Accordingly, an effective band width can be defined,

$$\nu_{\text{eff}} = \frac{1}{S(0)} \int_0^\infty S(\nu)d\nu, \tag{B.41}$$

which becomes very broad for small correlation times.

In stochastic dynamics, there are several stochastic processes which play a very important role [7, 13]: Wiener processes, Ornstein–Uhlenbeck processes and Poisson processes. They will be briefly revised below.

B.2.1   Wiener Processes and Brownian Motion

A *Wiener* (W) *process* [14] is a mathematical model that describes the Brownian motion undergone by small particles. This type of process presents the following characteristics:

1. It undergoes very rapid motions as, for example, it happens in low viscosity fluids and at high temperatures.
2. Ceaseless motion with very irregular trajectories (the velocity of Brownian particles is undefined).

If W_t is used to denote the position of the Brownian particle from some arbitrary point at $t_0 = 0$, then $W_0 = 0$. W-processes are Gaussian, i.e., all the finite dimensional probability distributions are Gaussian and have stationary independent increments. This means that the increments $W_{t_1}, W_{t_2} - W_{t_1}, \ldots, W_{t_n} - W_{t_{n-1}}$ are independent for $t_1 < \ldots < t_n$. The corresponding hierarchy of probability densities is given by

$$p(x, t) = \frac{1}{\sqrt{2\pi t}} e^{-x^2/2t} \tag{B.42}$$

and

$$p(x_1, t_1; \ldots; x_n, t_n) = p(x_1, t_1)p(x_2 - x_1, t_2 - t_1) \cdots p(x_n - x_{n-1}, t_n - t_{n-1}). \tag{B.43}$$

Although the probability density of a given increment is Gaussian, this stochastic process is not stationary itself. Furthermore, expectation values satisfy the following properties:

1. $E\{W_t\} = 0$.
2. $E\{W_t W_s\} = \min(t, s)$.
3. $E\{W_t^2\} = t$.

Wiener sample paths are continuous functions with probability one, but nowhere differentiable. In other words, $(W_{t+h} - W_t)/h$ is Gaussian distributed, but when h goes to zero the Gaussian distribution diverges. Moreover, the limit in the mean square,

$$\lim_{n \to \infty} \sum_{k=1}^{n} (W_{t_k} \quad W_{t_{k-1}})^2 = t \quad s, \tag{B.44}$$

is almost surely, where $s = t_0^{(n)} < \ldots < t_n^{(n)} = t$ is a sequence of partitions of the interval $[s, t]$, such that the size of each partition goes to zero for $n \to \infty$.

B.2.2 Ornstein–Uhlenbeck Processes and Brownian Motion

In *Ornstein–Uhlenbeck* (OU) *processes* [15], the role of the stochastic process is assigned to the velocity of the Brownian particle. Thus, the particle position (which is no longer a W-process) can then be obtained by integration. OU-processes are stationary, with $E\{X_t\} = 0$, and their correlation functions are given by decreasing exponentials,

$$E\{X_t X_s\} = \frac{\sigma^2}{2\gamma} e^{-\gamma |t-s|}, \tag{B.45}$$

where γ is the damping rate. Like W-processes, OU-processes are also Gaussian and do not have independent but correlated increments. The hierarchy of probability densities is given by

$$p(x) = \sqrt{\frac{\gamma}{\pi \sigma^2}} e^{-\gamma x^2 / \sigma^2} \tag{B.46}$$

and

$$p(x_1, t_1; \ldots; x_n, t_n) = p(x_1) p(x_2, x_1; t_2 - t_1) \cdots p(x_n, x_{n-1}; t_n - t_{n-1}), \tag{B.47}$$

with

$$p(x, y; \Delta t) = \sqrt{\frac{\gamma}{\pi \sigma^2 (1 - e^{-2\gamma \Delta t})}} e^{-\gamma (y - x e^{-\gamma \Delta t})^2 / \sigma^2 (1 - e^{-2\gamma \Delta t})}. \tag{B.48}$$

As mentioned above, the integral of an OU-process,

$$Y_t = \int_0^t X_s ds, \tag{B.49}$$

with $Y_0 = 0$, renders the Brownian particle position. This integration has to be understood as *realization-wise*, i.e., as almost surely. Moreover, since X_t is a continuous function of x with probability one, the integral is well-defined—the integral over a Gaussian process is another Gaussian process. Thus, $E\{Y_t\} = 0$, while the covariance $E\{Y_tY_s\}$ is different from zero. The W-process is recovered in the limit $\sigma \to \infty$ and $\gamma \to \infty$, with σ^2/γ^2 constant, usually equal to 1.

B.2.3 Poisson Processes

In order to describe discrete steps, *Poisson* (P) *processes* [16] in terms of discrete variables with independent increments are considered, rather than continuous W-processes. Considering all sample paths start from zero at time zero ($X_0 = 0$), the hierarchy of probability densities are Poisson distributions,

$$p(j, t) = \frac{(\lambda t)^j}{j!}e^{-\lambda t}, \tag{B.50}$$

with $j = 0, 1, \ldots$ and zero values for $j < 0$, and

$$p(j_1, t_1; \ldots; j_n, t_n) = p(j_1, t_1)p(j_2 - j_1, t_2 - t_1) \cdots p(j_n - j_{n-1}, t_n - t_{n-1}). \tag{B.51}$$

The increments of a P-process are also stationary, although the P-process itself is not stationary in a strict sense, since

1. $E\{X_t\} = \lambda t$
2. $E\{X_tX_s\} = \lambda \min(t, s)$

P-processes do not have almost surely continuous sample paths and are not almost surely differentiable functions.

B.3 *Markov Processes and Noise*

In general, if the power spectrum has a finite effective frequency band, the process is said to keep *memory* of its past. This is often described by means of a *colored noise*. However, there are many physical situations of interest where the fluctuations of a surrounding or *environment* are very fast. This gives rise to very broad effective frequency bands, which may cover frequencies even higher than those characterizing the system, described by some stochastic process X_t. In the limit $\tau_{corr} = 0$ or *zero memory*, the subsequent values of the stochastic variables describing the process at each time are independent, i.e., they are completely random. In these cases, the changes induced in the system by the environment will essentially depend on the strength of the latter fluctuations, increasing as such fluctuations become larger. For example, consider an OU-process. If $\tau_{corr} \to 0$ and $\sigma \to \infty$, the power spectrum is constant (flat) and therefore its corresponding correlation function becomes a Dirac δ-function. Actually, this δ-function can also be obtained from a Gaussian function whose variance goes to zero, i.e.,

$$\delta(t - s) = \lim_{\sigma^2 \to 0} \frac{1}{\sqrt{2\pi\sigma^2}} e^{-(t-s)^2/2\sigma^2}. \tag{B.52}$$

A δ-correlated process, with a flat spectrum, is called a *white noise*, since all frequencies are present and contribute equally, as in the case of white light. Since the OU-process is a Gaussian process, in the no-memory limit this process is known as *Gaussian white noise*. White noise can also be interpreted as the time-derivative of a process with stationary independent increments. In this sense, Gaussian white noise would be the time-derivative of a W-process; similarly, the same can be said for a P-process. However, neither the W-process nor the P-process are differentiable in the mean square sense. Thus, one can write

$$\xi_t = \dot{W}_t, \tag{B.53}$$

with $E\{\dot{W}_t\} = 0$ and $E\{\dot{W}_t \dot{W}_s\} = \delta(t - s)$.

Markov processes are stochastic processes with their time-evolution only depending on the present time, thus displaying a very short memory—i.e., the past history is rapidly forgotten. The hierarchy of joint probabilities can be reconstructed from just two distribution functions. It can be shown that a system is Markovian if the fluctuations are white, whereas non-Markovian systems are ruled by colored noise. Accordingly, the mathematical condition for a process to be Markovian is

$$P(X_t \in B / X_{t_m} = x_m, \dots, X_{t_1} = x_1) = P(X_t \in B / X_{t_m} = x_m). \tag{B.54}$$

This condition holds for all ordered set of times $t_1 < t_2 < \dots < t_m < t$, for all Borel sets B and all $x_1, x_2, \dots, x_m \in \mathbb{R}^d$. Thus, the probability of the event $X_t \in B$, conditioned on m previous events, only depends on the latest event $X_{t_m} = x_m$,

$$p(x, t | x_m, t_m; \dots; x_1, t_1) = p(x, t | x_m, t_m). \tag{B.55}$$

This conditional probability is also called *transition probability* or *propagator*,

$$T(x, t | x', t') = p(x, t | x', t'), \tag{B.56}$$

which satisfies the following properties:

1. $\int T(x, t | x', t') dx = 1$ (normalization).
2. $\lim_{t \to t'} T(x, t | x', t') = \delta(x - x')$.

When the propagator only depends on the time-difference $t' - t$, it is called *homogeneous*. In this sense, for example, the W-process is a homogeneous process in time, although it is not stationary.

The so-called *Chapman–Kolmogorov* (CK) *equation* for the propagator is

$$T(x_3, t_3 | x_1, t_1) = \int T(x_3, t_3 | x_2, t_2) T(x_2, t_2 | x_1, t_1) dx_2, \tag{B.57}$$

which in its differential version reads as

$$\frac{\partial T(x, t | x', t')}{\partial t} = \mathcal{L}(t) T(x, t | x', t'), \tag{B.58}$$

where $\mathcal{L}(t)$ is a linear operator generating infinitesimal time translations,

$$\mathcal{L}(t) \equiv \lim_{\Delta t \to 0} \frac{1}{\Delta t} \int T(x, t + \Delta t | x', t) dx' - \delta(x - x'). \tag{B.59}$$

For a homogeneous Markov process, $\mathcal{L}$ is time-independent. In this case, (B.59) can be formally solved to give

$$T_\tau(x|x') = e^{\mathcal{L}\tau} \delta(x - x'), \tag{B.60}$$

for $\tau \geq 0$. The one-parameter family $\{T_\tau / \tau \geq 0\}$ represents a dynamical semi-group [17], since τ is restricted to nonnegative values. This fact is related to irreversibility, which is the characteristic feature of any stochastic process. In other words, the exponential operator is not invertible in the total space of all proba-bility distributions. The simplest Markov process is given by a deterministic process: $\dot{x} = f(x)$, with $x(t) \in \mathbb{R}^d$ and $f(x) \in \mathbb{R}^d$. Consider the phase flow associated with such a differential equation is denoted by $\Phi_t(x)$ (see Chap. 1) and the phase curve with time is obtained for a fixed x and initial condition $\Phi_0(x) = x$. The corresponding propagator is [17]

$$T(x, t|x', t') = \delta(x - \Phi_{t-t'}(x')), \tag{B.61}$$

By appealing to the semigroup property $\Phi_t(\Phi_s(x)) = \Phi_{t+s}(x)$, it can be shown that the CK equation is satisfied by the propagator (B.61). The differential CK equation for a deterministic process is the Liouville equation,

$$\frac{\partial T(x, t|x', t')}{\partial t} = -\sum_i \frac{\partial \left[f_i(x) T(x, t|x', t') \right]}{\partial x_i}, \tag{B.62}$$

where only the initial conditions are assumed to be random. The time-evolution of this equation describes the deterministic drift.

Sometimes it is necessary to describe instantaneous jump processes. The corre-sponding differential CK equation is then given by the master equation

$$\partial_t T(x, t|x', t') = \mathcal{L}(t) T(x, t|x', t')$$
$$= \int \left[W(x|x'', t) T(x'', t|x', t') - W(x''|x, t) T(x, t|x', t') \right] dx'', \tag{B.63}$$

where $W(x|x', t)$ is the *transition rate* accounting for the instantaneous jump from the state x' at time t to the state x. The total rate for a jump at time t, $\Gamma(x', t)$, is obtained by integrating $W(x|x', t)$ over x,

$$\Gamma(x', t) = \int W(x|x', t) dx. \tag{B.64}$$

A more standard form for (B.63) is

$$\frac{\partial p(x, t)}{\partial t} = \int \left[W(x|x', t)p(x', t) - W(x'|x, t)p(x, t) \right] dx', \qquad \text{(B.65)}$$

where $p(x, t)$ is the probability density. The so-called Kramers–Moyal expansion of the master equation can be obtained from a different rewriting of (B.63) after performing an appropriate Taylor series expansion. This leads to a partial differential equation of infinite order.

The continuity condition for Markov processes arises when the probability for a transition, during an increment of time with size larger than a given small amount, decreases more rapidly than the increment of time as it goes to zero. Diffusion processes satisfy this condition, but they are not deterministic. The corresponding differential CK equation is the Fokker–Planck equation expressed in terms of the probability density,

$$\frac{\partial p(x, t)}{\partial t} = -\sum_i \frac{\partial \left[g_i(x, t)p(x, t) \right]}{\partial x_i} + \frac{1}{2} \sum_{i,j} \frac{\partial^2 \left[D_{ij}p(x, t) \right]}{\partial x_i \partial x_j}, \qquad \text{(B.66)}$$

where g_i and D_{ij} are the first and second moments of the jump distribution (the drift and diffusion coefficients, respectively). The D matrix is known as the *diffusion matrix*, which is symmetric and positive semidefinite. The Fokker–Planck equation (B.66) can be seen as a truncation of the Kramers–Moyal expansion to second order. Moreover, it can also be recast as a continuity equation,

$$\frac{\partial p(x, t)}{\partial t} + \sum_i \frac{\partial J_i(x, t)}{\partial x_i} = 0, \qquad \text{(B.67)}$$

after defining the probability current density

$$J_i(x, t) = g_i(x, t)p(x, t) - \frac{1}{2} \sum_j \frac{\partial \left[D_{ij}p(x, t) \right]}{\partial x_j}. \qquad \text{(B.68)}$$

Similar equations can also be written for the propagator.

Piecewise deterministic processes, arising from the combination of deterministic and jump processes, are also very important in many applications of the theory of open systems. The corresponding differential CK equation is known as the *Liouville master equation*, which in terms of the propagator, reads as

$$\frac{\partial T(x, t|x', t')}{\partial t} = -\sum_i \frac{\partial [g_i(x, t)T(x, t|x', t')]}{\partial x_i}$$

$$+ \int [W(x|x'')T(x'', t|x', t') - W(x''|x)T(x, t|x', t')]dx''. \qquad \text{(B.69)}$$

The so-called *waiting time distribution*, $F(\tau | x', t')$, is the probability for the next jump to occur during the time interval $[t', t' + \tau]$ starting from x' at time t'. Its general expression is given by

$$F(\tau | x', t') = 1 - e^{-\int_0^\tau ds \Gamma(\Phi_s(x'))}. \tag{B.70}$$

If one has a pure jump process, the drift is zero and the waiting time distribution is an exponential function.

If the propagator $T(x, t | x', t') = T_{t-t'}(x - x')$ is invariant with respect to space-time translations, the stochastic process is a *Lévy process* [18, 19] (homogeneous both in space and time). These two requirements allow to directly work with the integral CK equation. Lévy processes are also stable if by performing a time scaling the new process can be expressed as the original one multiplied by a scaling factor. This property is called *self-similarity* or *fractality*.

B.4   Stochastic Differential Equations

In order to describe piecewise deterministic processes in terms of random variables instead of probability densities or propagators, a stochastic differential equation (SDE) is necessary to account for their time-evolution. For example, in the case of a Gaussian white noise, the time-evolution of a one-dimensional random variable X_t is governed by the SDE

$$dX_t = f(X_t)dt + bg(X_t)dW_t. \tag{B.71}$$

In integral form, this equation reads as

$$X_t = X_0 + \int_0^t f(X_s)ds + b \int_0^t g(X_s)dW_s, \tag{B.72}$$

where the first integral can be understood as an ordinary Riemann integral, while the b coefficient of the second one is related to the diffusion coefficient. This second integral, on the contrary, is problematic due to the intrinsic properties of the W-process. While the Riemann integral is independent of the different evaluation points along the interval $[0, t]$, the same does not hold for the stochastic integral. To obtain an unambiguous definition of the stochastic integral, a choice of the evaluation points has to be previously decided. Thus, if

$$\tau_i^{(n)} = (1 - \alpha)t_{i-1}^{(n)} + \alpha t_i^{(n)}, \tag{B.73}$$

with $0 \leq \alpha \leq 1$ and $t_0 = t_1^{(n)} < \cdots < t_n^{(n)} = t$, there are two reasonable choices (which lead to desirable features):

1. The *Itô integral*, if $\alpha = 0$.
2. The *Stratonovich integral*, if $\alpha = 1/2$.

Although the latter conserves the ordinary rules of integration, the Itô integral is much more appealing from a mathematical viewpoint, for it retains the important properties of the W-process—indeed, for a white noise it is the only reasonable choice. The integral of the quantity $W_s dW_s$ within the interval $[0, t]$ can be used to illustrate how each one of these integration schemes works. In the case of Itô integration,

$$\int_{t_0}^{t} W_s dW_s = \frac{1}{2} \left[(W_t^2 - W_{t_0}^2) - (t - t_0) \right],$$ (B.74)

while in the case of Stratonovich integration, the integral becomes

$$\int_{t_0}^{t} W_s dW_s = \frac{1}{2} \left(W_t^2 - W_{t_0}^2 \right).$$ (B.75)

An important mathematical result establishes that any stochastic integral based on a different choice of α can be expressed as the sum of the corresponding Itô integral plus an ordinary Riemann integral.

Only a certain class of stochastic processes can be used in the Itô integral, the so-called *non-anticipating stochastic processes*, G_t. These processes have the property that G_t is only known from the past history of the W-process up to time t. That is, G_t is independent of all increments of the W-process for times greater than t. Thus, the stochastic process $Y_t = \int_{t_0}^{t} G_s(\omega) dW_s(\omega)$ is an Itô integral if:

1. Y_t is a non-anticipating process.
2. Y_t has almost surely continuous sample paths.
3. If $\int_{t_0}^{t} E\{G_s^2\} ds < \infty$, then $E\{Y_t\} = 0$ and $E\{Y_t Y_s\} = \int_{t_0}^{min(t,s)} E\{G_u^2\} du$.
4. $E\{Y_t | \mathcal{F}_s\} = Y_s$ (the martingale property) with $s \leq t$; $\mathcal{F}_s$ with $s \geq 0$ is a filtration or family of sub-σ algebras of $\mathcal{F}$ if $\mathcal{F}_s \subseteq \mathcal{F}_t \subseteq \mathcal{F}$, with $s \leq t$.

The Stratonovich SDE is similar to the corresponding Itô one, but with the difference that integration follows the usual rules of the Riemann integral, as said above. Thus, given a SDE, it is always possible to change of integration scheme or interpretation, Itô or Stratonovich. For additive noise ($g(X_t) = $ constant), there is no difference between both schemes. However, for multiplicative noise ($g(X_s) \neq $ constant), there are important differences. Both integrals lead to mathematically consistent calculus. However, the question on which form is the correct one to describe physical systems has led to a very long controversy [20] and, in the end, the last word always comes from the comparison with the experiment. In any case, the standard Langevin equation can always be expressed in terms of a SDE given by (B.71).

The Itô stochastic calculus is based on the fact that second order terms have to be retained. Thus, whereas in ordinary calculus $(dt)^2 \to 0$ and $dt\, dW_t \to 0$, in stochastic calculus $(dW_t)^2 = dt$. For example, given $f = f(W_t, t)$, the Taylor series expansion of f leads to

$$df(W_t, t) = \left(\frac{\partial f}{\partial t} + \frac{1}{2} \frac{\partial^2 f}{\partial W_t^2} \right) dt + \frac{\partial f}{\partial W_t} dW_t. \tag{B.76}$$

A stochastic process X_t obeys an Itô SDE,

$$dX_t = a(X_t, t)dt + b(X_t, t)dW_t, \tag{B.77}$$

whenever

$$X_t = X_{t_0} + \int_{t_0}^{t} a(X_{t'}, t')dt' + \int_{t_0}^{t} b(X_{t'}, t')dW_{t'} \tag{B.78}$$

for all t_0 and t. If the solution of the Itô SDE is unique, then it is Markovian. If X_t fulfills the Itô SDE and f is a functional of X_t and t, then

$$df(X_t, t) = \left[\frac{\partial f(X_t, t)}{\partial t} + a(X_t, t)\frac{\partial f(X_t, t)}{\partial X_t} + \frac{1}{2}b^2(X_t, t)\frac{\partial^2 f(X_t, t)}{\partial X_t^2} \right] dt$$
$$+ b(X_t, t)\frac{\partial f(X_t, t)}{\partial X_t}dW_t, \tag{B.79}$$

which is known as *Itô formula*. It can be proven that if a and b depend explicitly on time, the SDE defines a diffusion process. The stochastic equation is always linear in dW_t. A generalization to more than one dimension is quite straightforward.

Following the similar prescription, in general, given a function $f(\mathbf{x}(t), t)$ of a vector stochastic variable $\mathbf{x}(t)$, a second-order series expansion in increments leads to

$$df(\mathbf{x}, t) = \frac{\partial f(\mathbf{x}, t)}{\partial t}dt + \nabla f(\mathbf{x}, t) \cdot d\mathbf{x} + \frac{1}{2}\nabla^2 f(\mathbf{x}, t)(d\mathbf{x})^2. \tag{B.80}$$

If $\mathbf{x}(t)$ satisfies the Itô SDE

$$d\mathbf{x}(t) = \mathbf{a}(\mathbf{x}, t)dt + bd\mathbf{W}(t), \tag{B.81}$$

then

$$df(\mathbf{x}, t) = \left[\frac{\partial f(\mathbf{x}, t)}{\partial t} + \mathbf{a}(\mathbf{x}, t)\nabla f(\mathbf{x}, t) + \frac{1}{2}b^2\nabla^2 f(\mathbf{x}, t) \right] dt + b\nabla f(\mathbf{x}, t)d\mathbf{W}(t). \tag{B.82}$$

In this diffusion process, the probability distribution is described by the Fokker–Planck equation

$$\frac{\partial F(\mathbf{x}, t)}{\partial t} = -\nabla \cdot [\mathbf{a}(\mathbf{x}, t)F(\mathbf{x}, t)] + \frac{b^2}{2}\nabla^2 F(\mathbf{x}, t). \tag{B.83}$$

In some applications it is also very useful to define the so-called *mean forward* and *backward derivatives* of a given function $f(\mathbf{x}(t), t)$ [21],

$$
\begin{aligned}
D_+ f(\mathbf{x}(t), t) &= \frac{\partial f(\mathbf{x}, t)}{\partial t} + \mathbf{a}_+(\mathbf{x}, t)\dot{\nabla} f(\mathbf{x}, t) + \frac{1}{2}b^2\nabla^2 f(\mathbf{x}, t), \\
D_- f(\mathbf{x}(t), t) &= \frac{\partial f(\mathbf{x}, t)}{\partial t} + \mathbf{a}_-(\mathbf{x}, t)\dot{\nabla} f(\mathbf{x}, t) - \frac{1}{2}b^2\nabla^2 f(\mathbf{x}, t),
\end{aligned}
\tag{B.84}
$$

respectively, where $D_+\mathbf{x}(t) = \mathbf{a}_+(\mathbf{x}, t)$ and $D_-\mathbf{x}(t) = \mathbf{a}_-(\mathbf{x}, t)$. If $\mathbf{x}(t)$ defines the position of a given particle and is symmetric under time reversal, i.e., $\mathbf{x}(t) = \mathbf{x}(-t)$, the corresponding Itô SDEs are

$$
\begin{aligned}
d\mathbf{x}(t) &= \mathbf{a}_+(\mathbf{x}, t)dt + bd\mathbf{W}(t), \\
d\mathbf{x}(t) &= \mathbf{a}_-(\mathbf{x}, t)dt + bd\mathbf{W}_-(t),
\end{aligned}
\tag{B.85}
$$

where $E[d\mathbf{W}_-(t) \cdot d\mathbf{W}_-(t)] = |dt| = -dt$ for the reverse time flow. The Newtonian velocity is then defined as the mean velocity (or flow velocity) of a Brownian particle,

$$
\mathbf{v}(\mathbf{x}, t) = \frac{1}{2}\left[\mathbf{a}_+(\mathbf{x}, t) + \mathbf{a}_-(\mathbf{x}, t)\right].
\tag{B.86}
$$

If both derivatives are not equal, their difference defines the vector field

$$
\mathbf{u}(\mathbf{x}, t) = \frac{1}{2}\left[\mathbf{a}_+(\mathbf{x}, t) - \mathbf{a}_-(\mathbf{x}, t)\right],
\tag{B.87}
$$

which is called the *osmotic velocity* [21]. Similarly, a mean acceleration can be defined as

$$
\mathbf{A}(\mathbf{x}, t) = \frac{1}{2}\left[D_+\mathbf{a}_-(\mathbf{x}, t) + D_-\mathbf{a}_+(\mathbf{x}, t)\right].
\tag{B.88}
$$

One of the central aspects of stochastic dynamics is the concept of *noise-induced transition*. Under some conditions, the influence of environmental fluctuations can be far from being negligible. Indeed, the effect of external noise may depend on the system state. A transition takes place at points of the parameter space (mean value of the external noise, its variance, its correlation time, etc.) where the functional form of the mapping from the sample space into the state space changes qualitatively—e.g., the number and location of extrema of the stationary probability density [5]. This probability density is usually written in terms of a "probability" potential which is subject to topological analysis. It can be shown that only when the external noise is multiplicative, new potential wells or states can be created and some of the existing ones will be destroyed. In this regard, it is of particular interest the so-called *stochastic resonance mechanism* [22].

B.5 Stochastic Processes in Quantum Mechanics

In quantum mechanics, the general probability theory has to be applied in a rather different way, within the context of Hilbert spaces $\mathcal{H}$ and noncommuting algebras of operators, variables and observables or measurable quantities. Details on the mathematical notions involved in the formulation of this theory and its statistical interpretation [23, 24], and more particularly in the theory of measurement [17, 25], will not be accounted for here. Only those elements which are more closely related to the issues considered throughout this monograph are briefly discussed. To simplify the notation, both the random variable and its Hermitian operator will be denoted by the same symbol A, being easy to discern when the latter plays the role of a variable or an operator, respectively.

The basic tool when dealing with statistical mixtures is the statistical or density operator ρ. This operator has to be understood as describing a statistical ensemble consisting of a large number, N, of identical quantum systems prepared in the following way: M sub-ensembles each one described by a normalized state vector $\psi_n(n = 1, \cdots , M)$ formed by N_n elements. Accordingly, the density operator reads as

$$\rho = \sum_n w_n |\psi_n\rangle \langle \psi_n|, \tag{B.89}$$

where the statistical weight of each state is given by $w_n = N_n/N$, with $N = \sum_{n=1}^M N_n$. The density operator is Hermitian or self-adjoint, positive and with trace equal to one. A further discussion on the density operator and its properties can be found, for example, in [24, 26].

An observable via its spectral family leads to a real random variable, which describes the probabilities for all possible measurement outcomes. The sample space is the real axis and the algebra of events $\mathcal{B}$ is given by the Borel sets of the real axis. Thus,

$$E\{A\} = \text{Tr}\,\{A\rho\} \equiv \langle A\rangle, \tag{B.90}$$

where Tr stands for the trace operation. Within this context, the associated variance, which renders information about the dispersion of A, reads as

$$\sigma(A) \equiv (\Delta A)^2 = \text{Tr}\,\{A^2\rho\} - (\,\text{Tr}\,\{A\rho\})^2 \equiv \langle A^2\rangle - \langle A\rangle^2. \tag{B.91}$$

That is, according to this expression, there are no dispersion-free measurements.

Observables are the result of measurements and therefore display fluctuations around some average value. These average values can be obtained from pure or mixed state.[2] In the case of a pure state, if it corresponds with an eigenvector of A, the fluctuation is zero. However, those values have nothing to do with the typical

---

[2] The distinction between pure and mixed states is usually characterized by the definition of the convex linear combination in the convex set of density operators in the Hilbert space. Thus, ρ is a

errors associated with the measuring devices. In this regard, Heisenberg's uncertainty principle relates very precisely the fluctuations of two non-commuting operators, stating that the corresponding observables cannot be simultaneously measured with an infinite precision. Usually, after a measurement the system is projected in an eigenstate of the measuring device (ideal measurement). However, not always this is the case. Indeed, within the context of the quantum theory of measurement [25], Aharonov et al. [27] introduced the definition of *weak measurement*. This concept is based on assuming that the coupling between measuring device and observed system can be set so weakly that the uncertainty associated with a single measurement is still very large compared with the separation between the observable eigenvalues. Hence, after a weak measurement the system is not left in an eigenstate of the observable, but rather in a superposition of the unresolved eigenstates. This procedure is useful for the amplification and detection of weak effects [28–30], e.g., a direct detection of the photon wave function [9] or the associated photon trajectories [31].

In the case of dissipative and/or stochastic dynamics, the operator $\delta A = A - \langle A \rangle$ is also called a *fluctuation*, and it can be shown that $(\Delta(\delta A))^2 = (\Delta A)^2$ (see (B.9)). When one is interested in the average moments $\langle A^n \rangle$ (n being an integer) of a given operator A, sometimes it is more convenient to evaluate the so-called characteristic function of A, $\langle e^{i\xi A} \rangle$, where ξ is a real parameter—note that the nth moment arises from the partial differentiation with respect to $i\xi$ of nth order and evaluated at $\xi = 0$. Moreover, the Fourier transform of the characteristic function of A gives the probability distribution function or diagonal matrix elements of the density operator in the representation in which A is diagonal. In this regard, for example, the Wigner, Poisson and exponential distribution functions can be expressed as the Fourier transform of a given characteristic function [26]. When a set of observables commute, a characterization by joint probability distributions is also possible. For the particular case of the Wigner distribution, which is given in terms of the phase space coordinates (e.g., the position and momentum of a particle), see for example [26].

The noise in quantum mechanics has to be treated with some special care. For example, in the quantum Langevin equation the noise is a quantum operator and the noise (symmetric) autocorrelation function is a complex quantity because, in general, it does not commute at different times. At very low or zero temperatures, the noise is still correlated at very long times [32, 33]. Furthermore, it is not proportional to a Dirac delta function in time. Thus, we have the situation that, although there is no memory in the standard Langevin equation, the quantum process is not Markovian [34].

When dealing with open quantum systems, the total Hamiltonian is often expressed as the sum of the Hamiltonian accounting for the quantum system of interest (S), the Hamiltonian accounting for the environment (B) and their coupling. A measuring device can also be seen as an environment [17]. Many times, though, only the

---

convex linear combination of ρ_1 and ρ_2 when

$$\rho = \alpha\rho_1 + (1 - \alpha)\rho_2,$$

with $\alpha \in]0, 1]$. If ρ describes a pure state, then $\rho = \rho_1 = \rho_2$.

dynamics of the system of interest is relevant, which is obtained after tracing out over the environment (or bath) degrees of freedom. This gives rise to a *reduced system dynamics*, which is no longer unitary in time. Quantum Markov processes represent the simplest description of open (quantum) system dynamics. However, the extension of Markov processes to quantum mechanics requires the definition of a quantum dynamical semigroup [17]. Thus, consider a general dynamical map $V(t)$ which describes the transformation over time

$$\rho_S(t) = V(t)\rho_S(0),$$

where $\rho_S \equiv \mathrm{Tr}_B(\rho)$ is the *reduced density operator* or *reduced density matrix*. This map represents a convex-linear, completely positive and trace-preserving quantum operation. A quantum dynamical semigroup is defined as follows:

1. If $V(t)$ defines a continuous, one-parameter family of dynamical maps by varying t (with $t \geq 0$).
2. The Markov approximation for the homogeneous case is assumed.
3. The semigroup property $V(t)V(s) = V(t+s)$, with $t, s \geq 0$, holds.

This gives rise to a first-order linear differential equation for the reduced density operator, the so-called *Linblad equation*,

$$\dot{\rho}_S(t) = \mathcal{L}\rho_S(t), \tag{B.93}$$

where the generator $\mathcal{L}$ of the semigroup is a super-operator, since its action over operators yields another operator.

Coherence is a key issue in different branches of quantum mechanics. The opposite effect, decoherence, it appears when a quantum interference pattern is destroyed or suppressed. In a certain sense, it could be said that decoherence leads to the appearance of a classical world in quantum mechanics. Environment induced decoherence [25] is omnipresent and, in general, it is a short time phenomenon. When measuring, the dynamics of the system tends to be decoherent. In the weak coupling approximation between the system and environment, the trace operation carried out to obtain the reduced density matrix is key to the existence of any master equation. However, this operation is questionable when the system and environment are entangled at all times, including the initial state. The role of the initial conditions has also be widely discussed in the corresponding literature.

In order to define the state vector as a random variable in Hilbert space, it is necessary a new type of quantum-mechanical ensembles, which are not fully characterized by a density matrix [17]. This new ensemble is formed by M statistical ensembles of the type described above to define the statistical operator. Each one of the M ensembles consists of N_n elements prepared in a normalized state $|\psi_n\rangle$, in such a way that $N = \sum_{n=1}^{M} N_n$. This is the new sample space, where w_n is the measure corresponding to the ensemble characterized by the state vector $|\psi_n\rangle$.

Different types of measurements can be defined from different probability density functionals, $P[\psi]$. For example, for a subset A, such that $A \in \mathcal{A}$,

$$\mu(A) = \int_A D\psi D\psi^* P[\psi].\tag{B.94}$$

If $P[\psi] = \delta(\psi - \psi_0)$, one obtains the Dirac measure. The volume element in Hilbert space can be chosen to be the Euclidean volume element [17]. The expectation values are then obtained from

$$E\{F[\psi]\} = \int D\psi D\psi^* P[\psi] F[\psi]\tag{B.95}$$

for a functional $F[\psi]$—e.g., $F[\psi] = \mathrm{Tr}\,\{A\rho\} = \langle\psi|A|\psi\rangle$. Taking into account this new ensemble, the density operator can then be represented as

$$\rho = E\{|\psi\rangle\langle\psi|\} = \int D\psi D\psi^* P[\psi]|\psi\rangle\langle\psi|.\tag{B.96}$$

The time-dependence of the state vector leads to stochastic processes $|\psi(t)\rangle$ in Hilbert space, where the probability density functional is $P[\psi, t]$. The dynamics of the open quantum system can then be described by a differential stochastic equation for $|\psi(t)\rangle$ instead of a master equation. This is the idea behind what it is termed as *unravelling* of the master equation. The evolution in time of $P[\psi, t]$ can be expressed as

$$P[\psi, t] = \int D\phi D\phi^* T[\psi, t|\phi, t_0] P[\phi, t_0],\tag{B.97}$$

where T gives the conditional transition probability. Nevertheless, a similar prescription as before can be followed, which allows to describe the time-evolution of the wave function as a diffusion process. This process is governed by a SDE,

$$d|\psi(t)\rangle = -iK\{|\psi(t)\rangle\}dt + bM\{|\psi(t)\rangle\}dW(t),\tag{B.98}$$

where K is a drift operator and M is related to a certain diffusion operator. As can be noticed, here the noise is multiplicative, since the coefficient accompanying $dW(t)$ is not a constant (it depends on the wave function itself). This noise may lead to transitions, namely *noise-induced transitions*, which is an important aspect of the stochastic dynamics not fully developed yet.

References

1. Chandrasekhar, S.: Stochastic problems in physics and astronomy. Rev. Mod. Phys. **15**, 1–89 (1943)
2. Papoulis, A.: Probability, Random Variables and Stochastic Processes. McGraw-Hill, New York (1965)

3. van Kampen, N.G.: Stochastic Processes in Physics and Chemitry. North-Holland, Amsterdam (1981)
4. Risken, R.: The Fokker–Planck Equation. Springer, Berlin (1984)
5. Horsthemke, W., Lefever, R.: Noise-Induced Transition.. Springer, Berlin (1984)
6. Shiryaev, A.N.: Probability. Springer, Berlin (1995)
7. Wolfgang, P., Baschnagel, J.: Stochastic Processes. From Physics to Finance. Springer, Berlin (1999)
8. Margalef-Roig, J.: Curso de ocesos Estocásticos. CSIC, Madrid (2000)
9. Lundeen, J.S., Sutherland, B., Patel, A., Stewart, C., Bamber, C.: Direct measurement of the quantum wavefunction. Nature **474**, 188–191 (2011)
10. Gardiner, C.W.: Handbook of Stochastic Methods. Springer, Berlin (1983)
11. Radon, J.: Theorie und Anwendungen der Absolut Additiven Mengenfunktionen. Sitzungsber. Akad. Wissen. Wien **112**, 1295–1438 (1913)
12. Nikodym, O.: Sur une Généralisation des Intégrales de M. J. Radon. Fundamenta Mathematicæ **15**, 131–179 (1930)
13. Ross, S.M.: Stochastic Processes. Wiley, New York (1996)
14. Wiener, N.: The average of an analytical functional and the Brownian movement. Proc. Natl Acad. Sci. USA **7**, 294–298 (1921)
15. Uhlenbeck, G.E., Ornstein, L.S.: On the theory of Brownian Motion. Phys. Rev. **36**, 823–841 (1930)
16. Cox, D.R., Isham, V.I.: Point Process. Chapman & Hall, London (1980)
17. Breuer, H.-P., Petruccione, F.: The Theory of Open Quantum Systems. Clarendon Press, Oxford (2006)
18. Lévy, P.: Processus stochastiques et mouvement brownien. Gauthier-Villars, Paris (1965)
19. Applebaum, D.: Lévy processes—from probability to finance and quantum groups. Not. Am. Math. Soc. **51**, 1336–1347 (2004)
20. van Kampen, N.G.: Itô versus Stratonovich. J. Stat. Phys. **24**, 175–187 (1981)
21. Nelson, E.: Derivation of the Schrödinger equation from Newtonian mechanics. Phys. Rev. **150**, 1079–1085 (1966)
22. Gammaitoni, L., Hänggi, P., Jung, P., Marchesoni, F.: Stochastic resonance. Rev. Mod. Phys. **70**, 223–287 (1998)
23. Ballentine, L.E.: The statistical interpretation of Quantum Mechanics. Rev. Mod. Phys. **42**, 358–381 (1970)
24. Ballentine, L.E.: Quantum Mechanics. A Modern Development. World Scientific, Singapore (1998)
25. Wheeler, J.A., Zurek, W.H.: Quantum Theory and Measurement. Princeton University Press, Princeton (1983)
26. Louisell, W.H.: Quantum Statistical Properties of Radiation. Wiley Classics Library, New York (1973)
27. Aharonov, Y., Albert, D.Z., Vaidman, L.: How the resuslt of a measurement of a component of the spin of a spin-1/2 particle can turn out to be 100. Phys. Rev. Lett. **60**, 1351–1354 (1988)
28. Ritchie, N.W.M., Story, J.G., Hulet, R.G.: Realization of a measurement of a "weak value". Phys. Rev. Lett. **66**, 1107–1110 (1991)
29. Wiseman, H.M.: Grounding Bohmian mechanics in weak values and bayesianism. New J. Phys. 9:165(1–12) (2007)
30. Mir, R., Lundeen, J.S., Mitchell, M.W., Steinberg, A.M., Garretson, J.L., Wiseman, H.M.: A double-slit 'which-way' experiment on the complementarity-uncertainty debate. New J. Phys. 9:287(1–11) (2007)
31. Kocsis, S., Braverman, B., Ravets, S., Stevens, M.J., Mirin, R.P., Shalm, L.K., Steinberg, A.M.: Observing the average trajectories of single photons in a two-slit interferometer. Science **332**, 1170–1173 (2011)

32. Ingold, G.-L.: Path Integrals and their Application to Dissipative Quantum Systems. Lecture Notes Physics, vol. 611, pp. 1–53. Springer, Berlin (2002)
33. Gardiner, C.W., Zoller, P.: Quantum Noise.. Springer Complexity, Berlin (2004)
34. Ford, G.W., Lewis, J.T., O'Connell, R.F.: Quantum Langevin equation. Phys. Rev. A **37**, 4419 (1988)

Index

A. S. Sanz and S. Miret-Artés, *A Trajectory Description of Quantum Processes.*
I. Fundamentals, Lecture Notes in Physics 850, DOI: 10.1007/978-3-642-18092-7,
© Springer-Verlag Berlin Heidelberg 2012

PGSTL